The Bureau of Rec
Origins and Growt
Volume 1

William D. Rowley

Bureau of Reclamation
U.S. Department of the Interior
Denver, Colorado
2006

For sale by the Superintendent of Documents. U.S. Government Printing Office
Internet: bookstore.gpo.gov Phone: toll free (866) 512-1800; DC area (202) 512-1800
Fax: (202) 512-2250 Mail: Stop SSOP, Washington, DC 20402-001

ISBN-10: 0-16-075226-4
ISBN-13: 978-0-16-075226-1

TABLE OF CONTENTS

LIST OF ILLUSTRATIONS

CHAPTER 2

CHAPTER 3

CHAPTER 4

CHAPTER 5

CHAPTER 6

CHAPTER 7

INTRODUCTION

The Bureau of Reclamation has a grand tradition, and this volume tells the fascinating story of that tradition from Reclamation's beginnings to the end of World War II. It has been my privilege to work in Reclamation for 38 years—more than one-third of its history—and I have enjoyed every minute of that time.

The story of Reclamation is deeply entwined in the development of the American West in the twentieth century and beyond. Reclamation was established in 1902 by President Theodore Roosevelt to "make the desert bloom." Reclamation projects have been the seed for many of the modern American West's large agricultural and metropolitan centers. The Columbia Basin of Washington, the Central Valley of California, the Central Arizona Project, and the Shoshone and Platte River Projects areas of Wyoming are just a few of the many irrigated areas that are vital parts of the economic backbone of the West. Boise, Spokane, Las Vegas, Casper, and El Paso are some of the Western cities that have grown up around Reclamation projects.

Today, Reclamation provides one out of five western farmers with water for 10 million irrigated farmland acres. These farmlands produce sixty percent of the nation's vegetables and twenty-five percent of its fruits and nuts. We are the largest electric utility in the seventeen western states (operating 58 hydropower plants) and the nation's largest wholesale water supplier, administering 348 reservoirs with a total storage capacity of 245 million acre-feet. Nearly 30 million people all over the West depend on Reclamation projects for their municipal, industrial, and domestic water supplies.

Throughout its history, Reclamation has been an innovator in the engineering and science of dam design and construction, hydroelectric power production and delivery, water delivery, conservation, and multipurpose uses of water. Reclamation's masonry dams represent a distinguished lineage from East Park to Pathfinder, from Buffalo Bill to Arrowrock and Owyhee, and from Hoover to Grand Coulee, Shasta, Friant, and Morrow Point. Reclamation's embankment dams follow an equally distinguished lineage and include Belle Fourche, Anderson Ranch, and San Luis.

Reclamation's history is rich, filled with colorful personalities and the unique character of the West. It is marked by feats of engineering and economic wonder that have resulted in water development, resource management, and resource preservation. This book traces Reclamation's story to the end of World War II. I hope that you find this study as useful and informative as I do and that you will be as eager as I am for the concluding volume, which will cover World War II to the present.

John W. Keys III
Commissioner
Bureau of Reclamation

PREFACE AND ACKNOWLEDGMENTS

In 2002, the Bureau of Reclamation celebrated a century of service to the people of the United States. In that century the Bureau of Reclamation "grew up with the country," the American West. Precious water resources, managed to a large degree by the Bureau, now serve the West's growing modern society of a hundred million people. Water resources include water for agriculture, cities, industry, transportation, recreation, and the production of hydroelectric power from major western rivers. The Bureau of Reclamation built much of the water resource infrastructure that was absolutely necessary to tap the rich possibilities of the western environment.

The story that runs through the chapters of this first volume addresses the origins and implementation of what was to be called "reclamation" under the auspice and financing of a government service bureau. These pages look at the origins of Reclamation as an ambitious experiment or innovation in government land policies that from the beginning of the Republic sought to dispose of the Public Domain. Western lands without access to water seemed of little value, but once water and land could be joined, land became valuable. In many places the Bureau of Reclamation became the agency of this transformation. Arid public lands that obtained water rights became attractive land for settlement and private acquisition. The entire effort of "federal reclamation by irrigation" was a vast experiment ranging from the construction of minor to enormous multipurpose water projects that ultimately did far more than provide water for irrigation agriculture.

The story is not always a neat progression. The first chapter deliberately chooses to highlight important projects, events, and personalities that run the gamut of Reclamation's history in the twentieth century. More detailed and chronologically organized chapters follow. Likewise the story is not always a neat progression of one success to another. Grand crusades often announce high hopes and plans, but subsequent events can bring disappointments and deflated ideals that only unexpected turn of events rescue. History is a fickle subject that eludes iron clad laws of development and predictability, and the history of government reclamation confirms this. Conceived in the throes of a romantic movement to revitalize rural life and small farms, the Bureau of Reclamation by the mid-twentieth century became a chief arbiter of western water and power resources for a modern urban industrial society in the American West.

This work follows in the steps of previous histories of the program sponsored by public agencies: George Wharton James, *Reclaiming the Arid West: The Story of the United States Reclamation Service* (Washington, D.C.: Government Printing Office, 1917), Institute for Government Research, *The U.S. Reclamation Service: Its History, Activities and Organization* (New York: D. Appleton and Company, 1919), and Michael Robinson, *Water for the West* (Chicago: Public Works Historical Society, 1976). To the Senior Historian for the Bureau of Reclamation, Brit Allan Storey, must go the credit for envisioning the project, obtaining the support for it within the Bureau of Reclamation, and cheerfully directing the book's production. Commissioner John Keyes has been particularly supportive, taking a keen interest in the project since its beginning.

My colleagues afield, Professors Donald J. Pisani of the University of Oklahoma and Donald C. Jackson of Lafayette College contributed immensely in their roles on the advisory board for the planning and writing of this bureau history. A project of this scope, which will include a subsequent volume on the latter half of the twentieth century, could not have been undertaken in these time limits without drawing upon the excellent scholarship that has been performed in water and reclamation history in the last thirty years. But, of course, the ultimate responsibility for the soundness of the book and its interpretations rests with the author. Many others contributed to the work, not the least of whom was Carol Storey, who traveled with spouse Brit and the author to dams and irrigation sites and joined in the endless talk about water, water rights, dams, concrete, land classifications, canals, ditches, diversions, kilowatts, generators, exciters, and turbines.

A hearty thanks is gratefully extended to various librarians and archivists who facilitated the research with their cooperation, advice, and direction. They include Patrick Ragains, Business and Government Information Librarian at the University of Nevada, Reno; Sharon Prengaman, Librarian Technician at the University of Nevada, Reno; long-time friend Dr. Milton O. Gustafson, Chief, Civilian Reference Branch, now retired, at the National Archives and Records

Administration in College Park, Maryland; Carol Bowers, Reference Supervisor in the research room at the American Heritage Center on the campus of the University of Wyoming in Laramie; Eric Bittner, archivist, at the National Archives Branch in Denver where Record Group 115 is housed; Shelly C. Dudley, Senior Historical Analyst, Salt River Project in Phoenix, Arizona; John Douglas Helms, Historian for the Natural Resources Conservation Service in Washington, D.C.; Michael J. Brodhead, Historian, Office of History, U.S. Army Corps of Engineers, Alexandria, Virginia; and the always helphul staff at the Bureau of Reclamation's library in Denver.

Also, thank you to my colleagues in the Department of History at the University of Nevada, Reno for supporting my commitment to this project. Finally, my thanks go to Andrew Gahan, Ph.D. candidate in history at the University of Nevada. He pursued complicated research questions in the library, traveled to archives, and worked a productive summer internship in the History Office of the Bureau of Reclamation in Denver in service of this project.

William D. Rowley
University of Nevada, Reno

Professor William D. Rowley holds the Grace A. Griffen Chair in Nevada and the West in the History Department at the University of Nevada, Reno. He is a longtime student of western agricultural and resource history. His books include, *M. L. Wilson and the Campaign for the Domestic Allotment Plan* (University of Nebraska Press, 1970); *Reno: Hub of the Washoe Country* (1984); *U.S. Forest Service Grazing and Rangelands: A History* (Texas A & M Press,1985); *Reclaiming the Arid West: The Career of Francis G. Newlands* (Indiana University Press, 1996). From 1974 to 1990 he served as Executive Secretary of the Western History Association. More recently he has served on the governing board of the Forest History Society and the Executive Committee for the American Society for Environmental History and the Agricultural History Society. He was editor-in-chief of the *Nevada Historical Society Quarterly* from 1991-2004. Professor Rowley teaches graduate seminars and undergraduate courses in American environmental history, history of Nevada, history of California, and the history of the westward movement.

BUREAU OF RECLAMATION PREFACE

It has been my great pleasure to shepherd this book from idea to reality. This required selling the idea within Reclamation — a simple task, really, because of the pride of Reclamation's staff in the work it has done and continues to do. That pride is manifest in visible accomplishments across the seventeen western Reclamation states in the form of dams, powerplants, diversions, canals, and ten million acres of irrigated crops, about one-third of the West's irrigated cropland. That pride is also manifested in less tangible ways such as the economic activity clustered around the water resources made available to about one-third of the West's inhabitants.

William D. Rowley, a professor of history who holds the Grace A. Griffen Chair in the Department of History at the University of Nevada, Reno, was a logical choice for this work. His 1996 book, *Reclaiming the Arid West: The Career of Francis G. Newlands* introduced him to the topic of Reclamation. It was a great pleasure to learn that he was interested in this project.

I have been pleased to stay almost completely out of the intellectual creation of this history of Reclamation. Instead, Reclamation provided two peers to assist Professor Rowley in creation of this book: Dr. Donald J. Pisani, a professor of history and occupant of the Merrick Chair of Western American History at the University of Oklahoma, and Dr. Donald C. Jackson, a professor of history at Lafayette College in Pennsylvania. Both these historians are distinguished authors on topics related directly to the history of Reclamation. We have discussed the general outline of this work and provided editorial suggestions on draft chapters. In addition, Reclamation provided a student assistant to Professor Rowley to assist in his researches – Andrew Gahan, a Ph.D. candidate at the University of Nevada, Reno. While we have supported this work, because of the nature of the intellectual process and historical method, the selection of facts and their interpretation are the author's and do not necessarily represent the official views and policies of the Bureau of Reclamation — and they may not be cited as such.

I, with the able and indispensable assistance of Andrew Gahan and David Muñoz, have spent most of my energy on the project providing editorial assistance to Professor Rowley as we prepared the manuscript for publication.

The illustrations for the book are largely found in Reclamation's records at the Rocky Mountain Branch of the National Archives in Denver. Eric Bittner, Marlene Baker, and Rick Martinez were particularly helpful as I pulled these together.

Last, certainly not least, the design and graphic skills of Charles Brown, the volume designer, Bonnie Gehringer, Bill White, and Cindy Gray in Reclamation's Denver Office, Visual Presentations Group (D-8012), have been indispensable to this project. They have been called upon to scan hundreds of images and clean them up, redraw crude maps, locate and correct electronic files as old as six years, correct hundreds of textual and other errors, and generally push this project to completion. The assistance they have provided makes it possible for me to look forward to working with them on volume 2 of Professor Rowley's history of Reclamation.

Reclamation has been extremely supportive of this project in terms of funding and staffing. I wish to acknowledge that support at many levels and in many different forms, including review of the manuscript for factual error as it was prepared for publication. In particular, Commissioner John W. Keys III; Bob Johnson, regional director in the Lower Colorado Region; and Roseann Gonzales, the director of the Office of Program and Policy Services, have been very supportive.

Professor Rowley has highlighted numerous topics in Reclamation's history that are often overlooked or ignored. As a result he has made significant contributions to our understanding of the bureau. I look forward to volume two. It will cover the era of Reclamation's history after World War II, an era generally less visited by historians, and I expect it to shed new light on Reclamation's history.

Brit Allan Storey
Senior Historian
Bureau of Reclamation

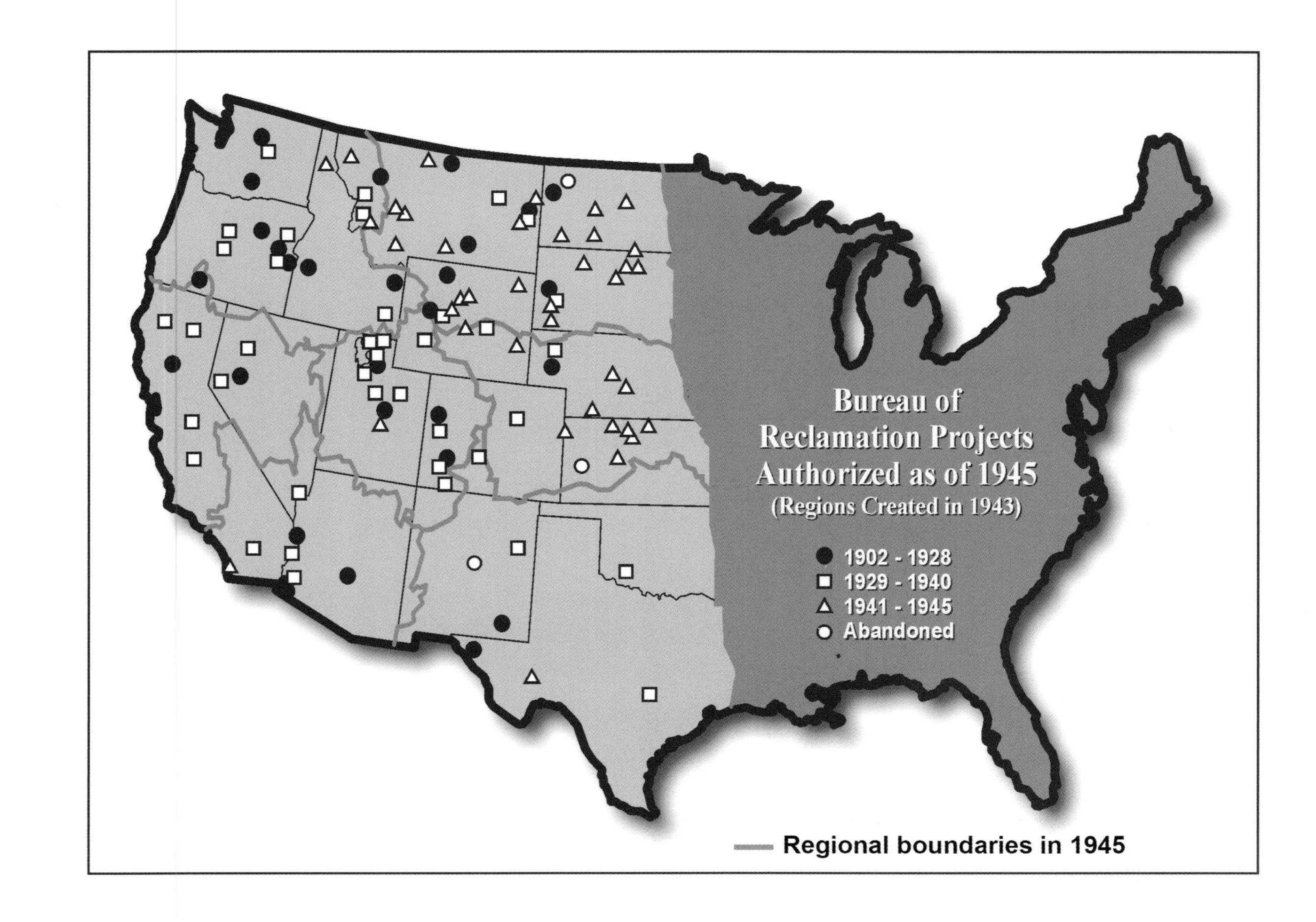
Bureau of
Reclamation Projects
Authorized as of 1945
(Regions Created in 1943)
1902 - 1928
1929 - 1940
1941 - 1945
Abandoned
Regional boundaries in 1945

CHAPTER 1:

ACHIEVEMENTS AND ACHIEVERS

Desert and rugged mountains mark much of the American West where nearly a hundred million people live with all the trappings of modern civilization – cities with skyscrapers, suburban homes with swimming pools, and farms with impressive production figures. Yet it is "a land of little rain," where deserts reach out "to argue with the sea." Population numbers and natural resource demands, especially for water, defy the environmental limitations of the region. Nineteenth-century western boosters declared the land would one day teem with millions. Foremost in a long line of optimistic, pro-growth westerners, William E. Smythe believed the West's water problems could be surmounted through the building of water storage and delivery systems. Filled with hyperbole and "hot air," his major book, *The Conquest of Arid America,* asserted, in a 1905 edition, "There is room for one hundred million people in the States and Territories between the Missouri River and the Pacific Ocean."[1]

While Smythe had the numbers right, his vision of an irrigated Eden missed the mark. Today's western population lives mostly in sprawling urban, industrial-commercial centers amidst a desert and semi-desert environment, defying earlier visions of a rural society of small farmers. Regardless of what patterns population growth assumed, the demand for water and its rational use became inextricably tied to the future of the West. The miracles of water made it all possible. Native

1.1. William E. Smythe was an early advocate of reclamation by the federal government.

Americans of the Southwest adapted to the arid climate by irrigating maize, pepper, squash, and bean crops. Early Mormon settlers in the Valley of the Salt Lake discovered that water must be brought to the land even before a plow could break its hard crust. By the 1850s these hardy religious pioneers at the base of the Wasatch mountains had developed a complicated system of small storage and diversion dams in the mountains that fed irrigation ditches in the valley. They demonstrated that with some adjustments nineteenth-century American agriculture was possible in the desert and mountain West. Of course, it required the construction of reservoirs, main canals, and ditch diversions along with community cooperation and effective water laws. Just as water resource development made possible ancient Native American desert communities and mid-nineteenth-century Church of Jesus Christ of Latter Day Saints (Mormon) agriculture, it became the key to modernizing and integrating the western arid region into the nation in the twentieth century.[2]

1.2. The prehistoric Hohokam culture, in the area of the Salt River Project, constructed very large main canals, as shown by this 1907 photograph of a horseback rider in a prehistoric canal near Mesa.

Water development meant far more than irrigation. In the twentieth century, the Bureau of Reclamation responded to the call for multipurpose water development.[‡] Its vast network of irrigation projects, dams, reservoirs, canals, and aqueducts supplied rural and urban water users, and, most of all, supplied an industrializing society the many-faceted resource of hydroelectric power. In the hundred years from the enactment of the national Reclamation Act in 1902 to the bureau's Centennial celebration in June 2002,

‡ A note on terminology. From 1902 to 1923 the U.S. Reclamation Service implemented the reclamation program in the West. In 1923 the name was changed to the Bureau of Reclamation. The term *Reclamation Service* refers to the period 1902 to 1923, but the term *Reclamation* is used interchangeably to refer either to U.S. Reclamation Service or the Bureau of Reclamation.

dams meant water for the West and power to turn the wheels of industry. Commissioner John W. Keys's keynote address at the Centennial gathering on the Colorado River before the massive structure of Hoover Dam declared that, "By the early 1940s, Reclamation had expanded into true multiple-use projects. Look at the great cities out here in the West – Boise, Salt Lake City, Spokane, Phoenix, Las Vegas – all built around Reclamation Projects."[3] First earmarked for agriculture, Bureau of Reclamation water now serves western society in the twenty-first century far beyond the original rural vision. The great metropolises of the western deserts, Phoenix, Los Angeles, Las Vegas, and Tucson, belied the original dream. With leisure time and ample income, urban populations not only enjoyed the immediate utilitarian benefits of water and power, but also the aesthetic and recreational values offered by the new desert lakes the great dams created. From the early years of the century through the 1960s, western dam building provided an infrastructure necessary for the utilization of the West's limited water resources. The results were water security for large populations and hydroelectric power for home and industry.

By the end of the twentieth century, the bureau's much-celebrated accomplishments in water engineering – dams, power, flood control, irrigation, navigation, and recreation – came from the works of a previous generation. They were an inheritance that turned rivers into engines of enterprise. The great dams (Hoover,[‡‡] Grand Coulee, Shasta, Glen Canyon, and San Luis) became

1.3. Grand Coulee Dam in 1946, a few years after completion of construction.

[‡‡] As will be explained, Hoover Dam has been known by various names, including Boulder Dam, and, rarely Boulder Canyon Dam. To avoid confusion, generally it will be referred to as Hoover Dam in this study, even during the period 1933-1946 when it was widely known as Boulder Dam.

national monuments to economic security for a region now fully integrated into the nation. While wild rivers disappeared into a series of lakes altering ecosystems of land and water, the undertakings occurred in a spirit of enterprise: what a previous generation considered the "greater good." Often, the great and proud works of one generation become the subject of critical analyses in the next. And this is perhaps as it should be as one generation gives way to another.

Overshadowing the arguments, the great dams stand astride the major rivers of the West. The water pours forth to agricultural lands and, increasingly, to a "higher economic use" in the growing cities. These services occur while Bureau of Reclamation hydroelectricity helps underpin the power needs of the region. A survey of the first half of the twentieth century points to achievements in dam building, storage facilities, and the production of hydroelectricity. Later in the century, policy changes occurred that reflected altered political goals, and more recently the rise of environmental concerns in a democratic society.

The history of the Bureau of Reclamation offers insight into the organization's transformation along with the changing needs and attitudes of American society during the twentieth century. Commissioner Keys emphasized transitions at the Centennial: "From the beginning, Reclamation has served the values and needs of the American people. And when those values and needs changed, so did we."[4] The early Reclamation Service cast its net as widely as possible when it launched projects in many western territories and states. But it labored under the various mandates and restrictions in the Reclamation Act: i.e., the 160 acre limit on farms in its projects eligible for government water, the prohibition against "Mongolian labor" in building the projects, and the requirement to operate in accord with state water laws. Hindsight lauded many of these undertakings but judged others harshly.

The Salt River Valley: An Early Success

The Salt River Project was one of the five original projects approved by the Department of the Interior in 1903. It ranks as one of Reclamation's most successful undertakings. In 1996, the Salt River Project celebrated the rededication of the Theodore Roosevelt Dam on the Salt River system in the mountains east of the sprawling metropolis of Phoenix, Arizona. The occasion was the completion of extensive renovation work by Reclamation on the dam that provides the key reservoir for a population of almost three million in the greater Phoenix metropolitan area. The dam was strengthened to withstand a probable maximum flood in accordance with the Dam Safety Act of 1984 and raised by seventy-seven feet to a height of 357 feet, providing nearly 300,000 acre feet of additional water storage – a 20 percent increase in storage capacity. Construction also included a 1,080 foot-long $21.3 million, steel-arch bridge across Roosevelt Lake to divert traffic from the narrow one-lane road across the top of the dam. Speeches at the rededication ceremonies praised Reclamation for the successes of its water storage facilities on the Salt River. The dams and reservoirs provided a stable supply of water that underwrote the expansion of agriculture and guaranteed water resources for the growth of central Arizona.[5]

1.4. Renovated Theodore Roosevelt Dam and the new bridge to remove traffic from the top of the dam. 1996.

Upon completion in 1911, Theodore Roosevelt Dam, commonly known as Roosevelt Dam, was the pride of the Reclamation Service and its Director Frederick H. Newell. The highest dam built to that date by the Service, it was the largest stone masonry arch dam in the world at 280 feet above its lowest foundation plus the additional seventy-seven feet added between 1984 and 1996. In addition to storing water for an already existing irrigation community in the valley more than sixty miles to the west, the Reclamation Service, according to Director Newell, designed the dam's massive structure to project an image of permanence and security. Based upon dam designs originating in nineteenth-century France, there was little that was innovative or experimental in its appearance, and purposely so. The huge stone and mortar edifice announced a conservative monument presumably designed to last forever and never fail. Engineers met rampant public fears when assigned to dam rivers throughout the United States. Memories of the over 2,200 lives lost in the Johnstown, Pennsylvania, dam failure and flood in 1889 persisted in the public mind. In response, Newell desired that the new dams of the Reclamation Service inspire public confidence in his bureau's engineering expertise. Roosevelt Dam became an early symbol of strength, endurance, and the wisdom of conservative dam technology that promised a prosperous future to the citizenry of the Salt River Valley. Reclamation historian Donald Pisani and dam historian Donald Jackson noted that, as the first major dam constructed by the federal government prior to the high dam era of the 1920s and 1930s, it was important that the public perceive Roosevelt Dam as permanent. A tried and tested technology was preferable to technological innovation in dam design and construction.[6]

1.5. Theodore Roosevelt Dam during construction using cyclopean blocks of local stone. 1909.

Many factors worked to promote the success of the Reclamation Service in the Salt River Valley, not the least of which was an already-thriving irrigation community on private lands. In 1903, the Salt River Valley Water Users Association contracted with the Reclamation Service to repay construction costs in ten years. Although extended to twenty years and more after 1914, the Association eventually met the obligations. With cost overruns, the Reclamation Service completed Roosevelt Dam by 1911. Fortunately, power revenues helped retire the debt. The experience confirmed the organizational abilities of a local community to contract for the building of a water storage facility and for the equitable and economical distribution of its water and power resources. Most importantly, the availability of water raised property values throughout the valley, a key factor in the growth of central Arizona in terms of both agriculture and urban wealth. "Salt River Valley," in the judgment of the project's major historian, "would become one of the most successful federal reclamation projects ever constructed."[7]

1.6. The importance of water user organizations in the West is shown in this 1914 portrait of the office of the Salt River Valley Water Users Association.

The Salt River Project proved the most successful of the five projects launched by the Reclamation Service in 1903. The others were the Truckee-Carson (Newlands) Project in Nevada, the Milk River Project in Montana, the Sweetwater (North

1.7. Using Fresno scrapers for construction on the L Line Canal on the Newlands Project in 1905.

Platte) Project in Wyoming, and the Gunnison (Uncompahgre) Project in Colorado. Success on the Salt River Project was aided by its favorable climate and private landholdings eager to receive Reclamation water. As early as 1889, local residents and the U.S. Geological Survey had already identified a prime reservoir site just below the mouth of Tonto Creek on the Salt River. Originally the Reclamation Service designated the Salt River Project as "the most feasible" because of an ideal dam site for water storage and an established irrigation community. The valley's 20,000 residents demanded far more water storage than provided by the existing private irrigation companies. A drought beginning in 1898, followed by rampaging floods in 1901, underscored the inadequacy of private efforts. Newell saw an opportunity to do what private capital did not wish to undertake – a theme he often sounded to justify the work of the Reclamation Service. While private companies attempted to build the large dam, they soon backed away from the project. It was also beyond the collective action of the local community. Prominent Arizonians, Benjamin Fowler, Dwight B. Heard, and others knew that Congress contemplated an ambitious program of dam building and water storage in the western states. Local newspapers excitedly reported a campaign by George Maxwell to gain federal support for a program of national reclamation in the arid land states and territories. Representatives from the Salt River Valley, especially Fowler, were in Washington in the spring of 1902 offering advice on a draft Reclamation bill sponsored by Representative Francis G. Newlands of Nevada. And, most promisingly, the bill had the support of the new President, Theodore Roosevelt, who only recently assumed the presidency after the assassination of President William McKinley in September 1901.[8]

1.8. Pathfinder Dam, on the Sweetwater (North Platte) Project, was completed in 1909.

For residents of the Salt River Valley, the passage of the Reclamation Act in June of 1902 came at a critical moment. The new Reclamation Service, established under the Act, approved the building of the long sought after dam on the Salt River and oversaw its construction by private contractors. The Service operated and maintained the dam until the Salt River Valley Water Users' Association assumed these duties in 1917. Some members of Congress saw government aid to reclamation only in terms of the development of the public lands. Not so the leaders of the Salt River Valley. With most of the valley already in private ownership and served by irrigation ditches under the management of private companies, the Salt River Valley was a leading example of the progress made by private irrigation efforts in the West prior to 1900. Critical wording in the

1.9. "Slim" Pickins, a farmer in the Uncompahgre Valley of western Colorado, about 1914 went to the county fair to show off his crops grown with water received through the Gunnison Tunnel.

1.10. Theodore Roosevelt speaking in Phoenix during activities for the dedication of Theodore Roosevelt Dam on the Salt River Project, March 20, 1911.

Reclamation Act, attributed in part to a delegation from the Salt River Valley that was in Washington, D.C., during the critical months before passage, acknowledged that lands under private ownership could be included in government irrigation projects so long as those lands also shouldered the debt burden incurred for the project's construction. The words in Section 4 of the Reclamation Act "and upon lands in private ownership which may be irrigated by the waters of the said irrigation project," ensured private landholders a place in the history of federal reclamation projects and were particularly important to residents of the Salt River Valley.[9]

Named after President Roosevelt, the Theodore Roosevelt Dam proved the key to the expansion of an agricultural and urban-suburban population in the valley. Of all the early reclamation projects, this facility — the dam, hydroelectric powerplant, and canals — served the greatest number of people in both an agricultural and urban setting. The Salt River Project succeeded for four reasons: (1) an established irrigation community familiar with the problems of irrigating the land, (2) an excellent dam site that provided ample reservoir storage, (3) an effective water users' association with knowledge of financing and cognizant of urban growth and the potential for electrical power revenues in the community, and, finally, (4) a federal government that consented to build storage facilities that could serve privately-held as well as public lands.[10]

In the summer of 1909, a future Secretary of Agriculture, Henry A. Wallace, set out from Iowa to report on the conditions of the reclamation projects in the West. In the Salt River Valley he noted that a good number of the residents originally came to the valley as "health seekers." Doctors recommended the hot, dry climate for a variety of ailments, particularly tuberculosis. Wallace was a student at Iowa State Agricultural College and a columnist for his family's newspaper, *Wallaces' Farmer*. He was an astute observer of western irrigation projects, as his reports to the newspaper reflect. His first article announced that he was "On the Trail of the Corn Belt Farmer." He termed the migration of the corn-belt farmers into the western country a "second winning of the west." He was particularly interested in the new irrigation projects, but he also observed dry farming and farmers in the grain country, in fruit growing, and even cattle country.[11]

What Wallace found in the Salt River Valley was not a village of invalids but a thriving agricultural community that awaited completion of the big dam "as massive as the hills." Yes, some had come as refugees with failing health, but others recovered and joined those with sound bodies to hold and develop irrigation farms. Irrigation farming involved what Wallace called, "puttering work" — work that took time, patience, and persistence and could be handled by a man with a small number of acres. Not all were committed to small homesteads because it was generally realized that land stood on the verge of enormous increase in price once more water was available. As Wallace noted, "The value of the land here is going to be measured by the reading of the gauge at the Roosevelt Dam." The "puttering work," as Wallace termed it, raised crops of oranges, dates, sugar beets for a local sugar factory, and 4,000 acres of barley. He reported that farmers often cut four crops of alfalfa a year. In the warm climate, ostriches could be raised for their feathers. On dairy farms, the Holstein breed predominated because of its great milk yield.[12]

1.11. Irrigating an orange grove near Phoenix in March of 1908.

Not only was the great dam to provide irrigation water, but it offered "tremendous possibilities" in the development of hydroelectric power. The dam's hydroelectricity was a welcome dividend to many aspects of the project. Hydroelectric production did not detract from the quantity of irrigation water; low cost electricity meant

1.12. In 1906 Walter Lubken photographed this huge crop on a date palm near Mesa, Arizona. Salt River Project.

1.13. Ostriches on the Salt River Project, 1908

1.14. An apiary on the Salt River Project in 1914.

conveniences for home and farm, and revenues from the sale of power could be applied toward retiring the debt for the project's construction. In fact, power from the dam would enable residents to tap more underground wells with electrical motors, pumping water from the 200-foot level, serving to further expand irrigation acreage. Wallace could see only a bright future for the valley: "The available water supply of this valley ought to make it the Egypt of the Western Hemisphere." It was an extraordinary statement from an Iowan who represented interests not entirely friendly to the development of western irrigation because of the competition it posed for Midwestern agriculture. Wallace took satisfaction that the valley's future rested with specialized crops that would not compete with the basic corn-hog agriculture of Iowa. One of the favorite responses of Iowans to the question of how they liked life in Arizona was: "no more winters for me."[13]

California: From the Colorado River to the Central Valley Project

As with Arizona, the Bureau of Reclamation played a major role in promoting the growth and prosperity of other western states. California stands out. But Reclamation's success in California came later than in the Salt River Valley and then only by launching gigantic projects to serve a state with the largest agricultural (and as it turned out, urban) potential of any western state. By the end of the twentieth century, California developed the most comprehensive and complicated hydraulic societies in the American West. Its geography, population concentrations, diverse climates, and rich agricultural potential demanded the storage and transportation of water over great distances. Despite persistent demands on water resources in the Golden State, the Bureau of Reclamation (the name changed from Reclamation Service in 1923) arrived late. Although Reclamation acquiesced and eventually endorsed the first major water storage and transportation project — the city of Los Angeles's scheme to tap the water resources of distant Owens Valley — it was not a construction partner. In fact, it abandoned its plans in 1906-07 for an irrigation project in the Owens Valley in favor of sending the water to the Los Angeles Basin. By 1917, the Reclamation Service irrigated only 100,000 acres in California. Still, its limited participation in California was not for lack of trying, especially in the San Joaquin Valley. In spite of all of its hit-and-miss efforts in California, the Reclamation Service planted in that state three important ideas about its work and its potential offerings to the West: (1) farmers need not bear the entire cost of irrigation, (2) electrical power generated by storage dams could be used to subsidize agricultural water delivery, and, finally, (3) the bureau saw the prospects of serving new urban constituencies with water and power.[14]

Reclamation's future achievements in California rested upon launching the Boulder Canyon Project (Hoover Dam) and its invited participation in the state's Central Valley Project. Both were essential for the development of California's impressive waterscape that blended nature and human ingenuity — although some inevitably argued that nature's needs suffered at the hands of that ingenuity. As early as 1902, U.S. Geological Survey engineers, many of whom soon became members of the Reclamation Service, explored dam sites on the lower Colorado River. A future Director of Reclamation, Arthur Powell Davis, was among them. He concluded early on that a larger storage facility would be necessary for the lower Colorado. The construction of a series of high dams on the lower Colorado was a pet project of Davis throughout his career with Reclamation. It was a goal that neither he nor Reclamation would be able to realize during his leadership tenure from 1915-1923.[15] All the elements necessary to build the Boulder Canyon Project, that is to say Hoover Dam and the All-American Canal to the Imperial Valley, would not come together until late in the 1920s. Suffice

1.15. An Imperial Valley strawberry field, about 1920.

it to note at this point that the demands on the part of Los Angeles for water and power, and the immediate irrigation needs of the Imperial Valley, dominated the arguments to tame the Colorado River with high dams, storage facilities, and hydroelectric turbines.

A series of catastrophes suffered by private irrigation interests in the Imperial Valley kept alive Reclamation plans to dam and control the lower Colorado. By 1901, private irrigation interests in California diverted water from the river into the Colorado Desert, which developers for commercial reasons immediately renamed the Imperial Valley to make it more appealing to buyers. But violent floods eroded the headworks on diversion ditches from the Colorado River so extensively that the entire flow of the river broke into the Imperial Valley, causing immense damage and hardship. One major problem with the diversion canals from the Colorado was that they passed through Mexican territory for about fifty miles. In addition to problems at the source of its first diversion canal, the irrigation company opened a second diversion from the river in Mexico in 1904, partly to foil plans of the Reclamation Service to take control of the project under the argument that the water was diverted from a federal river.

Imperial Valley interests demanded what they called an All-American Canal to avoid sharing much of the water with an expanding Mexican farming community south of the border. Supporters of the All-American Canal believed that the Reclamation Service should support this effort, but, to their surprise, the new Director of the Reclamation Service, Arthur Powell Davis, balked. Consistent with his long held opinions about the development of Colorado River resources, he declared there could be no effective canal from the Colorado to the Imperial Valley unless high dams upstream brought the river under control. Davis's 1922 report on the Colorado River officially announced the key role Reclamation would play in utilizing the resources of the river. It projected a major dam on the river and a massive water reservoir to control the river for the multiple purposes of irrigation, water supply to urban areas, and the generation of hydroelectric power.[16] Both Joseph E. Stevens, in his book, *Hoover Dam: An American Adventure* (1988), and Norris Hundley, in his work, *The Great Thirst: Californians and Water, 1770s – 1990s* (1992), note that a major dam on the Colorado was "the centerpiece" to irrigation security in the Imperial Valley, flood control, new hydroelectric development, and an aqueduct to southern California population centers. Both credit Davis with a comprehensive vision that laid the foundation for the Boulder Canyon Project and eventually the spectacular accomplishments in the building of Hoover/Boulder Dam — "the primal dam" in the words of another author.[17]

While the Imperial Valley demanded security from floods and assurances that Mexican interests would not lay claim to more water, the city of Los Angeles exhibited a growing appetite for water and power. Los Angeles's Metropolitan Water District of Southern California, as well as the private power interests of California Edison, showed mounting interest in damming the lower Colorado. In comparison to the growth of other far western states, California's growth was phenomenal, especially southern California. States of the upper Colorado River Basin feared that California interests might soon dam the lower Colorado and assert a prior right claim for the utilization of the river's waters under western water law's concept of prior appropriation. Six other states shared the drainage system of the Colorado. While Reclamation was reluctant to commit itself to anything but a major dam on the river, it was clear in many quarters that an interstate agreement must be reached on the distribution of the river's resources (especially water and power) prior to the construction of any federal dam. Secretary of Commerce Herbert Hoover, a Stanford graduate in mining engineering and former head of the Food Production Control Program during World War I, assumed the major role in negotiating an interstate agreement.

In the same year (1922) that Davis reported on the necessity for a large dam on the Colorado, Hoover successfully negotiated the Colorado River Compact with six of the seven states within the Colorado River Basin. Although Arizona dissented, the consent of six of the states guaranteed eventual approval of the compact. States in the upper Colorado drainage agreed to divide the waters of the Colorado at this point before California's influence and population grew even greater. States on the lower Colorado River could not agree on a division. Their allocations were later determined by the Boulder Canyon Project Act and a Supreme Court decision. Losing no time after the appearance of Davis's report and the signing of the Colorado River Compact, Senator Hiram Johnson of California and Congressman Phil Swing, whose district included the Imperial Valley, introduced the Swing-Johnson Bill, or Boulder Canyon Bill, to Congress. This bill faced challenges and delays for six years until passage in 1928. By that time, Arthur Powell Davis was not on hand to oversee Reclamation's greatest undertaking up to this point. He left the Reclamation Service in 1923 just before a "Fact Finder's" investigation reported problems with the slow development of irrigation projects in 1924.

Under the new Director, Elwood Mead, the Boulder Canyon Project served to rejuvenate the Bureau of Reclamation. The onset of the Great Depression and the election of Franklin D. Roosevelt to the presidency in 1932 brought reform programs to combat the Great Depression in the form of the New Deal. Roosevelt's New Deal used the federal government to put America back to work and defeat the paralyzing effects of the Depression. Huge federal projects and programs became standard New Deal policies in the 1930s. The Hoover Dam project symbolized American determination to overcome the grip of the Great Depression as Roosevelt released emergency monies to speed its completion. Field investigations and preliminary design work had been carried on for decades. Preparations for construction began in 1930, and construction proceeded at full speed after Reclamation awarded the contract for the big dam to the newly formed Six Companies in 1931. It was a critical decision to put the building of the dam up for bid and place it in the hands of a private conglomerate. The construction superintendent of the Six Companies was a former Reclamation engineer, Frank Crowe. He put together the winning bid of over $40 million and supervised the construction. After the decision to release funds for construction, Crowe brought the project to completion by 1935, under bid and in less time than contract specifications.[18]

1.16. Frank Crowe is shown here, the middle Reclamation employee, at the ceremony for the placement of the first concrete at Arrowrock Dam, November 11, 1912.

Boulder Dam, as it was then called (officially renamed Hoover Dam by an act of Congress in 1947), was a monument to American engineering skill and might. It proclaimed victory over the rampaging Colorado River and provided the means to divert water for the benefit of southern California irrigation and urban water supplies. Hoover later made it possible for Nevada and Arizona to also divert water. Most importantly, the dam's electrical turbines offered the potential of recovering, by manyfold, the entire cost of the structure. The latter point particularly appealed to an economy-minded

Congress. Successful completion of the big dam stood as concrete testimony to America's determination to overcome the ravages of the Great Depression and reinvigorated the lagging spirit of Reclamation. Even before the onset of the Great Depression, the big dam on the Colorado exerted a powerful attraction over the engineering minds inside the Bureau of Reclamation. In the 1930s, detailed blueprints for this project soon translated into one of the world's largest concrete and steel structures which, in turn, transformed the Bureau of Reclamation. Thereafter, the Bureau of Reclamation increasingly discerned its future as a multipurpose organization that built dams, managed reservoirs, provided power, controlled floods, and delivered both agricultural and urban water.[19]

No more could the Bureau of Reclamation's vision focus exclusively upon the rural agricultural West. The big dam on the Colorado made Reclamation a major player in the transformation of the West from its marginal economic position in the nation into a region with important centers of industry and finance. The undertaking also gave the American construction community a new confidence in its ability to tackle gigantic projects. The "can-do attitude" served the nation well when the challenges of World War II called for huge construction projects ranging from West Coast shipyards to airplane factories to the development of the atomic bomb in the Manhattan Project.

Hoover Dam spelled security for southern California's Imperial Valley, power and water for Los Angeles, and even promised to make Nevada a viable state. During the early construction phases of this federal project in the Nevada and Arizona desert, the California legislature approved the Central Valley Project to bring water from the state's northern mountains to the Central Valley, as far south as Bakersfield and the base of the Tehachapi Mountains. To include southern California, the plan envisioned sending water from the Kern River into the Los Angeles Basin, with Kern River water users receiving replacement water from rivers far to the north, i.e., the Klamath River and/or the Trinity River. Conceived by Robert Bradford Marshall before the 1920s, the plan was multipurpose. As envisioned, it would irrigate 12,000,000 acres in the Central Valley, improve navigation to and in San Francisco Bay, stop intrusion of salt water into rich delta lands of the San Joaquin and Sacramento Rivers, provide flood control, and produce enough electrical power to industrialize northern California.

Marshall began his career with the U.S. Geological Survey (USGS) in 1889, during John Wesley Powell's ill-starred Irrigation Survey of the West. He found himself in California by the 1890s in various capacities as a geographer in charge of topographical work with the USGS and in the administration of national parks. By 1919, he devoted full time to the promotion of his "Marshall Plan." He believed the project should be paid for by state bonds that could be paid off with revenues from hydroelectric and water sales to farms and cities. The price tag, he estimated, was under a billion dollars.[20]

California's legislature moved toward approval of a Central Valley Project after over a decade of wrangling. Sectional politics, a battle between private and public power, fights among irrigation districts, drought, and finally the crises of the Great Depression pushed Californians toward approval of an elaborate water project. The comprehensive water plan passed the state legislature in 1933, but funding was delayed by the Depression and a referendum fight launched by private power interests against it. The referendum to stop the project was narrowly defeated by a vote of 459,712 to 426,109. While originating as a state effort, the law kept the door open to federal aid and even ownership of the project. Congress responded by authorizing the Bureau of Reclamation to begin construction on the Central Valley Project (CVP) in 1935 with an appropriation of $20,000,000. Work on Shasta Dam began in 1937, but World War II delayed its completion. Shasta Dam began producing electricity in 1944, but it was not officially opened until 1950. Reclamation's success in building the CVP gave it a commanding position in the water development of the West's most populous state.[21]

1.17. At a ceremony in 1938 marking the beginning of heavy construction on Shasta Dam, these notables posed for a portrait: left to right are Director Earl Lee Kelly of the California State Department of Public works; Commissioner John C. Page; Secretary of the Interior Harold L. Ickes; California State Engineer Edward Hyatt; Project Construction Engineer Walker (Brig) R. Young of Hoover Dam.

Beyond California: The Grand Coulee of the Pacific Northwest

A thousand miles from the Southwest, in "the far corner" of the nation, as Pacific Northwest writer Stewart Holbrook termed it, Grand Coulee Dam stands across the waters of the Columbia River as another monument to the accomplishments of the Bureau of Reclamation.[22] While not the largest river in the nation, the Columbia carried more water than the Colorado and promised to be the engine to unlock the agricultural and industrial growth potential of the Pacific Northwest. A series of events and personalities on the local and national scene coalesced to draw Reclamation into building what was, at the time, the largest man-made structure on earth. At the outset, many charged that the government was building a white elephant in a far off semi-desert environment. Others celebrated it as the "Eighth Wonder of the World." The story of Grand Coulee Dam reveals the gradual and cautious involvement of the Bureau of Reclamation. It did not rush to build Grand Coulee Dam or to develop the Columbia Basin Project. Circumstances, more than premeditated stratagem, drew Reclamation into the rising momentum of a local campaign for a high dam.

1.18. Grand Coulee Dam in 1948, soon after completion.

The Columbia River enters Washington State from Canada. On the eastern side of the Cascades, it flows through desolate country hundreds of miles from the state's main population centers in western Washington and ninety miles away from the one major city in eastern Washington, Spokane. The few agricultural communities in central Washington had little political power. With population concentrated around Puget Sound, the outpost city of Spokane adopted a chamber of commerce term, "Hub of an Inland Empire," to lessen the image of its isolation.[23] On the Columbia Plateau in central Washington, there was no burgeoning irrigation community with both agricultural and urban needs for water such as existed in the Salt River Valley. A major argument for not considering the Columbia Basin Project, as it came to be called, was the absence of any immediate demand to open the agricultural land of central Washington. Also, the power resources of the river were far beyond the capacity of the then existing market to consume. Arguments for the project, Reclamation officials hinted, seemed to seek their justification too far into the future. Put simply, in the agricultural slump that followed World War I, no irrigation or power project in central Washington could be justified. Some believed that this project would irrigate a million acres and require creation of new farming communities. Could this occur in a glutted agricultural market?

Before the construction of Grand Coulee Dam and its completion in 1941, the Columbia flowed unobstructed, "unused, and wasted" to the sea. To the generation that undertook the reclamation of the American West by the hand of government, or for that matter by private and community enterprise, rivers rushing or moving lazily to the sea represented wasted water and energy. In time, of course, the valuation of rivers and water changed. By the late twentieth century, observers noted that dam building on the Columbia had converted a free-flowing river, a phenomenon of nature, into an "organic machine." In a 1963 preface to an edition of an early twentieth-century book on the magnificence of the Columbia River, the writer anticipated the critiques of dam building on the river: "Today the Columbia is no longer the turbulent rapids-broken stream. It has become a series of dams and lakes backed up behind them. No longer only a waterway, it is now a vast power plant, a subject of political controversy, and an expression of the nation's, as well as the region's, changing economic and social modes." A tamed and harnessed river served the needs of industry and industrialized agriculture. To the citizens of the remote and stark country of eastern Washington at the beginning of the century, a dammed river with a reservoir to supply irrigation water and hydroelectric power presented a welcome path to a prosperous future for their sage-covered plateaus. These visions, while only partially realized east of the Cascade Mountains in Washington, have been more fully achieved in the cities of the Pacific Northwest. In the river itself, the once rich salmon runs have fallen victim to the dams and the progress they brought to the region. [24]

Just as the flow of the river was gigantic, the task of harnessing it was enormous. Political and economic barriers faced early boosters of damming the Columbia. Unlike the Salt River Project, there was no already established irrigation community to agitate for a project; additionally, the remoteness of the region and its small population presented little market for large amounts of electricity to be generated from a big dam. From before the beginning of their official campaign in 1918, advocates of damming the Columbia were lawyers, businessmen, professionals, editors, promoters, and politicians. Few farmers participated. In fact, farm organizations in Washington refused to support the concept of a big dam on the Columbia because they feared it would detract from other, smaller projects desired in other parts of the state.

Backers of irrigation in central Washington first rallied their cause to an area they called "the Big Bend of the Columbia," but then, in 1918, changed the name to the "Columbia Basin Project." These same promoters understood that the project was beyond the resources of local communities, state government, and certainly private enterprise. They also knew that Congress permitted the Reclamation Service to combine powerplants with its dams. The 1906 Town Sites and Power Development Act provided that sale of power from these dams could be applied toward the cost of the projects and the cost of irrigating lands. For Columbia Basin Project visionaries and promoters, power was uppermost in their minds. Its enormous value would make all other things possible for the project. [25]

All things possible or not, in the early decades of its career, the Reclamation Service had no intention of committing itself to the development of the Columbia River Basin. Too many drawbacks existed. On a visit to Spokane in May 1909, Director Newell declared, "It may not be in this generation or the next, but the time will come when an immense irrigation project will be carried through for the reclamation of the Big Bend country."[26] The only two Reclamation projects listed for Washington State in 1909 by the *Reclamation Record*, a monthly periodical of the Reclamation

Service, were the Yakima Project which was described as 66 percent complete, and the Okanogan Project, which was 96 percent complete.[27]

Undeterred, the faithful backers of a Columbia Basin Project pushed on. They included, among others, local lawyers Billy Clapp and James Edward O'Sullivan. The editor of the *Wenatchee Daily World*, Rufus Woods, backed a big dam on the Columbia that required pumping water up on to the Columbia River plateau for irrigation purposes. The Spokane Chamber of Commerce supported a rival "gravity" plan to divert the Pend Oreille River from Idaho into the Spokane River, and, by various aqueducts, to the 2.5 million acres to be irrigated on the plateau. The big dam plan on the Columbia was known as "the pumpers plan" and Spokane's as "the gravity plan." But the differences extended far beyond the method and means by which the water was to be delivered to the land. Deeply ingrained in the rivalry was the contested ground of public versus private power in the 1920s.

Washington became a particularly bitter battle ground on this issue. As the two proposals gained political momentum in the state, each represented the divergent interests of public and private power. Both foresaw the irrigation of land and the development of the agricultural potential of the Columbia River Basin. The Spokane, or gravity, plan had the backing of private power interests such as the Washington State Power Company that feared the high dam on the Columbia would produce an excess of power and devalue the price of power throughout the region. In its view, the high dam threatened the profit margins of private power. Despite the controversy, the opportunity for the development of something big in eastern Washington attracted the attention of local promoters in the 1920s.

Initially, Reclamation kept its distance from the Columbia River Basin proposals. Already facing severe criticism in the early 1920s for opening too many projects, Reclamation's biggest fear was creating a white elephant for itself with a venture on the Columbia. The Colorado River project was a different story, and that is where Reclamation concentrated its energies during the 1920s. Director Arthur Powell Davis did visit the Grand Coulee in 1920 amidst much local fanfare. And while he declared a dam feasible, the director refused to back any plan. Reclamation did not want to position itself in the battle over which plan for irrigation to endorse or become entangled in Washington State's public versus private power struggle. One source wisely summed up Reclamation's actions and attitudes toward the Columbia Basin Plan from 1918 to 1933 saying it "eschewed involvement" while acting interested. Absorbed in Colorado River affairs, Reclamation gave little attention to the Columbia, but it worried about the possibility of the U.S. Army Corps of Engineers (Corps) seizing the project and making it a larger project than the development of the Colorado. Under its "wily" new and seasoned Commissioner, Elwood Mead, Reclamation showed interest but not commitment. The political climate in Congress was not right for the undertaking of a new project beyond the major effort and monies committed to the Boulder Canyon Project.[28]

Although they frequently sent representatives to Washington, D.C., local dam and irrigation proponents for the Columbia got nowhere with Congress or with the Republican Administrations in the 1920s. Neither could the state legislature of Washington afford to bear the cost of a Columbia Basin Project alone, although it did make appropriations for planning and preliminary drilling to test the bedrock for a major dam. Outside of Congress, the White House, and Washington State government, there were appeals to the agencies themselves, the Bureau of Reclamation and the Corps. Both the Spokane Chamber of Commerce and the Columbia Basin League courted them shamelessly. Studies by the Corps eventually endorsed a big dam on the Grand Coulee, much to the disappointment of Spokane's proponents of the gravity plan to bring water from Idaho. But the Hoover Administration, and, especially, Secretary of Agriculture Arthur M. Hyde declared the project impractical in the face of the nation's agricultural surpluses. The advocates of the project were not farmers and Hyde could not "see how the project can possibly constitute a sound opportunity for any prospective settler." Taking its cue from the Hoover Administration, the Corps reversed itself and refused to endorse the Columbia Basin Project. This move, in the opinion of Paul Pitzer, author of a major work on the building of Grand Coulee Dam and the Columbia Basin Project, eventually cost the Corps the project.[29]

The election of Franklin D. Roosevelt to the presidency in the fall of 1932 spelled new policies and bolder implementation of public works construction programs already launched by the Hoover Administration. In his first "One Hundred Days" after inauguration in March 1933, Roosevelt requested from Congress and received a $500,000,000 appropriation to fund emergency projects. In a Portland campaign speech on September 21, 1932, the presidential candidate endorsed the principle of "public power" and said it could be the "yardstick" by which to measure the cost of power in the Pacific Northwest and prevent private power monopolies from overcharging the people. The new administration determined to wage war on the Depression in both the agricultural and industrial sectors. Roosevelt's agricultural advisors suggested several approaches to improving agricultural conditions: convincing farmers to grow less food and fiber, persuading farm families that many must leave their farms for urban settings, and relocating farm families from unproductive to productive lands. The Columbia Basin Project might serve as a destination for the relocation of down-and-out Great Plains farmers. This became a new rationale for the Columbia Basin Project. Building it would not add to the surpluses but serve to relocate some farmers to more productive and promising lands. An overriding consideration was the promise that electrical power would foster industrial development if a big dam arose on the Columbia. Detractors said the lack of an industrial base in the Pacific Northwest should discourage the undertaking, but backers of the big dam took the attitude that "if you build it, they will come."

Finally, in mid-October 1933, following the lead of the new administration, Elwood Mead declared, "I am convinced that Grand Coulee will have to be built as a federal reclamation project along the lines of Boulder Canyon." This meant that the State of Washington would relinquish to the federal government the task that it had already begun on a very preliminary basis. The Secretary of the Interior, Harold Ickes, on November 1, 1933, confirmed that it would be Public Works Project #9. Appropriations to begin work immediately flowed from emergency funds. Set in motion by presidential fiat, Congress eventually appropriated money on a pay-as-you-go basis, unlike the Boulder Canyon Project Act of 1928 that funded that entire project. The Great Depression, however, was the "step-parent" of Grand Coulee Dam and the Columbia Basin Project. It grew out of the emergency measures invoked by the New Deal to fight the Depression and achieve full employment.[30]

In 1934, Reclamation made the decision in favor of the high dam at Grand Coulee. Completed in late 1941 during World War II, the dam and its turbines turned out electricity for Pacific Northwest industries that included aluminum production, airplane factories, and expanded shipyards. The war, however, postponed irrigation on the project until 1948. The project was testimony to Roosevelt's belief in planning and controlled land use, and the belief that public power would keep power prices fair and low. True to this vision, the Pacific Northwest enjoyed low power prices for forty years, attracting industry and growth to the region. By 1990, estimates suggested that Grand Coulee's power alone brought 40 billion dollars in revenue, paying back the dam costs many times over. Like Hoover Dam, Grand Coulee Dam was a monument to what the American people could build even in the depths of the Great Depression. When faith in the future was failing in the 1930s, the very breadth and height of structures taming the unpredictable forces of rivers brought confidence. Critical voices faded as Grand Coulee's power bolstered the war effort. Only after the war did Reclamation confront the difficulties of developing the irrigation aspects of the project. As it turned out, the Columbia Basin Project was not built before its time, but rather in the nick of time for the power demands of war industries. By mid-century,

1.19. Grand Coulee Dam at night emphasizes the massive amount of electricity produced at the historic left and right powerplants.

the Pacific Northwest could only praise this technical achievement and inheritance from the depression decade and the era of big dams. Literature and documentary newsreel productions celebrating Grand Coulee Dam became standard fare in the school rooms of the Pacific Northwest.[31]

The Colorado-Big Thompson Project (C-BT)

From the Pacific Northwest to the Continental Divide, big projects gained approval of a New Deal Administration committed to putting America back to work. In Colorado, the most populous state to span the Rockies,

1.20. The Colorado-Big Thompson Project provides a supplemental water supply to farms on the East Slope of Colorado.

the Colorado-Big Thompson Project attracted the interest of the Interior Department and the Bureau of Reclamation. It was a project to transport water from the comparatively water-rich western slope of the Rockies to the population and agricultural centers in the drainage of the South Platte River over the mountains and to the east. As in the Columbia Basin, local advocates of water development saw in the crisis of the Great Depression an opportunity to press forward with their goals of federal aid.

In principle, the proposal violated advice against inter-basin water transference by nineteenth-century American scientists, John Wesley Powell and George Perkins Marsh. Some states even had laws against inter-basin water transfers, in part, because of the Owens Valley example in California. Notwithstanding these cautionary echoes from the past, Reclamation expressed interest. As early as 1889, the Colorado legislature ordered a survey of a canal from Grand Lake, on the west side of the Continental Divide, to South Boulder Creek on the east side. Some farmers already engaged in small scale diversions of water from the west side to the eastern slope. In 1905, the Reclamation Service commissioned engineer Gerald H. Matthes to survey a route connecting Grand Lake and the upper waters of the Colorado River (at this time named the Grand River from its source to the junction with the Green River) to points east in the watershed of the Big Thompson River and onto the South Platte River with an eye to expanding irrigated agriculture in the vicinity of Greeley, Colorado.[32]

The engineering challenges of transmontane water transfer in Colorado paled before the complications of local politics that pitted the West Slope of the Rockies against the East Slope. The project faced the opposition of Colorado's veteran West Slope congressman, Edward T. Taylor. Taylor claimed responsibility for changing the name of the Grand River to the Colorado to underscore its origins in Colorado and that the river was a part of the larger Colorado that drained vast areas of the western United States. The name change alone drew attention to the far-reaching consequences of diverting water from the normal flow of this great river of the West. Having served in Congress since 1909, Taylor determined to hold on to as much of the headwaters of the river as possible for his West Slope constituency.

As the voice of western farm and ranch interests, Taylor early opposed both federal interference with and regulation of western resource uses. He was a defender of the free and open range and a fierce critic of the Forest Service's grazing regulations on national forest lands. He moved to introduce leasing programs by the Department of the Interior for the grazing lands outside the national forests.[33] Yet, he did an about face on the issue at the beginning of the New Deal. As sponsor of the Taylor Grazing Act in 1934, he suddenly saw range deterioration and the dangers of a "dust bowl." He sought range regulation, but only under close control of local ranch interests. Proponents of schemes to transfer water from the western slope to the eastern slope in Colorado hoped the congressman would have a similar change of heart and allow water projects to go forward. Only Taylor's failing health began to open the doors for the project. Still, the congressman's strong voice in Colorado politics forced advocates of the Colorado-Big Thompson Project (C-BT) to battle him locally and nationally until their efforts finally succeeded in 1937.

1.21. Congressman Edward T. Taylor of Glenwood Springs, Colorado.

The Colorado-Big Thompson Project does not have a spectacular showpiece structure in the dimensions of either the Hoover or Grand Coulee Dams. The size of the dams was modest in comparison, but this transmontane diversion was complex and required a significant number of features.

They included: four reservoirs and dams, a powerplant, and pumping plant on the west side of the mountains; 13.4 miles of tunnel through the Continental Divide to carry water from Grand Lake and its associated reservoirs to the Big Thompson River on the east side; and a large system of reservoirs, powerplants, and canals to carry water down into the South Platte River where it could be delivered for irrigation and municipal supply. Reclamation stood willing to build the project under the leadership of Commissioner Mead, who had a long standing familiarity with Colorado's water needs. He had been a faculty member at Colorado State Agricultural College in the 1880s and had served as Assistant State Engineer in Colorado.

Formation of the Northern Colorado Water Conservancy District (NCWCD) provided Reclamation with a local water users' association to contract for the repayment of the project upon completion under an interest-free loan over forty years. The agreement would not have occurred had farmers been bound by the 160 acre limitation in the 1902 Reclamation Act. While the bureau often facilitated avoidance of this rule, i.e., on the Salt River Project, in 1938 Congress approved bills introduced by Colorado congressmen to exempt the C-BT Project from the "excess land provisions." Congress now declared the rule not applicable, "to certain lands that will receive supplemental water supply from the Colorado-Big Thompson Project." The legislation was crucial for the approval of the contract with the bureau on the part of landowners with the NCWCD. A historian of the district, Dan Tyler, writes: "Exaggerating the importance of this act is difficult. So many of the productive lands [owners' lands] in the District were in excess of 160 acres that if the Reclamation Act had been strictly adhered to, the Repayment Contract might never have been passed by the voters. The C-BT would not have been built." The move demonstrated a pattern of flexibility by Congress and the Bureau of Reclamation on the controversial "excess lands" provision of the original reclamation law.[34]

Still, Reclamation had to bide its time while the difficult political negotiations occurred within Colorado and the Interior Department. The transmountain tunnel's projected route beneath Rocky Mountain National Park raised criticisms within the Park Service, also located in the Department of the Interior. Secretary of the Interior Ickes noted that the tunnel would be constructed beneath the lands of Rocky Mountain National Park (both ends of the tunnel, however, were outside of the park). He hesitated to violate park boundaries when park officials raised objections. While encroachments upon national parks for water supplies and hydroelectricity works were not new, the incursions often produced controversy. The Rocky Mountain National Park Act of 1915 noted that the park "should not interfere in any manner with the development of water resources for irrigation purposes."[35] Still, when confronted with the tunnel proposal, park officials objected. They found a sympathetic ear in the Secretary of the Interior. Ultimately, Ickes put aside his doubts and yielded to the Bureau of Reclamation and the desire of President Roosevelt and Congress to push the project forward. He declared: "I cannot follow my own will in the matter before us. I have to follow the law and I tell you very frankly that between the Bureau of Reclamation and the Park Service, I am for the Parks, but I am sworn to obey the … " Ickes left the sentence unfinished. What he most likely meant was that he was sworn to obey not only the law but "the powers that be" in the administration and Congress.[36]

From the western side of the state, Congressman Taylor adamantly advocated a plan that would require "the acre-foot-for-acre-foot principle." This meant the construction of enough reservoir space on the West Slope to store water to replace the amount of water transferred to the East Slope of the Rocky Mountains. Reclamation and the East Slope objected to this principle, believing it too costly and far beyond what they deemed necessary to protect

1.22. Green Mountain Dam on the Colorado-Big Thompson Project. In order to obtain approval of the project, supporters had to agree to build Green Mountain Dam first in order to protect the water rights of the West Slope of Colorado.

the West Slope's water supply. Ultimately a compromise modified Taylor's approach in a way that was believed to provide the greatest use of West Slope waters without injury to its water resources. The compromise required that a survey be made of any major transmountain project to determine the effect of the project upon existing and future West Slope development. Based on this survey, "replacement" storage would be provided in an amount sufficient so that neither existing Western Slope rights, nor probable future West Slope development, would be adversely affected by the proposed transmountain diversion. Reclamation completed the required survey in early 1937. The East Slope/West Slope agreement required the construction and operation of Green Mountain Reservoir, a reservoir to provide "replacement capacity" on the West Slope. Work on the entire Colorado-Big Thompson Project began on the West Slope at Green Mountain Dam in November of 1938. Construction of Alva B. Adams tunnel did not begin until June of 1940. It was touted as the world's longest tunnel for irrigation purposes (13.4 miles). Labor disputes in the spring and summer of 1939 at the Green Mountain Dam construction site prompted Colorado's Governor Ralph Carr to call out the National Guard. The presence of troops defeated labor's demands against the company contracted to build the dam. As was the case with labor disputes during the construction of Hoover Dam, Reclamation stayed away from the conflict saying that the issues were between the union and the contractor.[37]

1.23. The Alva B. Adams Tunnel on the Colorado-Big Thompson Project carried water 13.4 miles under the Continental Divide and Rocky Mountain National Park. This photo shows the tunnel during construction in 1942.

After overcoming all these difficulties, the project faced further delay because of World War II. Arguments to suspend the project contended that it was not essential to the war effort. Proponents countered that the project would bring vast new acreage into production for wartime food and fiber. By delivering water to the South Platte Valley and Northern Colorado Water Conservancy District, important increases in sugar beet production could be expected. Proponents won the day, and construction on the Alva B. Adams Tunnel was completed on June 10, 1944. A "Boring Through" ceremony was held as the final 25 feet connecting the East and West Slope of the tunnel were blasted. Lining the tunnel with concrete took another three years because of labor shortages. The tunnel officially opened and delivered water on June 23, 1947, at another ceremony, fourteen years after approval of the first project funds by Congress in 1933.

Reclamation had completed a complicated water project in the upper Colorado River Basin that moved water across the Continental Divide. The transmountain water transfer carried enormous implications for the future growth of urban and rural Colorado on the east side of the mountains. In spite of many barriers presented by local interests and politics, the Colorado-Big Thompson Project ensured water resources for the growing population of the eastern slope in Colorado. This was done at considerable cost. It was agreed to compensate the basin of origin with a reservoir in exchange for diversions across the mountains, and a system of reservoirs and diversion structures were required to store West Slope water for diversion through the mountains. In 1957, the first full year of water delivery, 720,000 acres of irrigated land yielded $68.7 million in crops. By 1990, crop land was reduced to 630,000 acres as urban centers purchased water from agriculture. Still, the value of agricultural crops equaled $331 million. The transfer of water and land from agriculture to urban growth needs represented a significant change in water use. Some critics complained that the C-BT water storage and delivery system was never intended to support the growth of urban populations and that it amounted to a subsidization of growth in the Front Range of Colorado. Advocates of change welcomed the transfers of water to a higher economic use that a free market economy heartily endorsed. Other critics noted that more water was lost to evaporation from upper-basin reservoirs than was diverted out of the basin. [38]

By the 1950s, Reclamation had made its mark upon the West with the achievement of major projects: Boulder Canyon, the California Central Valley, the Columbia Basin, and the Colorado-Big Thompson, to name only four. They represented achievements in engineering, political will, and administrative leadership savvy. Above all, these projects laid the foundation for the major role Reclamation played in utilizing water resources in the post-World War II years for a dynamic western society — populous, economically vibrant, and capable of sustaining the process of modernization. In most venues, the gigantic water and power projects drew the admiration of the nation, if not the world.

The Achievers:

The Newell Era

Frederick Newell guided the early Reclamation Service as chief engineer (1902-1907)[39] and director until 1914. He earned a degree in engineering from the Massachusetts Institute of Technology in 1888 and began his career with the USGS under John Wesley Powell's leadership. Shortly after passage of the Reclamation Act in June 1902, he was appointed chief engineer and eventually director in 1907. Newell accepted the mandate of the Reclamation Act to build dams, water storage reservoirs, and water conveyance systems (canals and tunnels) to serve communities of small farmers. Reclamation became the last enlistee in a land policy designed to serve the social goals of settling families on the land. To this end, Congress earlier passed the Homestead Act of 1862, allowing 160 acres of free land to heads of families who took up land claims and resided on the land. But in much of the Far West, farms of this size without guaranteed water meant failure.

1.24. Frederick Haynes Newell, Director of the U.S. Reclamation Service, 1907-1914, Chief Engineer 1902-1907.

For Newell, the choices were clear: either build

water and land reclamation projects for the benefit of the people or give the arid lands over to "great stock ranches, controlled by nonresidents, and furnishing employment to only a few nomadic herders." Obviously the superior choice, the choice of "highest citizenship" as he put it, was to provide homes for independent farmers. In turn, this would prevent the unoccupied but reclaimable portions of the public domain from becoming a "speculative commodity" in the hands of land monopolists who would discourage small farm enterprise in the West. Government reclamation tried to make it possible for the small farmer to make a go of it in the arid lands of the West. In Newell's view it was legitimate to use land and water policy to achieve social goals. He wrote: "The development of water for irrigation is a matter of concern to all citizens of the United States, since they are the great landowners [of the public domain], and, as such, are, or should be, interested to see that their lands are put to the best uses."[40]

Congress pledged the Reclamation Act to "best uses" when it wrote into the Act the provision that government water should not serve more than 160 acres in one ownership. Yet, this did not prevent Newell from striking agreements with the Salt River Project that exempted it from the "excess lands" (lands in excess of 160 acres per individual owner) provisions of reclamation law. He also accepted the 1906 amendment to reclamation law that water storage facilities and their dams for reclamation projects could sell water and power to urban areas so long as the profits were used to develop the project and pay back project debt to the government.[41]

While Newell made compromises, his major focus was upon bringing water to the land. The Reclamation Service faced the tasks of surveying and choosing the projects, constructing hydraulic works, dealing with problems of settlers once on the land, settling complicated water rights issues, and negotiating repayment contracts. Newell became dismayed and distressed when settlers balked at their ten-year repayment schedules to the government for building the irrigation works required under the Reclamation Act. Much to his disappointment, many held lands not as farmers but as speculators – the very outcome he hoped government reclamation could avoid. The greatest problems on the reclamation projects, in his view, were "connected with human nature and its limitations" and not with the social and economic goals of the Reclamation Act. He explained away the high rate of farm failures on early reclamation projects as a natural weeding out of individuals who lacked the wherewithal to succeed on the land.[42]

That aside, the entire enterprise of government reclamation frustrated and disappointed Newell. He received public and congressional criticism because costs of building the projects were often woefully underestimated. In addition, planning often ignored fundamental problems of drainage and poor soils. Water users protested he was an aloof college-educated easterner who held western farmers in disdain. When he refused their demands for easier repayment terms, this confirmed their opinions. Newell concluded that the problems of "human nature" were far greater than the engineering problems of western reclamation. In his opinion, nineteenth-century western pioneers met the challenges of their generation with great courage and self-reliance, but the settlers of the following generation failed to live up to the opportunities offered them under the terms of the Reclamation Act. In the end, Newell left the Reclamation Service disappointed, while water users and Congress laid much blame on him for the failure of the reclamation projects to prosper. In the field of structural engineering, however, the Reclamation Service under Director Newell achieved preeminence as one of the world's great dam building organizations. At the same time, it pioneered the development of hydroelectricity in conjunction with the management of water for irrigation and urban uses.

Commissioner Elwood Mead

After Newell's departure in 1914 and under Arthur Powell Davis's leadership for most of the next decade, the Reclamation Service endured criticisms and congressional investigations. Officially renamed the Bureau of Reclamation in 1923, Reclamation made headway against its troubles after 1924 when Commissioner Elwood Mead assumed its leadership. Mead's opportunity to revive the good fortunes of Reclamation came with the Boulder Canyon Project. By the 1930s, Americans lived under the hardships of the Great Depression, but the Bureau of Reclamation, under Mead's leadership, grasped opportunities in these bad times. Opportunities came in the

1.25. Elwood Mead, Commissioner, Bureau of Reclamation, 1924-1936.

form of large projects with higher dams than ever before, made possible by new markets for hydroelectricity, namely Los Angeles. The Boulder Canyon Project early demonstrated how large projects put people to work in an economy desperate for jobs. Not only did the high dams provide an infrastructure for power and water delivery, some of the dams became monuments to the nation's productive forces, pathways to recovery, and the power of a nation capable of meeting the challenges of World War II. Mead died while still serving as Commissioner in 1936. His long career witnessed the transition of the Bureau of Reclamation into an era when it built truly massive dams (Hoover, Grand Coulee, and Shasta). They served multiple purposes, not the least of which was to help put many Americans back to work during the Great Depression.

Dominy's Drive and Energy

Floyd Dominy's arrival as Commissioner of Reclamation in the post-World War II period brought the high dam period to its zenith. His leadership of Reclamation from 1959 to 1969 was nothing less than "charismatic." Dominy began his career with Reclamation in 1946 and ended with his resignation during the first year of Richard Nixon's presidency. He came from central Nebraska, in what he called "the dry land prairie country," where his grandfather homesteaded in 1875. In 1932, in the midst of the Depression, he graduated with a B.A from the University of Wyoming and eventually found employment with New Deal agricultural relief work in Wyoming that took him to Washington, D.C., by 1939. He saw service with the U.S. Navy during World War II and returned to Washington where he went to work for the Bureau of Reclamation in the belief that new projects were on its agenda after the interruptions of the war. The choice proved a good one,

1.26. Floyd E. Dominy, Commissioner, Bureau of Reclamation, 1959-1969.

not only for his career but for the expansive growth of Reclamation in the postwar decades.[43]

Although Dominy took engineering courses in college, he did not obtain what was thought to be the prerequisite engineering degree to head Reclamation. While this was surprising in an organization whose tasks generally demanded engineering skills, it was not unprecedented. Michael Straus had a distinguished record of government service with a journalistic background and finally as Commissioner of Reclamation (1945-1953). Dominy's aggressive and can-do personality compensated for any lack of professional training leading to an engineering degree. In these years, it became increasingly important for the Bureau of Reclamation to represent itself before congressional committees to gain project approvals and increased budgets. Dominy was phenomenally successful. His appearances before congressional committees exuded spontaneity, authenticity, and a bulldog attitude that appealed to many congressmen. Consequently, Reclamation enjoyed one of its flushest periods in terms of financial support from Congress and a high tide of dam building through the 1960s. The dams and the power they produced became the major focus of the Dominy era in this period of Reclamation's history. He represented the final transition from the Newell era, committed to home building in rural settings, to a new era that embraced the power of the kilowatt to serve a fast growing urban West.

Both in symbol and reality, big dams helped pave the road to recovery from the Depression, contributed power for the subsequent war effort, and propelled the nation forward into the "Guns and Butter" economy of the Cold War. The wisdom of building the dams was also confirmed by demographic changes within the United States — following World War II, there was a large population shift to the western states. By the end of the 1940s, the dams were a celebrated part of the landscape despite earlier opposition by private power interests. They, too, had been beneficiaries of the big dams when they were invited to purchase public power and distribute it. The election of Republican President Dwight Eisenhower in 1952 gave new courage to the ambitions of private power and stiffened its opposition to public power projects. Dominy took little notice as he moved aggressively to work closely with Congress after his appointment as Commissioner by President Eisenhower in 1959. He played to the interests of members of Congress and their western constituencies who welcomed government projects and money in the form of dams, hydroelectricity, and reservoirs that provided recreational facilities. Dominy compromised with private power, sharing the tasks of transmitting and marketing power. It worked for Dominy and for Reclamation.

1.27. On the campaign trail, Dwight D. Eisenhower visited Hoover Dam on June 22, 1952.

1.28. Glen Canyon Dam under construction on April 9, 1963.

The crowning achievement of the Dominy era was Glen Canyon Dam and the 260 mile long Lake Powell reservoir behind it on the Colorado River. Authorized in 1956 as a part of the Colorado River Storage Project, the dam was completed in 1966. Congress approved it only after a long fight in which the President of the Sierra Club, David R. Brower, emerged as a major opponent of other dams on the Colorado River, most notably the Echo Park Dam proposal that would have flooded part of Dinosaur National Monument. The successful campaign against Echo Park Dam resulted in an unstated but *de facto* compromise agreement to build Glen Canyon Dam. In the process, however, Commissioner Dominy came to represent Satan personified to the nascent environmental community.[44]

The conservationist battle against dams on the Colorado might have blocked Glen Canyon had it not been for the close relationship Dominy enjoyed with key western Congressmen. Yet the building of Glen Canyon represented a turning point. The Sierra Club, under the skilled leadership of Brower, tapped into a growing sentiment among the public to preserve areas of natural beauty.[45] The wave of the future seemed to favor wild rivers

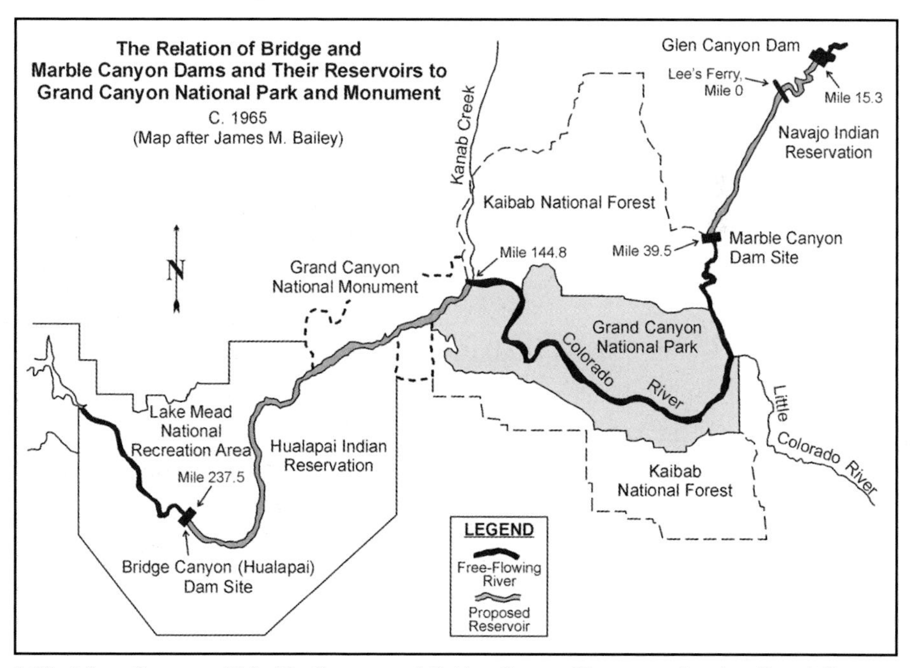

1.29. Map of proposed Marble Canyon and Bridge Canyon Dams as related to Grand Canyon National Park and Grand Canyon National Monument in the mid-1960s.

and pristine scenery over "constructed rivers" controlled and restricted by dams. The Dominy era, while celebrated by Reclamation for its accomplishments, closed with the Commissioner's failure to obtain two other dams on the Colorado, namely the proposed Marble Canyon and Bridge Canyon Dams, one below the Grand Canyon National Monument section of the river and the other on the river above Grand Canyon National Park. He agreed to partner with a coal fired electrical plant, the Navajo Generating Station near Page, Arizona, to pump water into an emerging Central Arizona Project rather than build more dams to produce the electricity for the project. In any event, the Dominy years revived the momentum begun in the New Deal

1.30. The Navajo Generating Station at Page, Arizona, is owned 24.9 percent by the Bureau of Reclamation to provide power to the Central Arizona Project. Courtesy of Carol DeArman Storey.

period to harness the forces of western rivers to provide power for industry and homes, recreation for urban populations, and expansion of irrigated agriculture.

Adjustments for a Different Future

After Commissioner Dominy, came a great divide in Reclamation's history. Some use the word "decline" and, others describe "a beleaguered Bureau of Reclamation." Clearly, the original mission of Reclamation "to go forth and build dams and storage facilities" came to an end. The environmental critique of Bureau of Reclamation work demanded accountability and assessment of project consequences, especially the big dams. After the passage of the National Environmental Policy Act in 1969 and the establishment of the Environmental Protection Agency in 1970, the Bureau of Reclamation's activities became subject to a host of regulatory reports and planning documents generally called environmental impact statements. In response, former Acting Commissioner and editor of *Reclamation Era*, William E. Warne, wrote in 1973 that, "Reclamation leaders have tried to prepare the way for an environmental tack." But the point should be made that Congress's growing lack of enthusiasm for funding new water projects presented a more formidable obstacle than the rise of environmentalism. With mounting expenses for the Vietnam War and pressure to sustain spending on domestic programs, Congress could not easily approve new western water projects. The combination of reduced funding and a growing environmental opposition to blocking free flowing rivers turned the Bureau of Reclamation on to a new course that demanded new outlooks, goals, and even a cultural change within the organization.[46]

That change did not come easily during the 1970s and 1980s. Reclamation, staffed largely with engineers, faced a future without dramatic construction projects on the order of Hoover, Grand Coulee, or Glen Canyon Dams to employ their skills and talents. It came as little surprise when the New York *Times,* on October 2, 1987, ran a modest announcement that the Bureau of Reclamation's dam building days were over. "Instead of constructing big water and power projects, the Federal agency will concentrate on managing the existing projects, conserving water and assuring good water quality and environmental protection," said the article. To many, this meant that the heyday of the Bureau of Reclamation resided only in its history, not the present or the future. President Jimmy Carter's attempt to block new western water projects suggested a limited future by the end of the 1970s.[47] It was not until the 1990s that a wide-ranging picture of the future of Reclamation emerged when President Bill Clinton's Administration appointed Commissioner Daniel P. Beard, who articulated new goals. He embraced the new constraints and saw a future full of possibilities for the importance and significance for the nation's largest water delivery and power management bureau.

Beard emphasized new constituencies for Reclamation. He painted a bright future while urging employees of the Bureau of Reclamation to stop talking about "the good old days" and look ahead to new agendas. His upbeat message did not disparage the past but simply said that the grand construction phase of reclamation history had passed. He avoided any hint at the darker ecological messages emerging in the current literature critical of great dams and their assaults on river systems. Not all were pleased with Beard's message to replace old visions with new ones, and certainly few could seriously entertain ideas that the great dam era should not have occurred. While the achievements of Reclamation were monumental, Beard believed it pointless to be stymied by a paralyzing sense of reverence or dismay for a bygone era. A new time was at hand. The Bureau of Reclamation must offer new services to a society whose changing values placed varied demands upon the resources of the natural environment.[48]

1.31. Daniel P. Beard, Commissioner, Bureau of Reclamation, 1993-1995.

As the mission of Reclamation changed to embrace new constituencies in the recreational and environmental community, the Commissioner acknowledged that the Bureau of Reclamation's original constituency, water users on projects, might feel neglected. Briefly, in the 1970s, even the name of the Bureau of Reclamation changed to "Water and Power Resources Service," reflecting the organization's struggle to redefine itself. The name attempted to denote its broader functions in the future rather than the limitations of its past implied in the term *reclamation*. By the 1990s, Commissioner Beard noted six future missions of the Bureau of Reclamation: (1) water conservation, (2) mitigation of environmental problems, (3) power and water for urban communities, (4) support for Native American water rights, (5) diversity awareness in the organization, and (6) water transfers. All suggested a smaller, more efficient "water management bureau with a more environmental mission."[49]

Each of these missions represented challenges to the Bureau of Reclamation. Making the future of the organization a concern in this critical time was, in itself, an accomplishment for Commissioner Beard. He believed the future was open-ended and called for new guidelines. An ability to adapt and respond testified to the long life of Reclamation. Achievers and achievements as well as shortcomings and unintended environmental consequences marked the history of Reclamation — all to be expected of an organization that made the leap from a champion of the nineteenth-century family farm to the builder of big dams for an industrialized urban society, and, by the beginning of the twenty-first century, to an arbiter and conservator of western resources. Commissioner Keys's 2002 Centennial Address sounded an enduring thread in Reclamation's history: overcoming water shortages and the Bureau of Reclamation's "collaboration … with all water users to leverage resources and make use of developed water supplies."[50]

Endnotes:

1 For these phrases, see Mary Austin, *Land of Little Rain* (Boston and New York: Houghton Mifflin, Inc., 1903); Walter Prescott Webb's much reviled article because he predicted limitations to growth in the West: "The American West Perpetual Mirage," *Harper's Magazine* 214 (May 1957), 21-35; William E. Smythe, *The Conquest of Arid America* (New York: Harper & Brothers Publishers, 1900); the literature on Smythe is extensive. For example, see Martin E. Carlson, "William E. Smythe: Irrigation Crusader," *Journal of the West* 7, 1 (1968), 41-7; also note David Wrobel reference to Smythe's population predictions in *Promised Lands: Promotion, Memory, and the Creation of the American West* (Lawrence: University Press of Kansas, 2002), 55.

[2] Leonard J. Arrington, *Great Basin Kingdom: An Economic History of the Latter-day Saints, 1830-1900* (Cambridge, Massachusetts: Harvard University Press, 1958).
[3] John W. Keys, III, "Address of the Commissioner Presented at the Bureau of Reclamation's Centennial Celebration at Hoover Dam on the Colorado River," June 17, 2002.
[4] Commissioner Keys, Address, June 17, 2002.
[5] "Arizona's Roosevelt Dam to be Rededicated," *U.S. Water News Online* (March 1996); "Roosevelt Dam Reaches New Heights," *Civil Engineering* 66 (May 1996), 10.
[6] Karen Smith, *The Magnificent Experiment: Building the Salt River Reclamation Project, 1890-1917* (Tucson: University of Arizona Press, 1982), 70; David McCullough, *The Johnstown Flood: The Incredible Story Behind One of the Most Devastating Disasters American Has Ever Known* (New York: Simon and Schuster, 1987); Donald C. Jackson, *Building the Ultimate Dam: John S. Eastwood and the Control of Water in the West* (Lawrence: University Press of Kansas, 1995), 189; Donald J. Pisani, *Water and American Government: The Reclamation Bureau, National Water Policy, and the West, 1902-1935* (Berkeley: University of California Press, 2002), 98.
[7] Smith, *Magnificent Experiment*, 2.
[8] Eric Rauchway, *Murdering McKinley: The Making of Theodore Roosevelt's America* (New York: Hill & Wang, 2003).
[9] Smith, *Magnificent Experiment*, 23; U.S. Department of the Interior, U.S. Geological Survey, F. H. Newell, "Reclamation Law," *Third Annual Report of the Reclamation Service, 1903-4* (Washington, D.C.: Government Printing Office, 1905), 27.
[10] U.S. Department of the Interior, U.S. Geological Survey, Louis C. Hill, "Operations in Arizona: Salt River Project," *Fourth Annual Report of the Reclamation Service, 1904-5* (Washington, D.C.: Government Printing Office, 1906), 64.
[11] Richard Lowitt and Judith Fabry, editors, *Henry A. Wallace's Irrigation Frontier: On the Trail of the Corn Belt Farmer, 1909* (Norman: University of Oklahoma Press, 1991), 1-3.
[12] Lowitt and Fabry, *Henry A. Wallace's*, 69, 60.
[13] Lowitt and Fabry, *Henry A. Wallace's,* 72, 104, 64.
[14] Frederick H. Newell to L.H. Taylor, August 13, 1906, writes that in view of the many difficulties existing in the Owens Valley, the Reclamation Service would not be initiating any projects there as planned, Record Group [hereafter cited as RG] 115, Entry 3, Box 18, National Archives and Records Administration [hereafter cited as NARA] Denver; George W. Woodruff, Acting Secretary of the Interior to Commissioner of the General Land Office, July 12, 1907, formally announced the abandonment of the Reclamation Service's Owen River Project to allow the city of Los Angeles to purchase government land for its Owens Valley aqueduct, RG 115, Entry 3, Box 18, NARA, Denver; the literature on Los Angeles's acquisition of water in the Owens Valley is extensive: see Remi A. Nadeau, *The Water Seekers* (Garden City, N.Y.: Doubleday & Co., 1950); Abraham Hoffman, *Vision or Villainy: Origins of the Owens Valley-Los Angeles Water Controversy* (College Station: Texas A&M University Press, 1981); William L. Kahrl, *Water and Power: The Conflict over Los Angeles' Water Supply in the Owens Valley* (Berkeley: University of California Press, 1982); Donald J. Pisani, *From the Family Farm to Agribusiness: The Irrigation Crusade in California and the West, 1850-1931* (Berkeley: University of California Press, 1984), 307-8, 334.
[15] Joseph E. Stevens, *Hoover Dam: An American Adventure* (Norman: University of Oklahoma Press, 1988), 19; Norris Hundley, Jr., *The Great Thirst: Californians and Water, A History* (Berkeley: University of California Press, originally published 1992, revised 2001), 204.
[16] Stevens, *Hoover Dam*, 18; Gene Gressley, "Arthur Powell Davis, Reclamation, and the West," *Agricultural History* 42 (July 1968), 241-57.
[17] Stevens, *Hoover Dam*, 18; Hundley, *Great Thirst*, 207-9; Jacques Leslie, *Deep Water: The Epic Struggle over Dams, Displaced People, and the Environment* (New York: Farrar, Straus and Giroux, 2005), 3-4.

[18] Al M. Rocca, *America's Master Dam Builder: The Engineering Genius of Frank T. Crowe* (New York: University Press of America, 2001), 183; Donald E. Wolf. *Big Dams and Other Dreams: The Six Companies Story* (Norman: University of Oklahoma Press, 1996).
[19] Theodore Steinberg, "'The World's Fair Feeling': Control of Water in Twentieth Century America," *Technology and Culture* 34 (April 1993), 402.
[20] Pisani, *From the Family Farm*, 398, 394.
[21] Pisani, *From the Family Farm*, 437-8.
[22] Stewart H. Holbrook, *Far Corner* (New York: Macmillan, Co., 1952).
[23] Kathrine G. Morrissey, *Mental Territories: Mapping the Inland Empire* (Ithaca, NewYork: Cornell University Press, 1997).
[24] Richard White, *The Organic Machine: The Remaking of the Columbia River* (New York: Hill & Wang, 1995); also note preface to a 1963 fourth edition of William Denison Lyman, *The Columbia River: Its History, Its Myths, Its Scenery, Its Commerce* (Portland, Oregon: Binfords & Mort, Publishers, 1963, originally published 1909), iv; Bruce Brown, *Mountain in the Clouds: A Search for the Wild Salmon* (New York: Simon & Schuster, Inc., 1982), 25.
[25] Paul C. Pitzer, *Grand Coulee: Harnessing a Dream* (Pullman: Washington State University Press,1994), 35, 23, 12, n. 22; Michael C. Robinson, *Water for the West: The Bureau of Reclamation, 1902-1977* (Chicago, Illinois: Public Works Historical Society, 1979), 27.
[26] As quoted from *Spokesman-Review*, May 20, 1909 in Pitzer, *Grand Coulee,* 13.
[27] "Progress of Work, October 1909," *Reclamation Record,* 1 (November 1909), 106-7; Christine E. Pfaff, *Harvests of Plenty: A History of the Yakima Irrigation Project, Washington* (Denver, Colorado: Bureau of Reclamation Technical Service Center, 2001), 27.
[28] Pitzer, *Grand Coulee*, 29, 23, 65.
[29] Pitzer, *Grand Coulee*, 65.
[30] Mead as quoted in Pitzer, *Grand Coulee*, 77, 80; "Title II of the NIRA authorized the creation of the Public Works Administration to create jobs while undertaking public works. The NIRA also gave the U.S. president unprecedented powers to initiate public works, including water projects. The Public Works Administration provided loans and grants to state and local governments and to federal agencies for municipal water works, sewage plants, irrigation, flood control, and waterpower projects. California's Central Valley Project, the Bonneville and Grand Coulee dams on the Columbia River, and Fort Peck Dam on the Missouri River were among the many major projects authorized under this legislation." United States Department of Agriculture, Economic Research Service, *A History of Federal Water Resources Programs, 1961-1970*, Miscellaneous Publication No.1379, by Beatrice Hort Holmes (Washington, D.C.: U.S. Government Printing Office, 1979); Kenneth D. Frederick and Roger A. Sedjo, editors, *America's Renewable Resources: Historical Trends and Current Challenges* (Washington, D.C.: Resources for the Future, 1991), 37.
[31] Pitzer, *Grand Coulee*, 358; Stewart H. Holbrook, *The Columbia* (New York: Rinehart and Co., Inc., 1956); Murray Morgan, *The Columbia* (Seattle, Washington: Superior Publishing Co., 1949); Murray Morgan, *The Dam* (New York: Viking, 1954).
[32] Daniel Tyler, *The Last Water Hole in the West: The Colorado-Big Thompson Project and the Northern Colorado Water Conservancy District* (Boulder: University Press of Colorado, 1992), 29, 31; Philip L. Fradkin, *A River No More: The Colorado River and the West* (Berkeley: University of California Press, 1996, paperback edition), 46; Fred N. Norcross, "Genesis of the Colorado-Big Thompson Project," *Colorado Magazine* 30, 1 (January 1953), 30.
[33] Fradkin, *A River No More*, 46.
[34] Smith, *Magnificent Experiment*, 54; Tyler, *The Last Water Hole in the West*, 95; Alfred R. Golzé, *Reclamation in the United States* (Caldwell, Idaho: The Caxton Printers, 1961), 67.
[35] As quoted in Tyler, *The Last Water Hole in the West*, 47.

[36] Hearing before the Secretary of the Interior on the Colorado-Big Thompson Water Diversion Project, Colorado, 1937 as quoted in Tyler, *The Last Water Hole in the West,* 88; T. H. Watkins, *Righteous Pilgrim: The Life and Times of Harold L. Ickes, 1874-1952* (New York: Henry Holt and Company, 1990), 582.
[37] Tyler, *The Last Water Hole in the West*, 108-22; U.S. Senate Document 80, 75th Congress 1st Session; *Congressional Record*, "The Colorado-Big Thompson Reclamation Project and the Northern Colorado Water Conservancy District," *Extension of Remarks of Hon. Lawrence Lewis of Colorado*, House of Representatives, Friday, April 25, 1941; Daniel Tyler calls "Green Mountain Dam, the West Slope's *sine qua non* for participation in the C-BT," *Last Water Hole in the West*, 95; also for discussion of labor issues in the construction of C-BT see Tyler, *Last Water Hole in the West*, 105-13.
[38] Tyler, *The Last Water Hole in the West*, 134-57, 461; Fradkin, *A River No More,* 113.
[39] From 1902 to 1907 Charles Doolittle Walcott, a palaeontologist and the director of the United States Geological Survey, was also the director of the U. S. Reclamation Service. He actively oversaw Reclamation and visited potential projects during that time, but, by 1907, he felt Reclamation was interfering with his Geological Survey responsibilities and moved to completely separate the two bureaus. In 1907, he resigned both directorships to become a successful, influential, and long-serving Secretary of the Smithsonian Institution.
[40] Department of the Interior, U.S. Geological Survey, *First Annual Report of the Reclamation Service from June 17 to December 1, 1902* (Washington, D.C.: Government Printing Office, 1903), 24.
[41] Smith, *Magnificent Experiment*, 54.
[42] U.S. Department of the Interior, U.S. Reclamation Service, *Twelfth Annual Report of the Reclamation Service, 1912-1913* (Washington, D.C.: General Printing Office, 1914), 3-5; Pisani, *Water and American Government* sums up this view by Reclamation officials of many on the projects, "farmers who failed lacked the 'pluck' and 'grit' of earlier generations of frontier Americans," 31.
[43] Floyd E. Dominy, Oral History Interview, April 6, 1994, Boyce, Virginia (Denver, Colorado: Bureau of Reclamation Oral History Program, 1994), 2, 7, 12.
[44] Mark W. T. Harvey, *A Symbol of Wilderness: Echo Park and the American Conservation Movement* (Albuquerque: University of New Mexico Press, 1994).
[45] See John McPhee, *Encounters with the Archdruid* (New York: Farrar, Straus and Giroux, 1971).
[46] Doris Ostrander Dawdy, *Congress in Its Wisdom: The Bureau of Reclamation and the Public Interest* (Boulder, Colorado: Westview Press, 1989), 1; William E. Warne, *The Bureau of Reclamation* (New York: Praeger Publishers, 1973), 218; Donald J. Pisani, "Federal Reclamation and Water Rights in Nevada," *Agricultural History* 51 (July 1977), 540-58.
[47] M. Glenn Abernathy, Dilys M. Hill, and Phil Williams, eds., *The Carter Years: The President and Policy Making* (New York: St. Martin's Press, 1984), 180-1; Marc Reisner, *Cadillac Desert: The American West and Its Disappearing Water* (New York: Viking, 1986).
[48] Patrick McCully, *Silenced Rivers: The Ecology and Politics of Large Dams* (London & New Jersey: Zed Books, 1996) chapter title: "Temples of Doom," 65-100; Paul R. Josephson, *Industrialized Nature: Brute Force Technology and the Transformation of the Natural World* (Washington, D.C.: Island Press, 2002); Donald Worster, *Rivers of Empire: Water and Aridity and the Growth of the American West* (New York: Oxford University Press, 1985) extends the criticism to Reclamation itself, 130; as does Reisner in *Cadillac Desert.*
[49] *Oregonian* (Portland), November 2, 1993, A-10 as quoted in Pitzer, *Grand Coulee*, 370; see also Commissioner Daniel Beard, Oral History Interview, Daniel Beard: 1993-1995 (Denver: Bureau of Reclamation Oral History Program, 1995).
[50] Commissioner Keys, Address, June 17, 2002.

1.32. A nineteenth-century *Harper's Weekly* lithograph of irrigated farms near Salt Lake City.

1.33. A ditchrider's home on the Belle Fourche Project, South Dakota, in 1921.

CHAPTER 2

VISIONS OF NATIONAL RECLAMATION

Introduction

President Theodore Roosevelt's signature on the national Reclamation Act (sometimes known as the Newlands Act) of June 17, 1902, confirmed western victory in a long campaign for federal aid to irrigation. It was not an easy sell. To win support in Congress, far western politicians and boosters engaged in both political log rolling and in elaborate historical, intellectual, and emotional arguments. Much of the intellectual argument hinged upon convincing Congress that western lands were different. They differed because in the country beyond the Missouri little rain fell except in the high mountains and in the windward marine climate of the Pacific Northwest. Periodic droughts compounded the general aridity of the interior West, where the open land and a pleasing climate seemed to beckon settlement and agriculture. But until more of the West's water could be brought to the land, the potential for agricultural settlement appeared limited. Watering the land meant diverting rivers, building dams, digging canals and ditches, and maintaining reservoirs. Although important progress, as Mormon settlement in Utah since 1847 demonstrated, had already occurred, national irrigation supporters wanted far more. They were not content with Mormon success and easy river diversion canals to scattered western valley farmsteads and ranches. And while some achieved success at dry land farming, western agriculture remained an uncertain enterprise without irrigation.

The Agrarian Myth in an Imperfect Land

Those who campaigned for a "New West" through the miracles of irrigation drew upon a rich archive of idealized and romanticized agrarian writing. J. Hector St. John de Crevecoeur in his *Letters from an American Farmer* (1783) focused upon the beauties, satisfactions, and rewards of rural farm life in the well-watered agricultural lands of New York and Pennsylvania. Here he saw the emergence of "this new man, this American," free and prosperous reaping the bounty of the land. Thomas Jefferson made land and small freehold farmers the hallmark of his political philosophy, which embraced the sturdy, independent yeoman farmers as the mainstay of democracy and the vitality of the republic. These writings, particularly Jefferson's, stamped American history and life with an agrarian mythology strong and enduring in its appeal. Historians and literary scholars have made much of "the agrarian myth" in various works. Little wonder that the champions of national aid to western irrigation embraced the myth to promote their cause. [1]

2.1. This 1908 Reclamation photograph from the Umatilla Project in Oregon, labeled "First crops under the project," projects the image of the Jeffersonian yeoman farmer ideal that strongly influenced Reclamation's early years. .

Scholars have pointed out the many flaws in an idealized agrarian view of the American past. In his work, *Democracy in America* (1834), Alexis de Tocqueville saw what he believed was the highly commercial nature of the American farmer, who regarded land and crops as commodities. The farm was less a crucible of democracy and hearthside contentment than a form of commercial enterprise. Nonetheless, the myth of an America wedded to the land served the cause of those who believed arid America offered the opportunity for a renewed agrarian experience. The special place of citizen farmers in the life of the republic possessed a long legacy and spawned an ideology for a new republic that stood on the verge of extending into millions of acres across a continent from sea to sea. [2]

Representations of nature in the West included sublime wilderness mountainous scenery, broad prairie lands, and the imperfections of an arid, desolate landscape sometimes designated "desert." Advocates of national irrigation focused upon aridity as a barrier to settlement and a prosperous future for the West as a full partner in national development. But if the West were anything, it was vast. To overcome this space, Congress embraced the transcontinental railroad to confront the endless miles of western distance and to bind one coast to another in the making of one nation. It provided loans and land grants in the National Railway Acts (1862 and 1864), which railroad companies used to build west to the Pacific after the Civil War. And to encourage western settlement as a complement to the transportation legislation, Congress enacted a liberal and, ultimately, free land policy in the Homestead Act of 1862, the Timber Culture Act of 1873, and the Desert Land Act of 1877, to name only the most outstanding pieces of free land legislation for the settlement of the West. Mechanized transportation, free land, new territorial and state governments, and policies that forced native tribes onto reservations all prepared the way for a new peopling of the West. Congress subsidized the western region by providing security, transportation, and a generous land policy to overcome the imperfections of these distant lands.[3]

The new challenges of expansion across the Mississippi River demanded knowledge of the region and all of its parts. Now that a railroad to the Pacific stood on the verge of conquering its great distances in the years immediately after the Civil War, Congress sought knowledge beyond geographical and topographical outlines – beyond what Lewis and Clark, Zebulon M. Pike, John C. Frémont, overland travelers, and the Railroad Surveys of the 1850s had earlier retrieved. Above all, questions persisted about the region's economic potential. Did it offer a prosperous future for settlement? This land of treeless plains, vast deserts, deep canyons, and spectacular mountain ranges yielded great wealth in precious minerals, starting with the California Gold Rush in 1849. What other mineral possibilities might exist and what of the land and the Native American tribes of the region? There was much to be learned and assembled about this region that was so different from the lands east of the Mississippi.[4]

Persuasive personalities lobbied Congress to commission new surveys of the West. Clarence King was among them. King was an admired, charming, well-spoken Yale graduate who began government service with Josiah Whitney's Geological Survey of California. In his classic autobiography, *The Education of Henry Adams*, Adams admiringly observed that King, " … had managed to induce Congress to adopt almost its first modern act of legislation." The result was a government survey under civil rather than military management. If one interprets Adams correctly, this was a significant step in the direction of establishing civilian government service bureaucracies. King emerged from the congressional session of 1867 as the civilian director of the 40th Parallel Survey, under nominal supervision of the War Department. King's survey paralleled the transcontinental railway, concentrated on geology, and was, according to Adams, "a feat as yet unequalled by other governments which had as a rule no continents to survey." [5]

King's effort and the three other authorized surveys from 1867 to 1879 became classic scientific works. Lieutenant George M. Wheeler headed another survey under the War Department. Two others under the Department of the Interior followed: Ferdinand V. Hayden conducted the United States Geological and Geographical Survey of the Territories and John Wesley Powell performed his famous work under the survey titled the United States Geographical and Geological Survey of the Rocky Mountain Region. King and Powell dominated the work of these early Geological Surveys. Powell's often quoted *Report on the Lands of the Arid Region of the United States, with a More Detailed Account of the Lands of Utah* (1878) emerged from his work on the Rocky Mountain region. An early historian of Reclamation, Alfred R. Golzé, claimed, "It prepared the way for the irrigation surveys of the Geological Survey in the following 23 years that led to the eventual passage of the Reclamation Act in 1902." The story is much more complicated than Golzé suggests, but his statement demonstrates the almost mythical standing Powell's arid lands report attained in the literature of reclamation.[6]

2.2. Clarence King, first Director of the U.S. Geological Survey. Courtesy of the USGS.

45TH CONGRESS, 2d Session. | HOUSE OF REPRESENTATIVES. | Ex. Doc. No. 73.

REPORT

ON THE

LANDS OF THE ARID REGION

OF THE

UNITED STATES,

WITH A

MORE DETAILED ACCOUNT OF THE LANDS OF UTAH.

WITH MAPS.

BY

J. W. POWELL.

APRIL 3, 1878.—Referred to the Committee on Appropriations and ordered to be printed.

WASHINGTON:
GOVERNMENT PRINTING OFFICE.
1878.

2.3. The cover of John Wesley Powell's famed *Report on the Lands of the Arid Region of the United States.*

Congress consolidated the surveys in 1879 under the U.S. Geological Survey (USGS), headed by King until 1881, and thereafter by Powell until 1894. Building upon the work of the initial surveys, the USGS accumulated excellent topographic and scientific information on the West. Advocates of western irrigation often noted that the USGS had an institutional knowledge and expertise that made it the ideal part of the federal government to undertake the task of planning for and even irrigating the arid regions.[7]

Internal Improvements and Other Arguments

Major western surveys, aid to western railroads, and liberal land policies established Congress as a major player in the economic development of the West. Promotion and subsidization of far western settlement was by no means Congress's first experience in aid to economic development. Early in the nineteenth century, the Federal Government funded river and harbor improvements to assist water transportation, commerce, and navigation on inland water ways and coastal harbors. Transportation (roads and railways) and harbor and river improvements all qualified for legitimate congressional support as internal improvements for the nation's welfare. Proponents of federal aid to arid land irrigation argued that it was also a legitimate internal improvement that the Federal Government could assume. Furthermore, Congress stood under some obligation to make water available for western public domain lands because, without water, the Homestead Act was unworkable in the arid West. Farmers in the humid Midwest and eastern Great Plains enjoyed free and cheap land, while nature's shortcomings denied the citizens of arid states the privileges of the Homestead Act. Congress could put nature's wrongs aright with a national irrigation program that qualified as an internal improvement.

After 1889, the cause of western irrigation as an internal improvement pressed upon Congress, when during a prolonged drought, new western states rapidly entered the Union — Washington, the Dakotas, Montana, and Wyoming, 1889; Idaho, 1890; Utah, 1896; Oklahoma, 1907; and Arizona and New Mexico, 1912. The special circumstances of the arid West demanded not only a revised land policy but, in the minds of many, justified a national program for aid to western irrigation. Mixed messages on the issue came from older states of the Far West along the Pacific Coast, and the Plains states of Nebraska and Kansas.[8]

The West's increased strength in the U.S. Senate gave it growing political clout by the late 1890s. Drought and economic depression added urgency to the western irrigation question and brought a stridency and emotionalism to the arguments for federal aid to western irrigation that often contrasted the favored lands of the East to the impoverished West. Generally, the eastern United States enjoyed adequate rainfall on rich agricultural lands, but western lands, although potentially rich, lay parched and poor. The picture presented the "haves" versus the "have-nots" dilemma. But what nature had ordained was difficult to dispute or even rectify. With the farmer migration into the plains after the Civil War, some argued that settlement itself would bring increased rainfall. Tree planting and plowing the land could make the arid portions of the Great Plains states – parts of the Dakotas, Nebraska, and Kansas – inhabitable and abolish forever the idea of the Great American Desert. Many embraced the idea that "rain follows the plow," as did scientists at the University of Nebraska; Ferdinand V. Hayden, head of the Geological Survey of 1871; the railroads; and land promoters. As settlement moved into the arid lands, newly planted trees and the plowed earth would offer up moisture to the atmosphere that would return in the form of rain to water crops. Indeed, the wet cycle of 1870s on the plains seemed to fulfill the promise, but when the cycle ended in the 1880s and 1890s, these promises proved spurious. By the 1890s, prominent scientists, John Wesley Powell among them, derided ideas that settlement would change climate.[9]

2.4. In camp during lunch on one of his geological surveys, Ferdinand V. Hayden is seated in the dark jacket, without hat, while his famed photographer and painter, William H. Jackson, stands on the far right. Red Buttes, Wyoming Territory, August 24, 1870. National Archives and Records Administration.

In any event, eastern congressmen did not readily accept the argument that the West was a deprived region by virtue of a stunted, ungenerous environment and, therefore, deserving of federal assistance for its agriculture. The West's story was not all hardship. It often enjoyed spurts of dramatic growth with the discovery of mineral wealth and the exploitation of grasslands and forests. Much more appealing to easterners was the point that irrigation might also be applied to their lands. Despite impressive average rainfall, periodic droughts often struck lands east of the Mississippi. This new "scientific farming," as some termed it, would place agriculture beyond the vicissitudes of nature even in the humid eastern part of the United States.[10]

Congress did encourage arid land redemption. While mainly directed toward mining operations, the important act of 1866 (sometimes labeled the first National Mining Act, anticipating those that followed in 1870 and 1872) permitted right of ways across the public domain for water ditches and recognized local customs and court decisions with respect to mining disputes and water use and rights. This meant that local jurisdictions could embrace prior appropriation water rights, abandoning traditional riparian/common law water doctrine, which granted only the owners of the banks of a water source access to it. An amendment to the act in 1870 strengthened these provisions. In effect, Congress adopted a local option approach (state courts and legislatures) in the realm of water laws, but on public lands, the Mining Acts of 1866 and 1872 fortified prior appropriation by granting rights of way across the public domain for the transportation of water. This facilitated river diversions for mining purposes, but also signaled to irrigation interests that water diversions could be used for agriculture under prior appropriation custom and law. These practices encouraged states that entered the Union in the following decades to abrogate riparian rights. The Timber Culture Act (1873) offered free land in return for tree planting to encourage climate change; the Desert Land Act (1877) offered 640 acres of land for the minimum price of $1.25 an acre if the settler irrigated a portion of it within three years; the Carey Act (1894) offered land to the states for irrigation enterprises. While most states failed to implement the Carey Act adequately, Idaho was an exception with a number of projects that achieved success, e.g., Twin Falls and American Falls. Land cessions to the states under the Carey Act were simply not enough to encourage projects except under the most favorable circumstances where private interests would take the initiative. Carey Act failures only prompted stronger demands for a comprehensive federal government initiative in irrigation that encompassed dam building, reservoirs, and the development of ditch systems to deliver water.[11]

The West as Problem

American westward expansion occurred faster than even Thomas Jefferson envisioned. After the Civil War, people attracted by mining, livestock agriculture, and trade moved into the Great Plains, the Mountain West, and along the streams of the Southwest. These were clearly not the verdant rolling hills of eastern America or the broad rich prairie soils of Illinois and Iowa. Won by purchase and conquest in the nineteenth century, the semi-arid and arid lands beyond the Missouri River raised barriers to American farmers far beyond the imaginations of Jefferson and those who romanticized an agrarian future for America. After the purchase of Louisiana in 1803, early western exploratory expeditions (Zebulon Montgomery Pike in 1805 and Stephen H. Long in 1820) characterized the area as the "Great American Desert," unfit for civilization. This nomenclature attached to the region well into the nineteenth century. During the debates over the Compromise of 1850, Daniel Webster argued there was no need to offend the southern interests by including a ban on slavery in the West: the Great American Desert itself barred the advance of the plantation system westward. He proclaimed that "an ordinance of Nature" stood in the way of the expansion of plantation agriculture and slavery.

For all the talk of spectacular scenery, amber fields of wheat, and boundless mineral resources in the American West, in much of the region crop agriculture required irrigation. At the end of the American Revolution, with its western border on the Mississippi River, the nation could boast of its supply of moisture from rain and snow, abundant woodlands, rich soils, and a myriad of rivers to facilitate transportation. The subsequent addition of a Trans-Mississippi West and, especially the Southwest yielded in the Mexican War settlement, gave to the United States a different, if not alien, landscape. Like Australia, this New West offered moisture and fertile lands on its coastal edge and in the lands adjacent to the Mississippi itself, but the interior presented formidable challenges to traditional agricultural settlement. Aware of this "outback," the Australian provincial government of Victoria sent Alfred Deakin to the United States in the 1880s to study initial American attempts at irrigation in arid lands. Deakin, a future Prime Minister of Australia, conferred with the young Elwood Mead, who dealt extensively with water and irrigation studies at Colorado State Agricultural College

2.5. Elwood Mead as he appeared while State Engineer of the Territory and State of Wyoming (1888-1899).

in Fort Collins. The meeting also introduced Mead to irrigation issues in Australia. Deakin made note of John Wesley Powell's observation that in the arid West, "only some 3 or 4 percent is irrigable at any price" — in an enormous land one-third the size of Australia. In many parts of the American West , the Australian reported "sand wastes" where there "is nothing to attract and everything to repel."[12]

In a rapidly modernizing America, material problems, whether in transportation, food production, mining, communication, or finance, prompted inventive technological responses. Reclamation was the ultimate technological fix or innovation to meet the problems created by arid western lands. It promised food production, thriving communities, and economic stability for the region. In 1849, the first blushes of the California Gold Rush seemed to foreshadow great riches with little effort, but the West, as a whole, was no Biblical paradise of milk and honey. Beyond the beckoning riches of California, the Willamette Valley, and forested lands surrounding Puget Sound, much of this country was a gamble for westward migrating Americans. Towns sprang up at rich mining strikes and just as quickly faded when ores played out. Large ranching outfits met disaster from bitter winters and drought. Attempts by private capital to construct irrigation projects often failed, and infrequent successes occurred only under the most favorable circumstances.

While not exactly free enterprise, industrious enterprise and community cohesiveness inspired by religious commitment in Mormon Utah set an example for how arid lands might be brought under cultivation by irrigation. The hard work and sacrifice in Utah stood in stark contrast to the seemingly easy wealth found in California. But Utah demonstrated it could be done. Utah proclaimed water a public resource that could be diverted from sources (rivers, lakes, springs) under a system of appropriation for the productive use of the water. Miners in California developed a "crudely evolved" appropriation doctrine. Mormons shaped their appropriation water law to suit farming needs and bring water from sources in the mountains to valley lands. Community and beneficial use of the water overruled private ownership or any monopoly on water in Utah. Equitable administration of the resource was essential and misuse or waste of water was subject to punishment by appointed water masters. While this system is similar to Mediterranean and Latin American practices, no direct link between Mormon thinking about irrigation practices and the outside world appears in the literature. This system required a high degree of community cooperation and willingness to accede to established authority in the distribution of a resource that was essentially under common ownership for the welfare of the community.[13]

John Wesley Powell admired the Mormon accomplishments. But his admiration was tempered by the recognition that the success of Mormon irrigation greatly depended on local community cooperation and organization, qualities little present in most of the West. The Greeley Colony in Colorado, under private capital development, also showed promising signs of a prosperous future, but by 1870 even this colony saw the limits and perils of extending into marginal lands. Local irrigation projects could succeed as demonstrated in the ditching of the Truckee Meadows by 1880 in western Nevada, but farther out in the Nevada desert, away from the waters of the eastern slope of the Sierra Nevada, the daunting challenges of irrigation required finances beyond local resources. The typical irrigation undertaking (in a valley overshadowed by nearby mountains — the vertical reservoirs of the Inland West) first diverted water from a natural water course into a main canal that ran along a "high line" above irrigable lands. Secondary canals or "laterals" diverted water from the main canal and these brought water into "corrugates" in the fields. Ditches then drained water from crops at one elevation to lower fields. The cost of construction of the main canal alone, which often must run for many miles to serve large valleys, could be prohibitive. [14]

There was an ongoing debate on how best to promote and shape a vision for economic stability and growth in the region. If yeomen farmers had a place in this vision, and they did, western reclamation could help make it possible far beyond the acres of land brought under the ditch by private and local enterprise. At the same time, new irrigated farmlands would secure the future of democracy. Pushing back the domains of aridity would keep alive the spirit of Manifest Destiny that had carried the nation all the way to the shores of the Pacific. These interior arid lands presented special challenges that required innovative policies for successful American settlement. Indeed, one historian of reclamation asserts that, "federal reclamation was the last stage of Manifest Destiny – the process of creating an integrated nation that stretched from sea to sea." [15]

Almost every American schoolchild learns the importance of the terms "Manifest Destiny" and "frontier" in the history of the United States. One represents an irrepressible force in nineteenth-century American expansion; the other a barrier and challenge awaiting transformation by the forces of civilization. The triumph of American destiny over frontier brings a victory that in and of itself legitimated the right of the American Republic to lay its claims all the way to the Pacific Coast and beyond, if necessary, for the security of the republic. *The Winning of the West*, as Theodore Roosevelt entitled his four volume history on the opening of the Old West (the trans-Appalachian lands) to American settlement, typified a literature filled with national pride, vindication of American expansion, and vindictiveness for all who stood in its way. Now the arid lands themselves became the final barrier, that "ordinance of nature" that must be confronted.[16]

Other Issues

In the 1890s, western drought, low agricultural prices, as well as industrial depression conspired to produce a tumultuous decade. These years saw the rise in agricultural states of a third party movement, the Populist Party, and, in the mining states of the West, an avid support for "free silver" and a Silver Party. Both ultimately demanded the coinage of silver money to combat what they viewed as a destructive deflation of the money supply caused by strict adherence to the gold money standard. The two major parties stood by the gold standard, except when William Jennings Bryan, of Nebraska, converted the Democratic Party to silver in 1896, but lost the cause in several bids for the presidency. The popularity of silver money in many western states temporarily removed the irrigation movement from the political spotlight.

In addition, by 1900, the conquest of the arid lands by federal irrigation competed with causes beyond the borders of the United States. After the Spanish-American War of 1898, the nation became an imperial power with newly acquired outposts, particularly the Philippines. Westerners, just getting over the defeat of their silver campaigns, now worried about foreign involvements undermining their bid for federal aid to western irrigation. Nevada's Congressman Francis G. Newlands, having dispensed with the silver issue, was one of many who wished that the government would stop irritating foreign lands and get on with the business of irrigating domestic lands. In short, the U.S. should concentrate on its "domestic empire" and avoid European style imperialism. After the beginning of the bitter Philippine Insurrection in 1900, Bryan, in his second unsuccessful presidential bid, decried the extension of Manifest Destiny beyond the nation's borders when he said, "Destiny is not as manifest as it was." At the same time, the Alaska

Gold Rush (1897-1900) inspired new confidence in western development. Once again, heart, drama, and romance in the Far West masked the constant complaints of failed mining economies, low silver prices, and drought-stricken agriculture.[17]

Beginning in 1897, the national economy started to recover from the depression years of the mid 1890s, but now, after the Spanish-American War, irrigation advocates feared that the costs of an overseas empire threatened an irrigation program for the West. Still, there was new hope. Just as the federal government had been instrumental in the "winning of the West" and vanquishing an old colonial power, so might it be the key to the final conquest of western deserts. An unprecedented intervention by government appeared necessary to transform these dry lands into gardens of productivity and homes for democracy. Science and engineering – the emissaries of progress – were critical to building water projects in the West, and government appeared to possess the organizational ability and capital resources to bring them into play. But doctrines of Social Darwinism and *laissez faire*, so fashionable among intellectuals and the captains of industry, reached even into the halls of Congress. They cast doubt upon government programs to aid any class or even a disadvantaged region of the nation. But a political pragmatism combined with an emotional enthusiasm on the part of irrigation advocates eroded the rhetoric of these heavy and dreary doctrines. The optimism of science, engineering, and planning, backed by the government, promised more for the building of communities in the West than did darker views about human mischief and avarice in the struggle for survival. Ideas aside, western states, although almost hopelessly diverse in their self interests, realized a new political power in the national government. Western politics, perhaps more than ideas, dictated that Congress could not for long ignore a comprehensive program of government sponsored irrigation. [18]

As a strong proponent of federally driven irrigation, Francis Newlands admitted that the decentralized, weak, nineteenth-century American government might not be up to the task. Despite the political upheavals of the 1890s, he and others looked to the national government as a source of energy and talent for the accomplishment of western irrigation. If the national government possessed the power to build a proposed Panama Canal, liberate Cuba, bring democracy to the Philippines, and tame the monopolistic forces of industry, surely bringing water to the arid lands of the West was possible. As few in his Democratic Party did, Newlands recognized the possibilities of government enterprise and its application to the water needs of the arid West.

Advice

Years before vast city populations and huge agribusiness empires demanded water, Congress sought the advice of some of the best scientific minds in America on the question of irrigating the western lands. In a letter to Congress in 1874, Vermont's widely traveled scientist and linguist, George Perkins Marsh, offered his wisdom on the irrigation of the West. Author of a renowned study of civilization's impact upon the environment, *Man and Nature: Or, Physical Geography as Modified by Human Action* (1864), Marsh was a man respected in the circles of knowledge and power. His letter covered a gamut of questions under the rather comprehensive title, "Irrigation: Its Evils, the Remedies, and the Compensations." The chief problems, in his opinion, were social and political — irrigation promoted land monopoly and the growth of large estates. In many instances, agriculture by irrigation left no middle class in the countryside, he wrote.

2.6. George Perkins Marsh reported to Congress on irrigation..

As American minister to Italy and widely traveled in the Middle East, Marsh studied irrigation in the Mediterranean countries. Beyond social and political problems, he noted "deleterious" miasmatic conditions created by waters draining from irrigated land and left stagnating in the soils. He drew attention to the economics of the heavy capital investment required to build storage and water delivery systems and to level and slope land for irrigation. Nevertheless, he concluded that the United States could learn from the mistakes of Europe as it undertook irrigation in its arid lands. While not generally an admirer of Mediterranean societies, Marsh believed Americans should take notice of the Spanish-Moorish system of community control of water supplies because it ensured the survival of the small farmer and prevented land monopolies. [19]

In the late twentieth century, environmentalist critics of western reclamation often seized upon the early Marsh letter as proof that scientists had doubts about the benefits of irrigation from the beginning – and about who would build the dams and control the water.[20] While these voices came from the late twentieth century, most studies of western irrigation in the years leading up to the national Reclamation Act of 1902 showed a mixture of caution and enthusiasm. One contemporary defender of managed water development, not unsurprisingly, came from California. William Hammond Hall, the first to hold the office of State Engineer in California, showed little of the caution recommended by Marsh toward irrigation schemes and dismissed the warnings against monopoly, pointing to Marsh's own recognition that the Moorish systems in Spain preserved the interests of the middle classes. He had conducted his own research into the subject and published his work, *Irrigation Development: History, Customs, Laws and Administrative Systems Relating to Irrigation, Watercourses and Water, in France, Italy, and Spain,* published in Sacramento in 1886. Hammond's and other investigations indicated that the search for information about irrigation reached far outside the halls of Congress. The search, in fact, moved beyond irrigation matters to dam construction, most always the key component in any irrigation project. In 1888, a classic work on the history and design of dam building appeared in the United States: Edward Wegmann's *The Design and Construction of Dams*. American dam engineers now had access to information on dams in Algiers, India, Italy, France, Spain, Belgium, and Britain. Only a few dams in the United States received notice.[21]

In 1873, Congress commissioned two additional reports. Nevada's Senator William M. Stewart, acting on behalf of his Southern Pacific Railroad clients in California, pushed a bill through Congress "to study irrigation in all of its aspects" in the San Joaquin, Tulare, and Sacramento Valleys of California. The report emphasized "the necessity for state planning and control over water resources; long-range comprehensive development of the agricultural potential of the Central Valley; and the cooperative use of state, federal and private resources to control nature and build an agricultural empire." The report was a remarkable piece of research on the part of an appointed commission of experts who studied the physical challenges of irrigation, proposed administration, financing sources, and the experience of other countries, most notably British India, with irrigation.[22]

In the second report, Congress asked Colorado River explorer and government scientist, John Wesley Powell, for a preliminary report on irrigation and the western surveys. Obscured by his more famous 1878 *Report on the Lands of the Arid Region of the United States, with a More Detailed*

Account of the Lands of Utah, this 1874 report to Congress mostly addressed the consolidation of federal surveys, but it did make reference to large streams that "can only be managed by cooperative organization, great capitalists, or by the general or state governments." The 1878 report was notably silent on the latter point. Those opposed to the possibilities of large irrigation development, based on these investigations, usually saw their self-interests at risk. For example, range cattle interests feared displacement by small farmers, and midwestern congressmen feared competition from new farms in the West. While Marsh's letter to Congress and portions of the California irrigation report brought to Congress's attention a wide array of irrigation problems around the world, the irrigation of the American West clearly presented challenges all its own. [23]

Americans found it difficult to compare their experience and environment with other nations. To do so would denigrate the "exceptionalism" of American history and frontier that made it so. Liberty and democracy had combined with frontier opportunity to create a nation among nations in the world. Surely, homegrown knowledge was more reliable than the experience of other nations. That is precisely what Congress assumed when it commissioned the various western surveys that "... laid the basis for the knowledge of the geology and natural history of the West." [24]

The legendary and popular American scientist, John Wesley Powell, was one of those knowledge builders. Powell, a disabled Union Civil War

2.7. John Wesley Powell pitched his first camp, during his 1871 expedition down the Colorado River, in the Willows near Green River, Wyoming Territory. Courtesy of the USGS.

2.8 John Wesley Powell's second expedition on the Colorado River at Green River, Wyoming Territory, in May of 1871. Courtesy of the USGS.

veteran, who lost an arm at the Battle of Shiloh, first gained fame with an account of his daring journeys down the Colorado River in 1869 and 1871 in his *Exploration of the Colorado River of the West* (1875). His *Report on the Lands of the Arid Region of the United States, with a More Detailed Account of the Lands of Utah* (1878), while published as a government document and not widely circulated, impressed the scientific community and subsequent generations. The accuracy of its detailed maps and suggested innovations in land policy for the Far West, as well as new approaches to water use and distribution in the arid lands, made the report the departure point for any serious discussion of western water and land policy. Overall, he suggested planned development of water and land resources. This meant a revamping of land policies to permit both large and small holdings within irrigation communities built around river basins or watersheds. While he emphasized planning, he believed it should occur at the local level and that federal land policies should be flexible enough to permit local communities to devise systems of landholdings that provided for large stock agriculture operations and small farm homesteads on the richer irrigable lands.[25]

Powell's reports typified an enthusiasm and caution for the western irrigation enterprise. His sober enthusiasm centered mostly upon the planning and foresight it would take to combine stock growing and diversified agriculture in the West. He was not a member of the "irrigation-at-any-price" school that developed as the campaign for federal irrigation intensified by the end of the century. He acknowledged that western America required new land laws and different patterns of settlement to adjust landholdings to river basins. Later, and most controversial from the point of view of western irrigation supporters, Powell spoke frankly before the 1893 Irrigation Congress in Los Angeles about the incontrovertible fact that only 3 to 4 percent of western lands were irrigable. More to the point, the best lands were already under irrigation and the West had very little water that was not already spoken for or appropriated. Powell's sober words sparked disappointment on the part of many who had regarded him as the foremost architect of a vast program for western irrigation. [26]

Misconceptions

In reclamation circles, Powell is revered as the early advocate, if not the originator of the vision of a federal presence in construction of irrigation works on a large scale. In his adulatory *Reclaiming the Arid West: The Story of the United States Reclamation Service* (1917), George Wharton James showed "… how Major Powell began the campaign of education, laying the foundations of the work Also how he trained some of the later leaders and workers while they were his subordinates in the United States Geological Survey." But this early hero of the irrigation movement was under no illusions about the challenges of the undertaking that had already discouraged many private associations. As a scientist, he was aware of the small percentage of the West susceptible to irrigation with the technology of his day. As his Los Angeles

RECLAIMING
THE ARID WEST

The Story of the United States Reclamation Service

BY
GEORGE WHARTON JAMES

WITH ILLUSTRATIONS

NEW YORK
DODD, MEAD AND COMPANY
1917

2.9. The cover page of George Wharton James's book on Reclamation.

speech indicated, he believed the best agricultural land was already in private hands and there was not enough water to fill the ditches already built.[27]

Often hailed as the document that made Powell "the Father of Reclamation," *Report on the Lands of the Arid Region of the United States, with a More Detailed Account of the Lands of Utah* was not exactly a call to arms for the irrigation movement. In fact, after 1874 Powell repeatedly shied away from endorsing either federal or state construction of irrigation dams and canals. Identification of Powell with the early movement for federal reclamation can probably be traced to the Reclamation Service itself. The *First Annual Report of the Reclamation Service* (1903) gave a brief history of the development of laws to promote the efficient use of water and land in the arid region, including the important Desert Land Act of 1877, which placed emphasis upon the irrigation of land by individual effort in return for generous land grants of 640 acres at $1.25 per acre. But the Reclamation Service maintained that Powell wanted more: "The need, however, for action on a larger scale by the General Government was first prominently set before the public by Maj. John W. Powell . He recognized clearly and stated definitely that the General Government must of necessity deal directly with the irrigation question." Later scholarly articles echoed this supposed early commitment of Powell to the cause of national reclamation.[28]

Instead, Powell's 1878 report saw the West in more complicated terms. No uniform water law or state administrative apparatus existed in the West.[29] State laws were as chaotic as the state boundaries that divided rivers and watersheds. He suggested that political units be reorganized around hydrographic basins; that states yield authority to these new units to resolve the problems of water sources in rivers flowing across state lines. According to Powell, the fundamental outlines of the landholding pattern must change for the West to practice irrigation. The rectangular survey established by Thomas Jefferson's orderly mind in the grid system under the Land Ordinance of 1785 must give way to a system that gave farmers access to water. The rectangular survey could not possibly adjust to these demands. The new land units must be expanded to at least 2,560 acres, or the equivalent of four square miles, in irregular shapes following topographical features instead of a grid system. The acreage was far more than the 160 acre homesteads deemed adequate under the Homestead Act. Larger land units would enable an agricultural enterprise in areas of less than 20 inches of annual rainfall (arid America) to include various resources: water, suitable irrigable land for crops, pasture land for grazing, and woodlands for fuels, fences, and shelter. Powell assessed the resources of the West and concluded that a new land distribution pattern was needed along with a political reorganization of the region to enable people to live upon the land and govern their own affairs.

What he called for, according to Donald Worster in his biography of Powell, *A River Running West*, was classifying the lands of the West according to their resources. Almost a decade and a half passed before Powell was called upon to undertake such a comprehensive land classification scheme in an Irrigation Survey of the West. After Powell's short-lived Irrigation Survey, the federal government did not again seriously entertain the idea of land classifications until the dust storms and farm surplus crises of the New Deal decade in the 1930s. As was the case in Powell's era, powerful interests feared that land use studies and classifications meant placing land under the "dead hand" of bureaucratic czars and land planning boards that could limit future economic development.[30]

While Powell was an important scientific mind of late nineteenth-century America, his reputation did not place him above controversy and criticism. In his biography of Powell, Worster notes "the deadly silence that fell around the Powell report of 1878." Its critics saw it as backward-looking, de-emphasizing individualism in favor of cooperation and community; much too committed to locking the West into a strictly semi-livestock agrarian future rather than the open ended future that a freewheeling land disposal system offered. Needless to say, the recommendations of the arid lands report were never implemented. And would it work, given the way land laws had been twisted and abused in the nineteenth century? In fact, the *Public Lands Commission Report* of 1880 paid little attention to Powell's 1878 report. It lauded the Homestead Act and expressed no concerns about its effectiveness in the arid lands. The Commission's report omitted Powell's suggestions for community-organized irrigation districts with small eighty acre farms, but endorsed his concept of large pasturage estates exceeding his suggested 2,560 acres, if necessary. Although Powell was a member of the land commission, he did not challenge its conclusions. He was in no position to upset current land legislation patterns or fly in the face of the powers that be in Washington. Powell had ambitions and was appointed successor to Clarence King as Director of the U.S. Geological Survey (USGS) in 1881.[31]

A Troubled Survey

Future troubles loomed for Powell. In 1888, Congress approved a hydrographic survey of the arid lands under the direction of Powell's bureau, the USGS. Again, Nevada's Senator Stewart spurred the passage of an irrigation study, and in 1889 he organized a tour by the Senate Committee on Irrigation of western irrigation sites. The legislation approving the survey declared that the arid region was "capable of supporting a large population thereby adding to the national wealth and prosperity." It asked Powell to identify dam and reservoir sites throughout the West and prospective lands to be irrigated in this grand survey. Many assumed that this survey was the foundation and certainly the doorway to a national reclamation or irrigation program for the West. This was not necessarily Powell's goal. As historian Donald Pisani and others have repeatedly emphasized, Powell never favored a national program. On a few occasions he expressed interest and support for federally constructed reservoirs but little more. "Powell usually argued that national work should end with comprehensive surveys of the arid lands," writes Pisani. Powell's antipathy to national control and development may be one of the factors that made trouble for him in his conduct of the Irrigation Survey.[32]

2.10. Senator William Morris Stewart of Nevada. Courtesy of the U.S. Senate Historical Office.

Several events brought the Powell Irrigation Survey to an inconclusive end. While Senator Stewart had been an early champion of the Irrigation Survey, he soon became Powell's nemesis. In the process he yielded the early mantel of irrigation advocacy to a newcomer in Nevada affairs, Francis Newlands, who usually commands the most attention of any Nevadan in the campaign for national irrigation. Stewart's attention, more so than Newlands, alternated between his interest in irrigation and the silver cause as panaceas for the woes of the Nevada economy. The droughts and harsh winters that plagued Nevada and the mountain West called for deliberate and

2.11. Senator Francis E. Warren of Wyoming. Curtesy of the U.S. Senate Historical Office.

immediate action on irrigation issues, but the seemingly slow pace of the Powell survey promised little. Other western politicians also pushed for national aid to irrigation. Senator Francis E. Warren of Wyoming was important. He ultimately wanted irrigation development to take place under the auspices of the states, with national aid taking the form of reservoir construction and land cessions to the states. Eventually, supporters of the "state party" came into conflict with those such as Newlands who opted for a federally controlled irrigation program.[33]

Senator Stewart had been associated with Nevada affairs since the early days of the Comstock Lode excitement in the 1860s. He was prominent in the state constitutional conventions of 1863-64 and was elected to a U.S. Senate term that ended in 1873, when he returned to California to work for Collis P. Huntington and the Southern Pacific Railroad. He came back to Nevada for another U.S. Senate seat in 1886 to serve primarily railroad interests in the nation's capital as Congress prepared to pass the Interstate Commerce Commission Act of 1887 to regulate the interstate railroad system. While the railroad wanted to protect itself from federal regulation, it had a growing interest in western irrigation. The Southern Pacific supported the convening of annual irrigation congresses in designated western cities beginning in 1891. Water for western lands meant an increase in land values for the Southern Pacific Railroad, the largest private landholder in Nevada with additional large holdings in the Lower Colorado River Basin in California. This helps to explain Senator Stewart's enthusiasm for irrigation – an enthusiasm usually tempered by the energies he devoted to other interests, including mining, free silver, and the protection of the railroad from either state or federal regulations.[34]

The form that reclamation would take was as yet unclear. By the late 1880s, Senator Stewart was convinced that it should occur but was without a clear program for it to follow. Newlands likewise was in the process of formulating his ideas, and Warren of Wyoming definitely wanted a state-based system. In sponsoring the 1888 Irrigation Survey bill, Stewart and others recognized the need to locate and reserve dam sites, reservoir sites, and irrigable lands. With this accomplished, the federal government might cede the lands to states. The states could then undertake irrigation projects and offer incentives for irrigators to form irrigation districts to construct dams, reservoirs, canals, ditches, and laterals. Or they could invite private companies to develop prime sites and lands. This state-based action plan largely grew out of the development of water plans in Wyoming. It appealed to Senator Stewart because he was very much a nineteenth-century American who could hardly imagine the federal government undertaking such an enterprise. These nineteenth-century attributes of mind marked him off from his colleague and early protégé in Nevada politics, Francis Newlands, who had only arrived in Nevada from California in 1888. What Senator Stewart wanted most for Nevada at the beginning of his resumed political career in the Senate were quick results on the irrigation front.[35]

2.12 Senator Francis G. Newlands of Nevada. Courtesy of the U.S. Senate Historical Office.

Powell's slow-paced Irrigation Survey could not deliver the goods. Nor did he demonstrate the enthusiasm for irrigation as a panacea for all the ills of the region that many boosters exhibited. He could only offer a prolonged study of the West that might take ten years to produce a detailed topographical map with a designation of irrigable lands, damsites, and reservoir sites. Stewart hated plodding studies. He was a man of action. As a no-holds-barred mining lawyer, he began his career in the frenetic gold rush society of California and, subsequently, earned riches in the legal wrangling over mining claims that consumed Nevada's rich Comstock Lode in the 1860s. Life for Stewart was about action and achievement not study, reflection, and the slow accumulation of factual knowledge that often fostered doubt, hesitation, and caution.

Stewart reacted to Powell's Irrigation Survey with disbelief and anger. In the eyes of Stewart, Powell lost his luster as one of America's most distinguished scientists and government servants and became a conniving, featherbedding bureaucrat. Something happened between Powell and Stewart when they toured the West in late summer of 1889 with the Senate Select Committee on the Irrigation and Reclamation of Arid Lands. Most notably, Powell refused to encourage the local western boosters who testified before the committee on tour. If he did anything, he sought to dampen their spirits and point to limitations on the acreage that could be irrigated in the future. As Professor Pisani notes, Stewart considered Powell a Cassandra about western irrigation:

> It was not simply that Powell exploded myths, not simply that he was a Washington bureaucrat, not simply that he proposed maps and surveys while most westerners demanded action, not simply that he was often inconsistent in public statements, not even that he was perceived as lordly and self-serving; his deepest sin was that he was a Cassandra who questioned dreams without offering others to take their place.[36]

As Stewart became suspicious of Powell's support for western irrigation, the Department of the Interior ordered a halt to land claims on the Public Domain. It had become alarmed over extensive filings upon lands that might prove irrigable under the new Irrigation Survey. The department feared that large land and water companies were making the filings in order to speculate in lands that would soon rise in value because of the Irrigation Survey. The solution was to invoke a provision attached to one of the appropriation bills for the Irrigation Survey that enabled the department to withdraw western lands from entrance and filing until completion of land classification. According to progress reports issued by Powell, this might take years. Meanwhile, prospects for western development would languish and give way to interminable planning. Little wonder that Senator Stewart and other irate westerners called for a halt to the Irrigation Survey. Indeed, Stewart personally led the fight to slash the Irrigation Survey's appropriations. Powell finally resigned as Director of the USGS in 1894 under a cloud of suspicion and disapproval from western interests and politicians.

Western politicians grew up in a climate of boosterism. They were understandably impatient and disappointed with the progress of the Powell Irrigation Survey. When Senator Stewart discovered Major Powell's true beliefs about the limited promises of western irrigation, he was bitterly disappointed — so much so that he vilified him, even though Powell was a respected scientist, explorer of the Colorado River, and disabled veteran of the Civil War. Stewart's vendetta brought to an end the Irrigation Survey that George Perkins Marsh had recommend as the first step toward the establishment of a national irrigation program almost two decades earlier.

The destruction of the Powell survey has not played well with later historians and writers. The prominent western literary figure, Wallace Stegner, in his Powell biography entitled *Beyond the Hundredth Meridian* (1953), saw Powell as the victim of rapacious western interests personified in the dark figure of Nevada's Senator Stewart. The Powell-Stewart brouhaha reflected the sharp division between the forces of good and evil characteristic of histories, novels, movies, and plays about the West. In the context of the times, it is understandable that western representatives could not accept a moratorium on western development and the virtual reversal of a century-long policy of the disposal and alienation of the public domain. In fairness, Powell was not totally the instrument of that abrupt change. He was the man caught in the middle and acquiesced in the moratorium on land entries to permit the survey so essential for national irrigation to go forward.[37]

Desperation and Disparate Voices

Westerners, with Senator Stewart the chief fulminator, responded with outrage when Powell's public statements deflated hopes for vast irrigated acreage. By the 1890s, western states schemed in one way or another to achieve economic growth and escape the grip of the decade's nationwide depression as well as a regional drought. While westerners suffered, they would not also suffer warnings at the 1893 National Irrigation Congress in Los Angeles from Powell and his hydrological

2.13 John Wesley Powell, Director of the U.S. Geological Survey from 1881-1894, including during the irrigation survey 1888-1892.

assistant, Frederick H. Newell, that irrigation presented but limited promises for the future of the West. Powell's discouraging remarks centered upon his estimation that irrigation should not be extended much beyond the lands already under irrigation and then only under the control of local communities. Control and planning was a rebuff to American individualism whether it came from the local level or the federal level as he had recognized in his *Report on the Lands of the Arid Region of the United States, with a More Detailed Account of the Lands of Utah* in 1878. Powell's pronouncements about the limits of western irrigation caused the movement to ridicule him and discount his previous investigations and explorations of the Colorado. Major Powell's reputation suffered so much that he failed to achieve (nor did he expect it) the fame accorded to the likes of Lewis and Clark and John Charles Frémont as renowned nineteenth-century explorers of the American West.[38]

Some suggested that the West's problems were particularly its own – stemming from its hostile, arid environment; distance from markets; and lack of manufacturing industries. Such thinking blamed the West's problems upon its own inherent disadvantages or the bad hand nature dealt it. Whatever the root cause, the western predicament presented a persisting problem for the nation. After 1890, new representatives from western states trumpeted the cause of national irrigation on the floor and in the committees of Congress. Nevada, while not a new western state, suffered from the sharp decline of mining with the depletion of the Comstock Lode. And by the late 1880s, no campaign for silver money could revive it. Nevertheless, its chief political figures, Stewart and Newlands, alternated between thinking free silver or irrigation would rescue the West and their impoverished state.

The deep economic depression of the 1890s and the energies diverted into the silver crusade stalled the drive for national reclamation, further clouding the bright future boosters promised. Stagnation and economic retrogression raised the questions of whether the region was stamped with a limited, parochial future constantly requiring outside help, but always left empty-handed by a reluctant federal government and exploitive corporations. Was it in danger of becoming a permanent backwater eking out a living "at the end of the cracked whip"? According to some, it was a victim, a plundered province with its wealth and property held largely by absentee capitalist overlords – be they cattle or timber barons or mining magnates.[39]

One of America's most prominent historians at the beginning of the twentieth century, Frederick Jackson Turner, refused to believe the West's problems were uniquely regional. In an 1896 article in the *Atlantic Monthly*, "The Problem of the West," Turner said the issue was not simply stalled regional development but a national paralysis. When western development was in trouble, the entire nation was in trouble, as were its chief ideals of democracy and opportunity. Turner believed that American access to free land in the West transformed institutions and insured freedom of opportunity and the growth of democratic ideals in the society. He created a "frontier" interpretation of American history. His famous essay, "The Significance of the Frontier in American History," delivered to the American Historical Association meeting at the Chicago World's Fair in 1893, observed that the latest census of the United States in 1890 revealed that open lands of the West were no more – the frontier was closed. How would opportunity and democracy fare without a frontier? [40]

2.14. Frederick Jackson Turner, historian of the American "frontier."

The arid lands promised a new frontier that offered opportunity as the nation faced economic distress, a rising tide of immigration, and poverty in the cities. The argument appealed to an "agricultural fundamentalism" popular in the rhetoric of politicians, academics, and crusading journalists who saw only beneficial social effects when people forsook cities for rural America. Water for the dry lands of the West, therefore, meant a new frontier, new challenges, and new opportunities for Americans. Despite the quick victory and acquisition of overseas possessions in the Spanish-American War, the decade of the 1890s proved an unsettling time. Economic depression, the rise of third party movements, and a continuing flow of immigrants into the United States looking for opportunities anywhere in society raised fears about the future. The American West offered a solution, a way out, an escape from the growing complexities and problems that the current economic trends of urban industrialism inflicted upon the nation. Redemption of lands through irrigation could make it happen. But did the redemption of arid lands and the building of western irrigation communities simply offer a refuge from modern life? Progressive westerners in the mold of Nevada's Congressman Francis Newlands rejected this view.

Irrigated lands represented progress and a new era of stability and growth for western states. He saw reclamation rescuing his own adopted state, poorest of the western commonwealths, from the uncertain fortunes of a mining and ranching economy.[41]

New Frontiers: Rural and Urban

A new irrigation frontier promised to defuse urban and industrial discontent evident in America during the depressed 1890s. The argument focused on the "safety valve corollary" of Turner's frontier thesis. The West and its cheap or free lands offered an escape. The poor and the exploited of America, instead of engaging in strikes and rioting in eastern cities, would move to vacant lands in the West. In the manner of a "safety valve," western lands siphoned off political and social discontent. Consequently, or so the theory went, the United States avoided the social upheaval and violence that threatened industrial societies elsewhere. Arid lands, once seen as a barrier to settlement, offered remedies for some of the ills of industrial society. But the safety valve, in this instance, could work only if the federal government renewed the frontier experience by launching a reclamation program.[42]

The call for government to assume greater social and regulatory responsibilities in response to "the will of the people" was becoming known as "Progressivism" in some circles. The conservation wing of the Progressive Movement sought to protect natural resources from the wasteful, destructive practices of nineteenth-century business – logging, grazing, and even mining. Land figured into the equation too. Good lands of the West must not be allowed to go to waste if their redemption only required the delivery of water, and they must not be monopolized by the few. This was merely an act of utilizing resources in a sound economical, efficient manner for the benefit of the people. Participation by government in reclaiming the western deserts met all the tests of a rising tide of reform called the Progressive Conservation Movement.[43]

But was this policy choice progressive, forward looking, and modern or backward, reactionary, and almost anti-modern? Historians disagree. With no little irony, Pisani asserts in his book *Water and American Government* (2002) that, far from "modernizing" the West and bringing it into the circle of national development, those who argued for the Reclamation Act of 1902 and its provisions to protect the small family farm froze their cause into traditional patterns of nineteenth-century individualistic farm making. Their visions were more closely related to "the agricultural model of 1800 or 1850" rather than being a sharp break with the past or leap into a progressive future. [44]

The drive for national irrigation gave new life to the century-old debate between Thomas Jefferson and Alexander Hamilton about the future of the United States. Jefferson, of course, believed the yeoman farmer on the land was the backbone of the republic; in contrast, Hamilton believed America's future lay in commerce, industry, and cities. While Hamilton's vision seemed to prevail, many resisted it. If this were the reality of progress, must the Jeffersonian vision be unprogressive and reactionary? As cities grew haphazardly amongst poverty and new foreign populations, many Americans preferred to believe that a better future lay with the land, that the nation had gone astray on dangerous paths to urban-industrialism that represented retrogression into poverty and dependence rather than self-reliance.

Revival of the Jefferson-Hamilton debate reflected a growing ambivalence about the city, the threat it posed to values, culture, manhood, womanhood, democracy, and prosperity. Both in the countryside and the city, the hard economic times of the 1890s increased doubts about the justice of an industrialized, urban society dominated by business interests. New farms in the arid West could offer an alternative to the uncertainties of urban-industrial life. While backers of irrigation for the West could only believe that watering the West kept alive the gift of free land and opportunity for rural frontier life to renew American democracy, Worster, in his 2001 biography of John Wesley Powell, *A River Running West*, believed that the rhetoric of the irrigation enthusiasts, which described reclaimed arid lands in the West as a safety valve or a refuge for the poor, missed the mark. Informed by the retrospect of an ever urbanizing America over the course of the twentieth century, Worster asserts that the poor of America were not seeking farms on which to make opportunity for themselves. Making desert farms was, at best, a losing gamble for them: "They did not look to the arid lands for opportunity; they looked to the unions, the eight-hour day, and the federal laws to protect them against exploitation."[45]

The Publicists

Still, the attraction of making a living from one's own farmstead exercised powerful appeals, according to articles appearing in the popular periodical press of the day – *Colliers*, *World's Work*, *Scribner's*, and *Pacific Rural Press*. Rural life appeared glamorized, where a living could be made from a farm without the supervision of an employer's critical eye and the dangers of work in factories, mines, or mills. The romanticized picture of rural life persisted despite a contrary narrative reflecting a harsh, bitter side in the novels of Hamlin Garland and others. Garland's novels, *A Son of the Middle Border* and *A Daughter of the Middle Border* (1917, 1921), made good cases for escaping the drudgery of farm life. Yet, while flight from the land might offer freedom and opportunity by some accounts, a good deal of the literature of the day dwelt on the dangers of city life; e.g., Theodore Dreiser's *Sister Carrie* (1901). Some believed that a compromise could be achieved in "a synthesis of the city and the country," as the American economic base shifted from agriculture to manufacturing. The diverse and contradictory literature, of course, reflected a nation undergoing a transition in its economic structure and living patterns as it moved along the road to urbanization and industrialization.

2.15. William Jennings Bryan was a fiery orator, free-silver, agrarian advocate..

For many, American farm life was crucial to the welfare of the society. The countryside's underlying ability to feed the cities was reassuring. Moreover, the land and its farms offered security, psychologically if in no other way, that individuals who failed in the cities might return to the land. Mixed in was a good deal of nostalgia for the bucolic life, but this grew dimmer as farm income fell behind that of factory workers. Beyond eroding romanticism, however, the ability to earn a living from the land had the mark of honest, worthy American manhood and womanhood about it. When the Democratic presidential nominee, William Jennings Bryan, denounced plutocrats in his famous "Cross of Gold Speech" at the 1896 Democratic Presidential Convention, he also declared: "Burn down your cities and leave our farms, and your cities will spring up again as if by magic; but destroy our farms, and the grass will grow in the streets of every city in the country." These words drew upon a long tradition of agricultural fundamentalism reaching beyond the borders and history of the United States that saw agriculture as the foundation for all other economic advances and even fundamental to the moral and spiritual welfare of a people.[46]

In contrast to the good souls of the countryside, people in the cities often survived more by wit, contrivance, and casuistry rather than honest work. Publicists and journalists seized upon the lingering myths of farm life, linking them to the health of the American republic. They perpetuated agrarian promises about the arid lands that rested more on the sentimental and emotional than on the hard realities of bringing lands under the ditch by community, state, or federal irrigation. As a new generation of Americans participated in this grand enterprise, they could reaffirm American values and be saved from the tedious, emasculating pursuits of urban life.[47]

The writings of journalist-publicist William Ellsworth Smythe (1861-1922) typified the extravagant, sentimental, and emotional promises of agricultural life. Smythe helped popularize the indecorous slogan: "Bring the landless man to the manless land." Beginning his journalistic career in Nebraska, Smythe helped organize the first annual National Irrigation Congress held in Salt Lake City in 1891. He popularized the cause of national aid for the irrigation of western arid lands and founded a new periodical, *Irrigation Age,* dedicated to making the Homestead Act workable in the arid lands. As he explained, the 1862 Act that granted 160 acres of free land proved unworkable beyond the 100th meridian, especially on western lands with less than twenty inches of rainfall a year. Climate and nature denied the interior Far West the benefits of the federal government's largesse in the Homestead Act legislation. Only the construction of federally aided irrigation works could extend the benefits of the Act to these lands and people. In fact, as historian

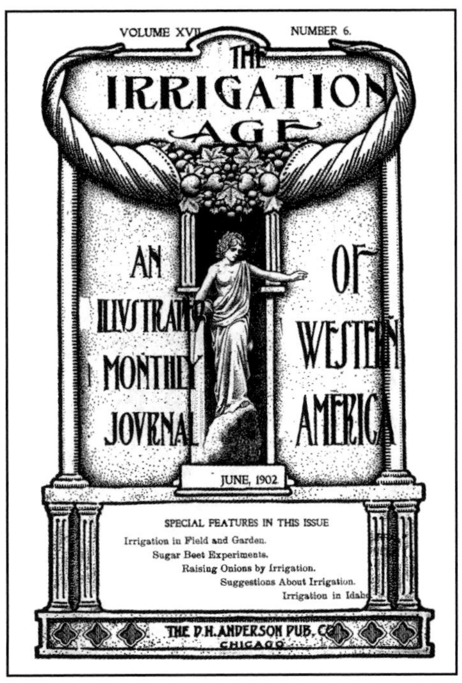

2.16. The June 1902 cover of *Irrigation Age,* a publication founded by William Ellsworth Smythe.

Pisani suggests, federal reclamation in this light came to be regarded as an "entitlement." Smythe also sang the glories of the agrarian society that could be built in the irrigated arid lands. That undertaking would revive American democracy and save individuals from the anonymity of urban life. A federal program to bring water to the arid lands would open a new American frontier.[48]

Smythe also saw opportunity in the irrigation cause for his own career as a journalist and publicist. The pages of the *Irrigation Age* ran numerous advertisements for agricultural implements and, more to the point, irrigation supplies such as ditch diggers, pipes, well equipment, valve works, and head gates. Not only would business suppliers flourish, but land prices would skyrocket, and the entire region would grow rich. All of this suggested that Smythe was more than an idealist. When the enthusiasm for western irrigation seemed to lag in the depressed 1890s, Smythe turned to the founding of small colonies where a cooperative spirit could prevail. Already, the trend seemed to favor ceding land to the states in the manner of the 1894 Carey Act, but even that move did not turn western acres into gardens. The inspiration for Smythe's faith, the annual irrigation congress, seemed increasingly an organization of the inland West with few connections to California or the Pacific Northwest. The irrigation congresses represented large numbers of land and water companies, lawyers, journalists, state officials, and businessmen rather than actual farmers and stockmen.[49]

Smythe propagated his ideas, and even acted them out in the colonies he supported, for almost a decade before he summarized them in his vision for the future of the West in the dramatically entitled book, *The Conquest of Arid America*, first published in 1900. In it, he recalled his attraction to the irrigation movement almost in terms of a religious conversion:

> Irrigation seemed to be the biggest thing in the world. It was not merely a matter of ditches and acres, but a philosophy, a religion, and a program of practical statesmanship rolled into one. There was apparently no such thing as ever getting to the bottom of the subject, for it expanded in all directions, and grew in importance with each unfoldment.[50]

On the verge of obtaining a national Reclamation Act in 1901, Smythe congratulated himself on launching the publication *Irrigation Age* ten years earlier "to preach the gospel to the heathen on both sides of the continent," a cause that "bore the white flower of civilization," he announced. Now at this moment he declared, "We are yet in the thick of events. Only the threshold has been crossed. At this time, rather more than ever before, there is need of preaching, teaching and doing." [51]

By this time, his work came to the attention of Representative Francis G. Newlands of Nevada. As a Californian and a Nevada politician since the late 1880s, Newlands had long been interested in irrigation. Although he was a beneficiary of the mining wealth of Nevada, he believed that mining was unable to provide a stable future to Nevada. Newlands underwrote Smythe's second book, *Constructive Democracy* (1905), which contained much starry-eyed hyperbole about the irrigated West. Farming the arid West offered both prosperity and the highest form of democratic government to the common man. The attachment of water rights to small freehold homesteads would favor social cooperation instead of cutthroat competition. The results promised economic, political, and social harmony.[52]

Another western irrigation enthusiast whose ideas attracted Newlands was George H. Maxwell, a California lawyer. Maxwell, with the backing of railroad interests, took up the cause of publicizing various irrigation undertakings in the 1890s. He had already devoted a great deal of his career as a water and mining lawyer to dealing with the problems of California's Wright Act that allowed for the formation of local irrigation districts with power of taxation and eminent domain.[53] From these battles, he concluded that irrigation must be dealt with from the level of the national government. After the release of a Corps of Engineers report by Captain Hiram Martin Chittenden entitled *Preliminary Report on Examination of Reservoir Sites in Wyoming and Colorado* in 1897, Maxwell became an advocate of federal reservoir building to aid western irrigation.

2.17. Hiram M. Chittenden. Courtesy of the U.S. Army Corps of Engineers.

The Chittenden report went so far as to support the federal government providing free water to western landowners financed by appropriations from the Rivers and Harbors bills that Congress passed almost annually. Not only did this major report

Read "Capturing Wealth from the Air" in this Number

The Homecroft Magazine

Maxwell's Talisman

A Journal of Construction and Social Education

with which has been incorporated

OPPORTUNITY

ONE YEAR 25 CENTS — JULY, 1908 — ONE COPY 3 CENTS

Every Child in a Garden

By LOUISE KLEIN MILLER

The Satisfactions of the Farmer's Life

By SOLON O. THATCHER

The Slogan of the Homecrofters

Every Child in a Garden—Every Mother in a Homecroft—and Individual Industrial Independence for Every Worker in a Home of his own on the Land

2.18. The cover of *Maxwell's Talisman: A Journal of Construction and Social Education*. July 1908. The swastika (gammadion figure; gammation figure; *crux gammata*; gammate cross) in the upper left-hand corner was a symbol commonly used to represent well being and harmony. It is unrelated to the later use of the swastika by the Nazis in Germany.

suggest the building of reservoirs, it also suggested building main line canals to serve new lands.[54] In 1898, Chittenden sent a letter to the seventh annual irrigation congress, meeting in Cheyenne, Wyoming, suggesting that if such work were to be undertaken, the Corps of Engineers, with its accumulated expertise in these matters, should be the obvious choice to conduct the enterprise. While this portended future conflict with any new bureau created to carry out federal irrigation programs, Maxwell and others were mightily impressed with an endorsement from a major figure in the well-respected Army Corps of Engineers. By 1900, Maxwell became a chief publicist for federal reclamation and founded the Homecroft Society to settle families upon the land in irrigated communities. He testified before Congress and published numerous articles about the benefits of federal development of western irrigation water resources for the benefit of not only the West, but also of the East as an outlet for "surplus labor" that should be encouraged to settle upon the land. "Arid America beckons to them with open arms," he wrote as he foresaw the new irrigated lands acting as a balm to social unrest caused by unemployment in the cities. He formed the National Irrigation Association in Wichita, Kansas, in 1897 with his talented associate Guy Mitchell, who served as its director and chief publicist. Mitchell's articles touting irrigation regularly appeared in national publications, including Maxwell's. Maxwell founded numerous publications with financing from various railroads, e.g., *The National Homemaker*, *California Advocate*, *Opportunity*, and *Maxwell's Talisman*, to publicize the reclamation of arid America.[55]

The Outcome

Congress came under increasing pressure to act. It faced pressure from several fronts: (1) the intellectual and historical arguments for western irrigation gained prominence in the press that sometimes took on sentimental and emotional tones from publicists, (2) railroad interests saw advantages for their companies both in serving new farming communities created by government financed reclamation projects and in enhancing the value of their own western lands, (3) new western state representatives used their votes and influence in Congress in favor of a government program, and, finally, (4) the severity of the economic depression in the 1890s and the proclaimed closure of the frontier combined to build a rationale for Congress to open new lands to create more opportunity in the hopes of avoiding further economic distress and social upheaval. But crop surpluses also caused midwestern farm groups to pressure Congress not to act.

Where private efforts failed or feared to venture, so its adherents proclaimed, government-sponsored reclamation, would open new horizons in the twentieth-century West. Federally financed dams, reservoirs, and main line ditches promised to spark regional development. Yet the dams offered more than water storage and flood control. They facilitated the electrical age by providing a source of power endlessly transformable for myriad tasks. Historically, dams directly harnessed waterpower to drive the mechanisms of industry – sawmills, grain mills, and large textile mills. Now they stood ready to produce a far more adaptable, pliant form of power. While imposing dam structures came to represent electrical power along with the capacity to store or conserve water, the more compact and mysterious mechanism known as the dynamo drew the attention of the curious at World's Fairs. The enthusiasm for irrigation came at the dawn of the electrical age. Few suspected that the two would join in an amicable marriage and march hand in hand into the twentieth century. Dams for water storage also became hydroelectric sites as electricity proved to be an unexpected bonus in the drive for arid lands reclamation.

2.19. Henry Adams, a historian who saw in the dynamo a new force that would shape the future..

Historian Henry Adams was a scholar of both American history and of Europe's Middle Ages. His career reached its apogee in the years after the Civil War when the American government presided over a far-flung nation attempting to cope with Reconstruction, a flood tide of immigration, and rapid western expansion. As a member of one of the founding families of the republic, he took a keen interest in public affairs.

2.20. In 1906 the powerhouse on the Strawberry Valley Project in Utah was about ready to operate – becoming the first permanent hydropower plant on a Reclamation project. For historian Henry Adams the dynamo symbolized the new era and the source of power for the modern age.

He had applauded the emergence of scientific surveys of the American West under civilian direction as forerunners of government service bureaus. Quick to observe new developments, he saw in the dynamo and electricity a power to create and revolutionize modern life, but Adams also admitted to an underlying trepidation about where this new power would take American society. Having studied the spirit and vitality of a pre-modern Christian Europe in the Middle Ages, he believed he saw in the dynamo a new force as powerful, in secular terms, as was the driving religious spirit of Christian Europe in the realm of the sacred. As a historian seeking larger causes and interpretations, Adams tried to identify "forces" that shaped the dynamics of an age. In his most widely read work, *The Education of Henry Adams*, he contrasted the sublime aesthetic achievements of the Middle Ages, moved by the power of religious conviction, in a chapter on the "The Dynamo and the Virgin," with modern society on the threshold of deriving its power from the electric dynamo. While the force of a previous era expressed itself in sublime spiritual expressions in art and architecture, he saw few of these promises in the stark, harsh, machine-centered aesthetics of the dynamo. Of course, what Adams felt reluctant to accept, and even disdained, was the cold modernist aesthetics of a new era entwined in the dynamo's steely wires. [56]

Unlike Adams, Nevada's Francis Newlands was no vacillating Hamlet on this score. He embraced the new age and its new power as well as the values of a new political movement, Progressivism. Not surprisingly, he declared reclamation the highest form of conservation as Progressives made the Conservation Movement a centerpiece of their cause. Under

2.21. With the dynamo came control panels such as this one at the Black Canyon Powerplant on the Boise Project. It came online in 1925.

reclamation, waste lands became productive, dams on rivers formed reservoirs to conserve water and hold back costly floods, and navigation improved. As an added dividend, the generation of electrical power was part and parcel of the management of water resources under a reclamation regime. Stamping out waste, whether in idle lands, in the prodigal use of forest resources, or in permitting rivers to rush wantonly and unused to the sea, was part of the Progressive Conservation Movement to which both Newlands and Theodore Roosevelt ascribed. In the words of Gifford Pinchot, Chief of the new Forest Service and principal advisor to President Roosevelt on conservation issues, this was "Wise Use," marked by rationality and efficiency, the opposite of the wastefulness common to nineteenth-century resource use. Conservation would move the people of the West beyond the reckless frontier stage of development into a modernizing, efficient society and economy.[57]

The Progressive reform movement of the first two decades of the twentieth century followed the protests of the Populists in the 1890s and assumed many of their causes, such as greater democracy in American political life and regulation of corporations and trusts. In addition, it directed government attention to social issues, including child labor, injured workmen, tenement housing, and the eight-hour work day. The application of expertise and science to the problems of society was also a Progressive innovation. Progressives saw themselves as the new wave of modern, up-to-date, urban reformers; a far cry from the shaggy, rural Populist agrarians who had failed to win basic reforms in the 1890s because, in the view of the Progressives, the Populists were oriented to the past and not the future. After thirty years of debate, the Progressives now pursued reform by coupling science and expertise with government to confront and solve many of the problems of modern America, including, if western irrigation supporters had their way, the reclamation of the West.[58]

2.22 The Boise River Diversion Dam Powerplant went into operation in 1912.

2.23. The Minidoka Dam Powerplant in 1911.

Irrigation advocates welcomed the Progressive Movement. They approved and identified with its reform programs that called for a stronger, more activist American government that could serve every region of the nation. If the federal government undertook reclamation of the West, it could also turn its attention to the reclamation of swamplands in the South and Midwest through drainage reclamation. The proposal sought to counter arguments that Congress should not fund a program that benefited only one section of the country. A more telling defense of a federal program for western reclamation grew out of debates over river and harbor appropriations. If Congress could fund harbor and river improvements on the two coasts, in the Mississippi Valley, and the Great Lakes, some western senators argued that the arid West was entitled to a share of Congress's largess in the form of dams, canals, ditches, and even dynamos. The time appeared almost at hand for the passage of important national legislation for the reclamation of the West at the beginning of the new century. Still, it would take a combination of favorable political circumstances to push Congress toward that objective. And this was by no means ensured with the reelection of President William McKinley in 1900.

Endnotes:

[1] J. Hector St. John de Crevecoeur, *Letters from an American Farmer* (London: Printed for Thomas Davies, 1783), 51; Henry Nash Smith, *Virgin Land: The American West as Symbol and Myth* (Cambridge, Massachusetts: Harvard University Press, 1950).
[2] Richard Hofstadter, *The Age of Reform: From Bryan to F.D.R.* (New York: Alfred A. Knopf, 1955); Alexis de Tocqueville, *Democracy in America*, trans. Harvey C. Mansfield and Delba Winthrop (Chicago, Illinois: University of Chicago Press, 2000), 529.
[3] David B. Danbom, *Born in the Country: A History of Rural America* (Baltimore, Maryland: Johns Hopkins University Press, 1995), 144-5; Claire Strom, *Profiting from the Plains: The Great Northern Railway and Corporate Development of the American West* (Seattle: University of Washington Press, 2003), 6.
[4] William H. Goetzmann, *Exploration and Empire: The Explorer and Scientist in the Winning of the American* West (New York: Knopf, 1966), 303-31.
[5] Henry Adams, *The Education of Henry Adams* (New York: Random House Modern Library edition, 1931, first published Massachusetts Historical Society, 1907), 312; Richard A. Bartlett, *Great Surveys of the American West* (Norman: University of Oklahoma Press, 1962), 142; J. Kirkpatrick Flack, *Desideratum in Washington: The Intellectual Community in the Capital City, 1870-1900* (Cambridge, Massachusetts: Schenkman Publishing Company, Inc., 1975), 86.
[6] Golzé, *Reclamation*, 21.
[7] Bartlett, *Great Surveys*, xiv; William D. Rowley, *Reclaiming the Arid West: The Career of Francis G. Newlands* (Bloomington: Indiana University Press, 1996), 101.
[8] Nelson A. Miles, "Our Unwatered Empire," *North American Review,* 150 (March 1890), 370-72.
[9] Henry Nash Smith, "Rain Follows the Plow: The Notion for Increased Rainfall for the Great Plains," *Huntington Library Quarterly* 10 (1947), 175-88; James C. Olson, *History of Nebraska* (Lincoln: University of Nebraska Press, 1955), 173-4; John Wesley Powell, "The New Lake in the Desert," *Scribner's Magazine,* 10 (October 1891), 467.
[10] Pisani, *Water and American Government,* 5, quotes the *New York Times*, May 28, 1871, October 29, 1873, and January 5, 1896, "beyond the reach of meteorological caprices."
[11] A. E. Chandler, *Elements of Western Water Law* (San Francisco, California: Technical Publishing Co., 1918), 3; Norman Smith, *Man and Water: A History of Hydro-Technology* (New York: Charles Scribners's Sons, 1975), 60; John T. Ganoe, "The Beginnings of Irrigation in the United States," *Mississippi Valley Historical Review* 25 (June 1938), 64.
[12] Alfred Deakin, *Progress Report: Irrigation in Western America, So Far as its Relationship to the Circumstances of Victoria* (Melbourne, Australia: Government Printer, 1885), 7-9; James R. Kluger, *Turning on Water with a Shovel: The Career of Elwood Mead* (Albuquerque: University of New Mexico Press, 1992), 58.
[13] Smith, *Man and Water*, 60.
[14] Donald Worster, *A River Running West: The Life of John Wesley Powell* (New York: Oxford University Press, 2001) notes Powell's admiration of the Mormon accomplishments in irrigation through community cooperation, but also their expanding population placed too many demands upon land and water that even cooperation could not accommodate, 354; for a fairly positive assessment, see Thomas G. Alexander, "Stewardship and Enterprise: The LDS Church and the Wasatch Oasis Environment, 1847-1930," *Western Historical Quarterly* 25 (Fall 1994), 340-64; Dan Flores, "Zion in Eden: Phases of Environmental History of Utah," *Environmental Review* 7 (Winter 1983), 325-44; for an excellent overview, see Marcus Hall, "Repairing Mountains: Restoration, Ecology, and Wilderness in Twentieth-Century Utah," *Environmental History* 6 (October 2001), 584-610; Alvin T. Steinel, *History of Agriculture in Colorado* (Fort Collins, Colorado: State Agricultural College, 1926); Stanley Roland Davison,

[14 continued] *The Leadership of the Reclamation Movement, 1875-1902* (New York: Arno Press, 1979), 136-7; Daniel Tyler, *Silver Fox of the Rockies: Delphus E. Carpenter and Western Water Compacts* (Norman: University of Oklahoma Press, 2003), 50-1; Donald J. Pisani, *To Reclaim a Divided West: Water, Law, and Public Policy, 1848-1902* (Albuquerque: University of New Mexico Press, 1992), notes the Greeley colony "eventually prospered," 81; Smith, *Man and Water* for a quick description of canals, laterals, ditches, and corrugates, 60.

[15] Anne F. Hyde, *An American Vision: Far Western Landscape and National Culture, 1820-1920* (New York: New York University Press, 1990); Pisani, *Water and American Government,* 273.

[16] Theodore Roosevelt, *The Winning of the West*, 4 volumes (New York: G.P. Putnam's Sons, 1889-96).

[17] Rowley, *Reclaiming the Arid West*, 109; Albert K. Weinberg, *Manifest Destiny: A Study of Nationalist Expansion in American History* (Baltimore: Johns Hopkins Press, 1935), as quoted in Patricia Limerick, *Something in the Soil: Legacies and Reckonings in the New West* (New York: W. W. Norton & Company, 2000), 310; Davison, *Leadership of Reclamation*, 248; James E. Wright, *The Politics of Populism: Dissent in Colorado* (New Haven, Connecticut: Yale University Press, 1974).

[18] George M. Fredrickson, *The Inner Civil War: Northern Intellectuals and the Crises of the Union* (New York: Harper & Row, Publishers, 1965), 210-1; Raymond H. Merritt, *Engineering in American Society, 1850-1875* (Lexington: University of Kentucky Press, 1969), 2.

[19] United States Congress, Senate, Letter of the Commissioner of Agriculture in answer to a Senate resolution of February 6, 1874, transmitting a paper prepared by Hon. George P. Marsh, on the subject of irrigation, "Irrigation: Its Evils, the Remedies, and the Compensations," 43rd Congress, 1st Session, Misc. Doc. No. 55.

[20] Lawrence B. Lee, *Reclaiming the American West: An Historiography and Guide* (Santa Barbara, California: ABC–Clio, 1980), 8, n17.

[21] Kevin Starr, *Material Dreams: Southern California through the 1920s* (New York: Oxford University Press, 1990) notes that, "Hall, in fact felt it necessary explicitly to refute Marsh in the preface to *Irrigation in California [Southern]",* 18; William Hammond Hall, *Irrigation in [Southern] California* (Sacramento: 1888); Gerald D. Nash, *State Government and Economic Development: A History of Administrative Policies in California, 1849-1933* (Berkeley: University of California Press, 1964), 189-90; Smith, *Man and Water*, 61; Edward Wegmann, *The Design and Construction of Dams Including Masonry, Earth, Rock-Fill and Timber Structures Also the Principle Types of Moveable Dams* (New York: John Wiley & Sons, 1904), fourth edition. Original edition published in 1888 as *The Design and Construction of Masonry Dams.*

[22] W. Turrentine Jackson, Rand F. Herbert, and Stephen R. Wee, "Introduction." *Engineers and Irrigation: Report of the Board of Commissioners on the Irrigation of the San Joaquin, Tulare and Sacramento Valleys of the State of California, 1873* (Fort Belvoir, Virginia: Office of History of the U.S. Army Corps of Engineers, 1990. Reprint published as "Engineer Historical Studies" Number 5), 45, n. 79.

[23] As quoted from U.S. Congress, House. J. W. Powell, "Geographical and Geological Surveys West of the Mississippi", House Report No. 612, 43rd Congress, 1st session, 10, in Davison, *Leadership of Reclamation*, 43, n. 12; Jackson, et al., *Engineers and Irrigation*, 18, 30-1.

[24] Bartlett, *Great Surveys*, 142.

[25] Deakin, *Irrigation in Western America* notes, that as an Australian, he began his investigations by reading Powell, 7-9.

[26] Davison, *Leadership of Reclamation*, 39; Bartlett, *Great Surveys*, 529.

[27] Brookings Institution, Institute for Government Research, *The U.S. Reclamation Service: Its History, Activities and Organization* (New York: D. Appleton and Company, 1919), 8;

[27 continued] George Wharton James, *Reclaiming the Arid West: The Story of the United States Reclamation Service* (Washington, D.C.: Government Printing Office, 1917), 13.
[28] *First Annual Report of the U.S. Reclamation Service,* 48; Henry Nash Smith, "Clarence King, John Wesley Powell, and the Establishment of the U.S. Geological Survey," *Mississippi Valley Historical Review* 34 (June 1947), 37-58; Wallace Stegner. *Beyond the Hundredth Meridian: John Wesley Powell and the Second Opening of the West* (Lincoln: University of Nebraska Press, 1953).
[29] Donald J. Pisani, "Enterprise and Equity: A Critique of Western Water Law in the Nineteenth Century," *Western Historical Quarterly* 18 (January 1987), 15-31.
[30] Worster, *A River Running West*, 347, 350; M. L. Wilson, "Land Utilization," Economics Series Lecture #25, April 16, 1932 (Chicago, Illinois: University of Chicago Press, 1932); Thomas G. Alexander, "The Powell Irrigation Survey and the People of the Mountain West," *Journal of the West* 7, 1 (1968), 52.
[31] Worster, *River Running West*, 359, 377-9.
[32] Richard J. Orsi, *Sunset Limited: The Southern Pacific Railroad and the Development of the American West, 1850-1930* (Berkeley: University of California Press, 2005), 243; Pisani, *To Reclaim a Divided West*, 143; Everett W. Sterling, "The Powell Irrigation Survey, 1888-1893," *Mississippi Valley Historical Review* 27 (December 1940), 434.
[33] Mary Ellen Glass, *Water for Nevada: The Reclamation Controversy, 1885-1902* (Reno: University of Nevada Press, 1964), 45; Pisani, *From Family Farm to Agribusiness*, 283-99; Orsi, *Sunset Limited*, 245-6; Donald J. Pisani, "State vs. Nation: Federal Reclamation and Water Rights in the Progressive Era," *Pacific Historical Review* 51 (August 1982), 268.
[34] Orsi, *Sunset Limited*, 243; Russell R. Elliott, *Servant of Power: A Political Biography of Senator William M. Stewart* (Reno: University of Nevada Press, 1983).
[35] For more on Wyoming Senator Warren's views, see William Lilley, III, and Lewis L. Gould, "The Western Irrigation Movement, 1878-1902: A Reappraisal," in *The American West: A Reorientation*, editor Gene Gressley, (Laramie: University of Wyoming Publications, 1966), 32.
[36] Pisani, *To Reclaim a Divided West*, 151-2.
[37] Pisani, *To Reclaim a Divided West*, 149, 152-3; for a viewpoint differing with Stegner on the Powell-Stewart blowup, see Alexander, "The Powell Irrigation Survey and the People of the Mountain West," 51-3; Thomas G. Alexander, *A Clash of Interests: The Interior Department and the Mountain West, 1863-1896* (Provo, Utah: Brigham Young University Press, 1977).
[38] Michael A. Bryson, *Visions of the Land: Science, Literature, and the American Environment from the Era of Exploration to the Age of Ecology* (Charlottesville: University of Virginia Press, 2002) see chapter "Powell and the American West," 99; Davison, *Leadership of Reclamation*, 217.
[39] William G. Robbins, "'At the End of the Cracked Whip:' The Northern West," *Montana: The Magazine of Western History* 38 (Winter 1988), 2-11.
[40] Frederick Jackson Turner, "The Problem of the West," *Atlantic Monthly*, 78 (September 1896), 290; Worster writes that the arid West's answer to the problem was, "give us water and we will make opportunity anew," *River Running West,* 528.
[41] Rowley, *Reclaiming the Arid West,* 67.
[42] For a discussion of the safety valve "corollary" to the Turner thesis see Ellen VonNordhoff, "The American Frontier as Safety Valve: The Life, Death, Reincarnation, and Justification of a Theory," *Agricultural History* 36 (Fall 1962), 123-49.
[43] Samuel P. Hays, *Conservation and the Gospel of Efficiency: The Progressive Conservation Movement, 1890-1912* (Cambridge, Massachusetts: Harvard University Press, 1959); Donald C. Swain, *Federal Conservation Policy, 1921-1933* (Berkeley: University of California Press, 1963), 3.

[44] Pisani, *Water and American Government* declares this the thesis of his book: " to see the Reclamation Act of 1902 and the events that followed as evidence of the persistence of 'frontier America' and traditional nineteenth-century values, rather than as the emergence of 'modern America,'" *xi*.

[45] Turner, "The Problem of the West," 290; Worster, *River Running West*, 528-9.

[46] As quoted in Hofstadter, *Age of Reform*, 35.

[47] Davison, *Leadership of Reclamation*, 256; Patricia Nelson Limerick, *Desert Passages: Encounters with the American Deserts* (Albuquerque: University of New Mexico Press, 1985), 77-94.

[48] Carlson, "William E. Smythe: Irrigation Crusader," 45; see also Lawrence B. Lee, "Introduction" in William E. Smythe, *The Conquest of Arid America* (Seattle: University of Washington Press, reprint 1969); Pisani, *Water and American Government*, 274.

[49] Davison, *Leadership of Reclamation*, 221, 232; Lawrence B. Lee, "William Ellsworth Smythe and the Irrigation Movement: A Reconsideration," *Pacific Historical Review* 41 (August 1972), 289-311.

[50] William E. Smythe, *Conquest of Arid America*; also quoted in Gordon E. Nelson, *The Lobbyist: The Story of George H. Maxwell, Irrigation Crusader* (Bowie, Maryland: Headgate Press, 2001), 34.

[51] William E. Smythe, "The 20th Century West," *Land of Sunshine* [predecessor of *Out West*], 15 (June-December 1901), 61-4.

[52] Rowley, *Reclaiming the Arid West,* 121.

[53] Nelson, *The Lobbyist*, 27.

[54] For a good summary of Chittenden's contentions that government aid to irrigation should be viewed as "an internal improvement" see Hiram M. Chittenden, "Government Construction of Reservoirs in Arid Regions," *North American Review*, 174 (1902), 245-58; Pisani, *Water and American Government*, 277.

[55] Warne, *The Bureau of Reclamation*, notes, "The Chittenden report came out strongly for a comprehensive system of storage and diversion works under a national policy of internal improvements [conducted by the Corps of Engineers] for the development of the region," 11; Pisani, *From the Family Farm to Agribusiness*, 292-3; Pisani, *Water and American Government*, 16-20; Donald J. Pisani, "George Maxwell, the Railroads, and American Land Policy, 1899-1904," *Pacific Historical Review* 63 (May 1994), 177-202; Robinson, *Water for the West*, 15.

[56] Adams, *Education*, 379-90; David E. Nye, *Electrifying America: Social Meanings of a New Technology, 1890-1940* (Cambridge, Massachusetts: MIT Press, 1990).

[57] Gifford Pinchot, The Fight for Conservation (New York: Doubleday, Page & Company, 1910); David Stradling, editor, *Conservation in the Progressive Era* (Seattle: University of Washington Press, 2004), 11-4.

[58] Hays, *Conservation and the Gospel of Efficiency*, 249, 265; Pisani, *Water and American Government*, 285.

2.24. Reclamation completed the Sun River Diversion Dam, Sun River Project, Montana, in 1915.

2.25. The first crop of oats on the Grand Valley Project in west-central Colorado in the fall of 1916.

CHAPTER 3

NATIONAL RECLAMATION BEGINS

Introduction

With the lifting of the depression after 1897 came rekindled efforts to achieve legislation authorizing national reclamation. President Theodore Roosevelt's signature on the national Reclamation Act of June 17, 1902, marked the success of a long campaign initiated decades earlier. The Act opened the way to the creation of the U.S. Reclamation Service within the Department of the Interior. Staffed by many former employees of the U.S. Geological Survey, the new Reclamation Service embraced an array of projects throughout the West. Under the leadership of Frederick H. Newell, the Reclamation Service seemed infused with the same spirit of change and achievement that pushed the nation into building the Panama Canal in that same decade. Indeed, Senator Newlands often compared the two. The U.S.

3.1. The Panama Canal's impressive Gatun Upper Locks in January 1912. Building the inter-ocean canal captured American attention and was often compared to the work of the U.S. Reclamation Service as it undertook challenging dam and irrigation projects in the arid West. Official Photograph of the Panama Canal Commission – Panama Canal Graphics Section. Provided courtesy of William P. McLaughlin.

Reclamation Service, like the Panama Canal Commission in the Department of War, demonstrated the ability and skill of government to initiate and build great projects beyond the resources of private enterprise.[1]

Although begun with a decisive air of optimism in the manner of the Panama Canal, by 1912 the Reclamation Service struggled to fend off widespread criticism of its western water projects. Meanwhile, the nation prepared to honor the Panama Canal Commission in 1915 on the completion of the Panama Canal as an engineering triumph at the Panama-Pacific International Exposition in San Francisco. Americans stood in awe of the

3.2. Situated on the Avenue of Palms at the Panama-Pacific International Exposition of 1915 in San Francisco, "The Fountain of Energy" celebrated the power of water to produce hydroelectric energy. Courtesy of the California Historical Society.

power of science, engineering, and technology as they visited the Panama-Pacific Exposition that affirmed the power of civilization to connect the commerce of two oceans by altering nature's very geography. The U.S. Reclamation Service's exhibit at the Exposition also demonstrated a nature modified to serve the needs of humankind. It featured an idyllic forty-acre farmstead "in a beautiful mountain rimmed desert valley" newly reclaimed against the background of what was labeled, "the highest dam in the world" on the Shoshone Project. In lectures illustrated with "motion films and colored stereopticon slides," the Reclamation Service told of its work in the West. The scenes, rural and bucolic, depicted its mission to reclaim the land and build a society in the West around hearth, family, and farm. The hydroelectric production that accompanied projects appeared to be a helpmate to domestic and farm chores. Nineteenth-century images of "turning land and water into an agricultural landscape" captured the purposes around which the original reclamation movement had rallied. And perhaps this backward vision toward the past was the crux of the difficulties in which the Reclamation Service found itself by 1915. Yet, criticisms aside, the reclamation program boasted of multiple projects, grand engineering feats, and strong congressional support within the West. It was unlikely that Congress would abandon an effort into which it had already invested almost $100,000,000.[2]

3.3. Completed in 1910, the 325-foot high Buffalo Bill Dam (Shoshone Dam), Wyoming, was the highest dam in the world for a time.

Breaking Political Barriers to National Reclamation

In 1900, President William McKinley's reelection confirmed the good fortunes of the Republican Party with the return of prosperity and the success of "the splendid little war" with Spain. Neither free silver nor anti-imperialism proved viable enough for the Democratic Party, under the leadership of the Nebraskan William Jennings Bryan, to wrest victory from the Republicans. After many close presidential elections, in 1896 Republicans commanded national politics with the decisive defeat of the Nebraska Democrat. Still, their leadership recognized dangers, as the politics of a Republican majority rested largely upon a stable, growing economy. Republicans barely escaped the challenges of the opposition party and third party populism in the depression and protests of the 1890s. With the South always solidly Democratic, the best strategy for Republicans was to undermine Democratic efforts outside of the South — this meant in the West. In both 1896 and 1900, Republicans succeeded, but the political arithmetic weighed heavily upon Republican minds. Should the economy relapse into depression, Democrats might once again tempt the West and South into an alliance.[3]

Nonetheless, the McKinley Administration was not overly worried. A confident and secure conservative Republican leadership was unlikely to embark upon a program of regional aid in the West if it felt that the region remained securely in the Republican column. In the first place, it was unnecessary; in the second place, Republicans in the Midwest and Northeast believed that a vast acreage of reclaimed western lands would threaten crop prices; in the third place, the proposal resembled previous western raids upon the Treasury, or "sops," granted to the West in special silver purchase acts (the Bland-Allison Act of 1878 and the Sherman Silver Purchase Act of 1890), the latter widely blamed for causing the panic and depression of the 1890s. Confident, self-satisfied McKinley Republicans were unlikely to move toward experimental legislation involving the general government in western irrigation schemes. Newlands, Maxwell, Smythe, and even Republican Senator Francis E. Warren from Wyoming could go through the motions of backing and writing various irrigation bills, but to what end? The economic upswing inspired confidence in the dominant party.

If western interests could be defeated on the inflammatory silver issue, surely reclamation, also a regional cause, stood little chance for approval from the Republican leadership. Earlier in the year

(January 26, 1901), Democratic Congressman Newlands had introduced such a bill in the House of Representatives. It met strong objections from Republicans who accused Newlands of proposing a raid on the public treasury, mainly to address the problems of his own impoverished state. The motive seemed all the more evident when Newlands supported only a national program and opposed state-based efforts that would have confined federal moneys under the direction of the states to the building of reservoirs and main canals. Since most western rivers flowed with no regard for state boundaries, Newlands explained, their waters should be subjected to national authority, regulation, and distribution for reclamation to succeed. Another strong voice for western irrigation, Wyoming's Senator Warren, championed a state-based program shaped by the ideas of Wyoming State Engineer Elwood Mead, who believed that irrigation efforts should come under the control of state water engineers and state water laws. Many worried that the Newlands approach threatened state water laws (that currently protected established interests, e.g., the livestock industry) and feared federal control of western waters. Predictably, by March Congressman Newlands's bill found itself bottled up in the Committee on Irrigation of Arid Lands. Senator Henry C. Hansbrough of North Dakota introduced a similar bill in the Senate that was defeated. Under the McKinley presidency, the drive for western reclamation legislation was an endless game going nowhere.

In the Senate, however, seemingly lost causes could find new life. In 1899, Senator Warren had resorted to the filibuster to turn the Senate toward consideration of an amendment to the Rivers and Harbors bill to build reservoirs and canals in arid states. He and fellow senator from Wyoming, Thomas Carter, argued that water improvements in the West should be considered part of the general flood control program. Rivers and Harbors projects never occurred in the arid states. Specifically, Warren argued for reservoirs in Wyoming on streams that fed the Powder River, a tributary of the Yellowstone and Missouri Rivers. These reservoirs in the headwaters, they argued, would help control floods and promote navigation on the upper Missouri River to Sioux City. The Warren Amendment called for two million dollars to build reservoirs and canals that could also be used for irrigation in a total appropriations bill of over thirty million dollars. If Senators Warren and Carter could obtain small western reservoirs funded by the federal government through Rivers and Harbors appropriations, the challenges and delays of enacting new reclamation legislation could be avoided. In addition, as was the practice with Rivers and Harbors projects, there would be no repayment responsibility because the appropriations served "the general welfare." While approved by the Senate, the amendment never made it out of a joint House-Senate conference committee. When the bill came back to the Senate for final approval, Warren launched a filibuster against it that threatened to kill the entire appropriation. He only gave up in the morning hours of March 4, 1899, before the close of the Fifty-fifth Congress. His filibuster threatened to derail a rich, $45,000,000, appropriation for many states. Warren's maneuver served notice that Rivers and Harbors appropriations might be held hostage to western interests in the future.[4]

Failure of the reclamation bill in the spring of 1901 offered an opportunity to once again invoke a filibuster and, this time, with telling effect. A set of circumstances enabled Wyoming Senator Carter to talk the Rivers and Harbors Bill of that session to death. He did this on March 3, 1901, just before President McKinley's second inauguration. In the fall of 1900, Senator Carter failed in his re-election bid in Wyoming. He was on his way out of the Senate, and the failure again of Congress to approve aid to western water development prompted him to carry out the filibustering strategy of 1899 to a deadly end on the Rivers and Harbors bill without fear for his own future standing in the Senate. The demise of the Rivers and Harbors bill, which included lucrative projects for many congressional constituencies, at the hands of one western senator, marked the beginning of the age of the "New West," according to historian Donald Pisani.[5]

Everything changed when the McKinley presidency ended abruptly in September 1901. McKinley's assassination raised the young and adventurous Theodore Roosevelt to the Presidency. How would he react to the proposals now stymied in Congress for national irrigation? The new president's first important message to Congress in December 1901 struck notes of action and innovation while addressing the major issues facing the United States in the first decade of the twentieth century. Here was a president who had lived in the West, at least briefly, and was acutely aware of its political importance to himself and to the Republican Party. Theodore Roosevelt was no ordinary Republican president. In fact, conservative factions within the Republican Party had shuddered at the possibility of a Roosevelt presidency should anything happen to McKinley. Notably for the West, Roosevelt took an immense interest in the issues of forest conservation and related issues of water supplies. As early as 1883, *The Nation* magazine noted proposals for the creation of a national forest preserve, "only valuable as a reservoir of moisture" for rivers in the arid lands that would help make "agriculture possible west of the 100th meridian by means of irrigation." [6]

Roosevelt thought likewise when he asserted, "The water supply itself depends upon the forest." He continued:

> The forests are natural reservoirs. By restraining the streams in flood and replenishing them in drought they make possible the use of waters otherwise wasted. They prevent the soil from washing and so protect the storage reservoirs from filling up with silt. Forest conservation is therefore an essential condition of water conservation.

The conservation portion of the message embodied the ideas of Pinchot and Newell on forests, water, and reclamation. Roosevelt admitted that the two had visited him in Washington even before he moved into the White House and, "laid before me their plans for National irrigation … and for the consolidation of the forest work of the Government in the Bureau of Forestry." Linking irrigation and forestry was part of a larger conservation package that eventually looked to multiple-use development of water resources in river basins. Such ideas were not new. Intellectuals and scientists employed in various Washington bureaus for almost two decades recognized the relationship of water, land, forest, and transportation. Often, they were members of Washington's noted Cosmos Club, a place that, according to one source, was "crucial in averting fragmentation within the scientific departments." Powell was an original member. Pinchot came later and credited another member, W. J. McGee, head of the Bureau of American Ethnology, for helping him mold his conservation ethic into a more comprehensive philosophy. McGee suggested a utilitarian premise for

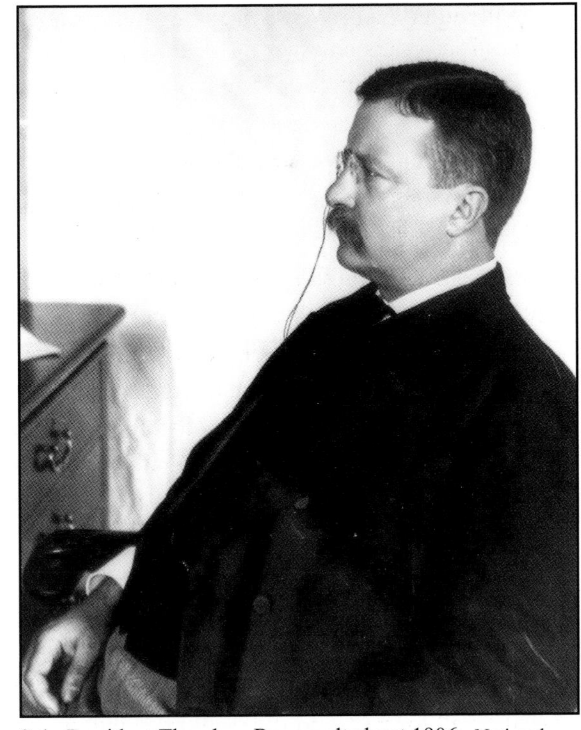

3.4. President Theodore Roosevelt about 1906. National Archives and Records Administration.

conservation based upon the principle of "the greatest good, for the greatest number, for the longest run" in the service of human society. The Irrigation Congresses pioneered a comprehensive approach to river resources with ideas about multiple use river development in their slogan: "Save the Forests, Store the Floods, Reclaim the Deserts, Make Homes on the Land," as did the publication *Forestry and Irrigation.* In the coming decades, multiple-use presented enormous challenges as rivalries emerged among the new government resource bureaucracies and as local government resisted federal decision making regarding western rivers, lands, and forests.[7]

Roosevelt made it clear that, "Forests alone could not regulate and conserve the waters of the arid region." Dam building must occur to create water storage reservoirs to conserve and provide streamflow into the dry summer months. Such monumental construction was beyond the powers and resources of private enterprise and even of state governments that could not address the problems of interstate water flows. He could only conclude that, "It is as right for the national Government to make the streams and rivers of the arid region useful by engineering works for water storage as to make useful the rivers and harbors of the humid region by engineering works of another kind. The storing of the floods in reservoirs at the headwaters of our rivers is but an enlargement of our present policy of river control, under which levees are built on the lower reaches of the same streams." If the waters that were available in the arid lands could be saved and used for irrigation rather than wasted, he also predicted that a hundred million people could live in the West, or, as he said, "a population greater than that of our whole country today." Making water available to western land was the key to giving it value and opening it to settlement. The congressional debates echoed this theme saying that it was "not in harmony with the progressive spirit of the age" that the American West "should remain practically a desert." Western congressmen also saw a more telling political argument for federal aid to western irrigation: if appropriations could be made for river and harbor improvements in Midwestern and coastal waterways, the water resources of the interior West could also be improved for irrigation purposes by federal dollars.[8]

Roosevelt lauded the creation of forest reserves in the 1890s and called for their more efficient administration under a Bureau of Forestry in the Department of Agriculture. He said, "The forest and water problems are perhaps the most vital internal problems of the United States." Not only did Pinchot and Newell claim they had influenced Roosevelt's thinking on these matters, but Elwood Mead believed that the President's words reflected some of his ideas related to the need to reform state water laws. Roosevelt's message to Congress endorsed reclamation but stepped gingerly through the thicket of western water laws and water rights by saying that national reclamation should occur, but distribution of water should be left to the settlers themselves under state laws. The policy of the government should be, Roosevelt insisted, "to aid irrigation in the several States and Territories in such a manner as will enable the people in the local communities to help themselves, and as will stimulate needed reforms in the State laws and regulations governing irrigation." Easily irrigable lands were already under the ditch, but there were vast areas of public land that could be opened to settlement through federal construction of reservoirs and main-line canals by the federal government that were impracticable for private construction.[9]

Roosevelt reiterated the longtime arguments of irrigation advocates when he said, "Settlement of the arid lands will enrich every portion of our country, just as the settlement of the Ohio and Mississippi valleys brought prosperity to the Atlantic States." And he believed that whatever the general government did for the "extension of irrigation" could also "improve, the condition of those now living on irrigated land." This opened the door for the benefits of national reclamation to flow to private landholders already in the West.[10]

At last there was presidential leadership on the issue of national reclamation. Key congressmen now faced the task of devising a bill acceptable in committee and both Houses. Representatives from western states quickly formed a joint House-Senate seventeen-member conference committee of one congressman or senator from each state and territory in prospective reclamation states, including Texas which, not having any public lands, was not covered in the Reclamation Act as passed in 1902. The committee included such prominent western members as Senators Newlands, Warren, and Fred Dubois of Idaho. It worked continuously through December 1901 and into January 1902 to produce an acceptable piece of legislation. But harmony was hard to achieve. Diverse and self-serving interests in each state and territory made it difficult to come up with a bill to satisfy all. Advocates of a centralized, national approach to reclamation faced the opposition of those who favored a decentralized, state-based, or local irrigation district approach. This meant that the centralizers, Newlands of Nevada and Henry Clay Hansbrough of North Dakota, with their advisor Newell, confronted the decentralizers, Warren of Wyoming and

John F. Shafroth of Colorado, with their advisor Mead. The two groups seemed at loggerheads, but the absence of Senator Warren during crucial negotiating sessions because of a family crisis gave the Newlands forces an opening. They came up with a revised Newlands/Newell bill that was introduced simultaneously in the House and Senate on January 21, 1902.[11]

Suffice it to say that the national Reclamation Act was on the president's desk for signature June 17, 1902. The president had embraced the Act, but in Congress some of the momentum behind its passage came from the not-so-veiled threat that western senators could again filibuster the Rivers and Harbors appropriations. While Roosevelt eagerly claimed major credit for passage of the Reclamation Act in his 1904 campaign, he did give Newlands credit years later in his 1913 *Autobiography*. He said Newlands fought hard for reclamation in Congress, but the President insisted that it was he who put reclamation at the top of his agenda when he took office.[12] In and outside of Newlands's own Democratic Party circles, the legislation became widely known as the Newlands Act, and some said aptly so because it brought "new lands" under cultivation. The most important features of the new law were the 160 acre limitation, use of revenues from public land sales to finance projects, provisions that project construction costs be paid back by water users, and the pledge by the federal government to respect state water laws and existing water rights. The Act laid the foundation for a powerful new federal presence in western water matters. It committed the government to undertake reclamation projects, not simply one project or a specific number, but feasible projects as determined by the Secretary of the Interior.

The Reclamation Act inaugurated one of the most important programs of internal improvement ever attempted by the federal government. In the Department of the Interior the work of reclamation fell under the administration of the U.S. Reclamation Service, which grew out of the work of the U.S. Geological Survey. Reclamation drew its financing from public land sales in the arid states. Originally, 51 percent of the revenue from federal land sales made within a state or territory had to be spent on reclamation within that state or territory. That was another way to accomplish the often-proposed policy of land cession to the states — earmarking the revenues on the land sales to projects within the state rather than to the general treasury. Ideally, the government would locate projects on unclaimed public domain. These lands were to be offered free under the 1862 Homestead Act, with the understanding that the claimant must reside upon them or nearby. The water was not free. Costs of building the water collection and delivery systems (dams and canals) must be repaid at no interest into a revolving Reclamation Fund over a ten-year period following an agreed upon date for completion of the construction. In addition, maintenance and operation costs were to be an ongoing expenditure for the project and paid by the landowner.

Reclamation and the Disposal of the Public Domain

The national Reclamation Act was one of several turn-of-the-century responses by Congress to the question of what to do with the unclaimed public lands in the West. The Reclamation Service and its irrigation projects became a new conduit for western settlement, just as the railroads and a generous land policy spurred western migration in the nineteenth century. For the Reclamation Service, the challenge was enormous, but Congress offered it the revenues from almost the entire remaining western public domain. And much was expected in return for the much that was given. It was the responsibility of the Reclamation Service to open wide these new doors of progress in the West.

Less progressive westerners held that the West should remain a gigantic grazing province, a sparsely populated region largely under control of big "outfits." These were highly capitalized absentee stock corporations employing transient "cowboy" labor upon an open and uninhabited free range. Such a future was unacceptable to ambitious western boosters. As Newell expressed it in *Irrigation in the United States* (1902), "It is unquestionably a duty of the highest citizenship to enable a hundred homes of independent farmers to exist, rather than one or two great stock ranches, controlled by nonresidents, furnishing employment only to nomadic herders." To establish a society of small independent yeoman farmers on the land, at best a romantic idea in the first place, proved a difficult task. [13]

Congress argued over the many possibilities for the disposition of western lands and a reform of its land policies in the 1890s. In the General Land Revision Act of 1891, Congress granted authority to the president to proclaim and designate forest reserves on the public domain. From 1891 to 1908, the presidents, and in particular Theodore Roosevelt, set aside millions of acres for forest purposes. In 1905, the forest reserves previously under the Department of the Interior became part of the National Forest system administered by the U.S. Forest Service, under the Department of Agriculture. The Forest Service took its place alongside the Reclamation Service as another public service bureaucracy created by Congress to utilize, conserve, and protect natural resources. The proliferation of federal government resource bureaus in the Progressive Era prompted one historian to observe that the period saw the emergence of "captains of bureaucracy" (especially Pinchot and Newell), just as the previous Gilded Age (1865-1900) saw the appearance of "captains of industry."[14]

As Congress moved ahead with forestry and reclamation policies on western lands, it also confronted decisions related to aesthetic and cultural resource issues. National parks, national monuments, and wildlife habitats or refuges drew the attention of Congress. It created Yellowstone National Park in 1872, Yosemite in 1890, Rainier in 1902, Mesa Verde in 1906, Glacier in 1910, and, finally, in 1916, the National Park Service. Included in the administration of parks were national monuments authorized by the Antiquities Act of 1906. On these lands, of course, Congress did not invite settlement, but rather provided destinations for a tourist industry that eventually assumed a major role in the western regional economy. Railroads, eager for passenger traffic, supported national parks as they supported, for similar reasons, national reclamation.[15] In reference to reclamation, however, Congress made it clear that the waters of these scenic and sacred national lands were not off limits; their waters could be dammed for utilization as need might occur. Yosemite National Park soon provided a flash point over this issue in one of the West's most noted conservation battles, when the city of San Francisco moved to dam the Tuolumne River in the Hetch Hetchy Valley in the park. Although the Reclamation Service was not involved in building the dam, it did support the intentions of San Francisco as consistent with its own policies of utilitarian conservation (the greatest good for the greatest number). It employed the same rationale when it acquiesced to Los Angeles's plans to drain the Owens Valley of water where Reclamation had planned a project.[16]

On the other hand, Congress appeared more interested in small farms than parks or cities. In the first two decades of the twentieth century, Congress surpassed its nineteenth-century record for homestead laws. For western Nebraska, Congress approved the Kinkaid Act in 1906, offering 640 acres for homesteading in the Sand Hills section of that state. In the same year, it opened national forests to homesteading, when the Forest Homestead Act granted homesteads of 160 acres to a claimant who successfully argued that a living could be made from the land through grazing and crop agriculture. In 1909, Congress approved the Enlarged Homestead Act, giving 320 acres of land to an individual. Finally, the Stock-Raising Homestead Act of

1916 made available 640 acres of non-irrigable free land. This flurry of land legislation, like the Reclamation Act, passed in the hope that more of the public domain would find its way into the productive hands of the nation's citizens or citizens-to-be and also to rectify inadequacies in earlier land legislation.

Even such continued largesse by the federal government in the twentieth century failed to dispose of all the lands in the West. By the 1920s, millions of acres of public domain still remained beyond management. The General Land Office exercised little oversight as it vainly waited for lands to be taken up under the complex of land laws Congress provided. It was an open question who should administer these lands, most of which were used for grazing. Various proposals to cede the lands to the states gained little support. On Indian lands, the 1887 Dawes Severalty Act vastly reduced the landholdings of the Indian reservations. The legislation, seen as a noble reform at the time, provided for the distribution of commonly held reservation lands to individual Indian farmers and families "in severalty" or under private property ownership. Much land fraud and deception resulted in the loss of millions of acres in reservation lands. Reclamation held some promise of improving the lives of Indian farmers if they wished to accept lands under the Severalty Act. Although severe abuses occurred in efforts to get water to Indian lands (including further reductions in those lands), Reclamation in its early days hoped that greater availability of water for agricultural purposes would make severalty legislation a workable land policy in association with some reservations.

At least the Reclamation Act held out the hope of ordered development on specified projects that could prove a model for the West to follow. Reclamation suggested that the future would probably take shape around some combination of crop agriculture, dairying, and stock operations using grazing lands within the National Forests and the remaining open range lands. The Reclamation Service saw its mission to foster progressive development of the many western states. Its leadership would be surprised and dismayed to learn that, a century later, many historians judged the goals of reclamation as backward looking, seeking to impose a small farmer agricultural system upon the West, while other economic forces, progressive or unprogressive as they may be, busied themselves building a different future.

Competitive Bureaucracies and Personalities

President Roosevelt turned to the U.S. Geological Survey (USGS), headed by Charles D. Walcott, to direct activities of the new Reclamation Service. The President's Secretary of the Interior, Ethan A. Hitchcock, made Walcott the first Director of the Reclamation Service in addition to his duties as director of the USGS in the Interior Department. Placing the Reclamation Service in the USGS, if only temporarily, was an apt choice, because the USGS already possessed experience in assessing western water resources in surveys after its founding in 1879. After the ill-fated work under Major Powell in the Irrigation Surveys of the late 1880s and early 1890s, the USGS had retained a core of trained, experienced, hydrological engineers in its longstanding work on the problems and resources of the arid lands. As Chief of the Division of Hydrography in the USGS, Frederick H. Newell headed up the new work as chief engineer until his 1907 appointment as Director of an independent Reclamation Service within the Department of the Interior.[17]

3.5. Charles Doolittle Walcott, Director of the U.S. Geological Survey (1894-1907), Director U.S. Reclamation Service (1902-1907). Courtesy of the USGS.

By August 6, 1902, Newell and Walcott were in the Far West scouting out reservoir sites for irrigation projects and confirming earlier site designations in Montana, Wyoming, Nevada, and California. By late September, Walcott returned to Washington, D.C., in time for Major Powell's funeral. The major had lived just long enough to see the passage of the national Reclamation Act. Newell, Powell's early protégé, however, saw himself at the beginning of an exciting and ambitious enterprise about which Powell, in his last years, had expressed misgivings. Newell shared those misgivings but, at this point in his life, was eager to get on with the building of a new, irrigated West whether from the local or national level. Newell began work, after graduation from the Massachusetts Institute of Technology, as a young surveyor with the USGS in a career with government that was to last for thirty years. By 1888, at age 27, he was appointed Chief of the USGS's Division of Hydrography. He worked closely with Powell until Powell resigned from the USGS in 1894, and thereafter he worked for Director Walcott.[18]

During these years, Newell's interest in federally assisted irrigation increased as campaigns by private organizations, politicians, and individuals intensified, e.g., the National Irrigation Congresses and, of course, the railroad-backed lobbyist, George Maxwell. By 1900, Newell related in his diaries, "I began to get acquainted with Francis G. Newlands of Nevada, discussing with him the opportunities for western development." Newell is often credited with counseling Newlands and drawing his attention to proposals of the early Irrigation Congresses for the cession of federal lands to the states as outright land gifts for financing irrigation. From the standpoint of Newlands and other advocates of federal reclamation, it was the national government itself that must become the active agent in turning the revenues from the sale of public lands into the financing of irrigation projects. Newell believed in home-making on western lands guided by federal

3.6. Frederick Haynes Newell and Charles D. Walcott during field work at a USGS/USRS camp on the Buffalo Fork of the Snake River near Jackson Lake, Wyoming. August 12, 1903. On April 23, 1904, Secretary of the Interior Ethan Allen Hitchcock authorized the Minidoka Project, for which Reclamation later built Jackson Lake Dam.

reclamation. Without guidance and oversight, he feared inevitable land monopolization in the West with little future for the small homestead farmer. Finally, Newell saw irrigation "to include the whole question of conservation and utilization of water in the development of the arid regions." He believed that the western lands should serve the higher purposes of conservation, which meant elevating them to use rather than idleness or even misuse under monopolistic grazing systems. And the technological abilities of the Reclamation Service would serve these greater causes.[19]

Newell appeared to be a happy choice for chief engineer of the newly independent Reclamation Service. In addition, the USGS was pleased to be the lead bureau to bring watered homesteads to the arid lands. This enabled it to build upon its accomplishments in surveying and classifying lands that began more than a decade earlier with the Irrigation Survey legislation in 1888. Others, however, were not so happy. The creation of a new bureau in the Department of the Interior to carry out the important work of reclamation, underwritten by funds from the sale of public lands, meant that other departments of the government were left on the sidelines in work that might last for a generation or more. Most evident, the Corps of Engineers, under the Department of War, took a back seat to the new activities in the West. If the Corps had experience in delivering water improvements to the East (river and harbor improvements: levees, jetties, dikes, navigation, and flood control improvements), why should it not play a role in western water improvements? After all, had it not been the Corps's engineer, Hiram Chittenden, who had suggested in a widely publicized pronouncement in 1898 that the federal government should undertake the building of reservoirs and canals for the delivery of water in arid lands of the West? He continued the argument in 1902 when he made the point that the West had a right to expect "internal improvement" appropriations for reservoir and irrigation systems in the same manner as the East Coast, the Mississippi Valley, and the Great Lakes states enjoyed appropriations for river and harbor improvements. Still, it was not clear that the leadership of the Corps wanted to build irrigation works or had the engineering staff to do so.[20]

Similarly, establishment of the Reclamation Service left the Department of Agriculture (USDA) out of the loop. By the late 1890s, alert officials in the USDA sensed the movement of forces in Congress toward support of irrigation in the West. This meant new appropriations and expansion of the government bureau designated for the task. Since the late 1880s, Elwood Mead's work in irrigation and water law in Wyoming had made him a candidate for an appointment with either the USGS or the Department of Agriculture. Looking toward authorization of a federal irrigation program in 1897, and after some prodding from Wyoming Senator Warren, Mead accepted an appointment in the Department of Agriculture as a consultant in irrigation studies. This soon led to a full time appointment in 1899 heading up Irrigation Investigations for the Office of Experiment Stations, also under the USDA.[21]

Mead's presence in Washington not only gave the Department of Agriculture a more prominent voice in irrigation matters, it divided the forces working for national irrigation legislation. His close association with Senator Warren gave him influence that reached all the way to President Roosevelt. Newell, Pinchot, and Newlands were not the only ones talking to the President on irrigation matters. In this role Mead came into competition with Newell, who, along with Newlands and Maxwell (with his railroad support), stood squarely for a federal approach to irrigation. Their ideas contrasted with the state-based approach that called for the state engineers in western states to take charge of the details of land irrigation with the federal government confined to ceding lands and building major reservoirs. While Mead and Warren lost out in the final phases of the legislative process, Mead did claim that he had some influence on the final Reclamation Act that offered protection to state water law and non-interference by the federal government in local water matters. Yet Mead and the Department of Agriculture believed that the Administration turned irrigation matters over to inexperienced men in the USGS and ultimately the Reclamation Service itself. Their criticism centered on the complaint that engineers, without the wider knowledge of farming and soils offered by the Department of Agriculture, were ill equipped to build agricultural communities. In private correspondence, Mead identified Newell as too much an engineer, a man who lacked knowledge of agriculture, soils, economics, water law, and rural life.[22]

Most observers agree that the expansion of federal government services at the turn of the century saw the creation of the "modern American state," but it was neither a smooth nor complete transition. The competition between Newell and Mead (or between the Department of the Interior and the Department of Agriculture) was only one example of what historian Donald Pisani calls an increasing fragmentation that occurred as new governmental bureaus clashed for control over the management of natural resources. Some guessed, as Mead certainly knew, that, if the work of the Department of Agriculture in irrigation had been "conjoined" with the Reclamation Service, much of the early trouble on the projects might have been averted. This still remains a topic of speculation.[23]

As the Reclamation Service secured its place and confined the activities of reclamation within the Department of the Interior, Mead found limited opportunities for his irrigation expertise in the USDA. In 1901, Mead had experienced a personal setback when he lost his right arm in a trolley car accident in Washington, D.C. This occurred shortly after he had agreed to teach a six week course at the University of California at Berkeley where he continued to be a consulting professor. His main employer, however, was the Department of Agriculture until 1907, but it became increasingly clear that the Department stood on the sidelines as major irrigation works proceeded under the Department of the Interior's Reclamation Service with Mead "shunted aside." Mead took his personal disability in stride and assured President Benjamin Ide Wheeler of the University of California that his work would go forward. And go forward it did. In the decades that followed, he finished his career as Commissioner of the Bureau of Reclamation and presided over the building of the Boulder Canyon Project. In 1935, he stood alongside President Franklin D. Roosevelt at the dedication of Boulder Dam on the Colorado River when the President reportedly said, "I came, I saw, and I was conquered." As for himself, standing there among the dignitaries, Mead could easily recall, that it was John Wesley Powell who, in the not-so-distant past, first saw the possibilities of a great dam on the river of the West. [24]

3.7. Alfred Deakin, who studied the arid lands of western America and later became Australia's Prime Minister. National Archives of Australia.

While Newell and others made their mark in building the early works of western reclamation, Mead sought out other venues for his talents even beyond his occasional lecture courses at the University of California. Mead had met Alfred Deakin, a future Prime Minister of Australia, while Deakin toured the United States to study American irrigation in the 1880s. In 1906, Mead was hired by the newly united Commonwealth of Australia to help the Province of Victoria promote irrigation projects and draft water laws. He remained in Australia until 1915. A year later, Secretary of the Interior Franklin K. Lane, in an odd twist of fate and policy, proposed that Mead take over as Director of the Reclamation Service, but he declined that opportunity until it again came to him in 1924.[25]

The Reclamation Act: Its Breadth and Struggle with Western Water Laws

Meanwhile, the great challenges of reclamation faced Newell. As Chief Engineer of Reclamation from 1902 to 1907, Newell lost no time in launching reclamation projects. President Roosevelt, always impatient for results, wanted reclamation projects to dot the land so that western states and territories would appreciate projects brought to them by his Administration. As he prepared for election to the presidency in his own right in 1904, he wanted the spotlight to shine on his achievements in reclamation in the new western states. Under political pressure, the Reclamation Service abandoned any idea of building one major demonstration project in order to learn from its mistakes. Instead, presidential politics quickly moved several projects ahead simultaneously for approval. The new Reclamation Service also faced pressures from members of Congress to build as many projects as possible as soon as possible, especially from members who had played important roles in the passage of the Reclamation Act of 1902. For instance, the designation "#1" on the Truckee-Carson Project in Nevada was no small reward and acknowledgement of Congressman Newlands's efforts on behalf of federal reclamation.

Besides the political context, the Reclamation Act itself presented problems. It proved to be no blessed Rod of Aaron that brought forth blossoms across the land to awe the people as in the Biblical tale (Numbers 17:8). While criticisms of the legislation are manifold, the Reclamation Act was clearly a breakthrough piece of legislation. It was a bold federal experiment in internal improvements. Whatever its criticisms and shortcomings, the Reclamation Act was not simply another piece of land legislation to speed the disposal and alienation of the public domain. It was a product of that strain of American land policy that sought to dispose of the public lands for specific social goals. In the process, the Act committed the federal government to building permanent works — dams, reservoirs, water delivery systems — and pledged it to maintain the infrastructure.

While the Reclamation Service labored under many of the idealistic vagaries of the Act, the next two decades saw important engineering accomplishments in terms of damming and controlling western rivers. The dams and the power they represented, both in terms of subduing rivers and providing hydroelectric energy, symbolized "progress" in the eyes of the public. The large structures represented the power of the nation to respond to the uneasiness about the future at the beginning of the century that the depression of the 1890s, immigration, political protests, and, now, imperialism presented. This is to say nothing about the Reclamation Service's purposeful choice of massive structure designs to reassure a public already skittish about dam safety after the 1889 Johnstown Flood in Pennsylvania, caused by a dam failure that killed over 2,200 people. The Reclamation Act, in a sense, reaffirmed the confidence asserted by an earlier historian of the nation, George Bancroft, who observed that here, in this nation with its "abounding harvests of scientific discovery

3.8. Frederick Haynes Newell about 1914.

3.9. Under political pressure Reclamation sometimes undertook too many projects, resulting in scenes such as this one on the Grand Valley Project, Colorado, in August of 1913. Reclamation photographer Henry T. Cowling labeled this photograph: "Waiting for water. Residence of B. B. Freeman … Mr. Freeman and family have been waiting nearly 6 years, having moved into this cabin in 1908."

… the wildest forces of nature have been taught to become the docile helpmates of man." Could anyone doubt that the American nation, with all the achievements of the nineteenth century behind it, would successfully face the challenges of the arid lands? After the impressive campaigns in support of government-aided irrigation by William Smythe, George Maxwell, western congressmen, the powerful railroads, and the manufacturers of farm and irrigation machinery, it appeared that the taming of mighty rivers could serve only the highest purposes of civilization. To expect Congress and its western constituency to turn away from this challenge would be a misreading of the nation's history.[26]

The title of the Reclamation Act identified the source of its funding and stated its purpose: "An Act appropriating the receipts from the sale and disposal of public lands in certain States and Territories to the construction of irrigation works for the reclamation of arid lands." These works included dams and reservoirs, but reclamation also involved settling people upon the land, building communities, and finding crops and markets. The social and economic side of reclamation proved far more difficult than building dams and water delivery systems. The Reclamation Act required that moneys received from the sale and disposal of public lands in the states of California, Colorado, Idaho, Kansas, Montana, Nebraska, Nevada, North Dakota, Oregon, South Dakota, Utah, Washington, Wyoming, and the territories of Arizona, New Mexico, and Oklahoma, other than those funds as set aside for educational purposes, should be placed in a "special fund in the Treasury to be known as the reclamation fund." The money was to be expended on surveys and the "construction and maintenance of irrigation works for the storage, diversion, and development of waters for the reclamation of arid and semiarid lands in the said States and Territories …"

The Secretary of the Interior was to designate and locate projects and withdraw from settlement all the lands under consideration for irrigation in order to guard against speculation. When appropriate, the Secretary of the Interior (really the Reclamation Service) could open the lands for settlement under the provisions of the homestead laws in tracts not less than 40 nor more than 160 acres, but the commutation provisions (freedom to sell the land before proving up) in the original 1862 Homestead Act did not apply. The Secretary of the Interior had the authority to limit the acreage per entry, "which limit shall represent the acreage which, in the opinion of the Secretary, may be reasonably required for the support of a family upon the lands." The Secretary also set the "charges which shall be made per acre upon the said entries ... and upon lands in private ownership which may be irrigated by the waters of the said irrigation project."

Herein were two of the most controversial items in the Reclamation Act: the 160 acre limitation and the requirement that project settlers pay back the costs of bringing water to the land. Ideally, payments to the Reclamation Fund would create a self-replenishing "trust fund" in the Treasury to provide capital for future projects. The Secretary set the number of annual installments, not exceeding ten years, for the payback. "The said charges shall be determined with a view of returning to the reclamation fund the estimated cost of construction of the project, and shall be apportioned equitably," said the Act. Private landholders within a project could share in project water, but not for tracts exceeding 160 acres to any one landowner, and the owner had to reside upon the land. This requirement, often referred to as the "excess lands" provision, produced a massive amount of literature debating the pros and cons of acreage limitation. Eventually, after many decades of evading this provision by both landowners and the Bureau of Reclamation, Congress, in 1982, all but lifted acreage requirements when they were changed to 940 acre limits. The original Act also declared that management and operation of reservoirs and the works necessary for their protection would remain with the Government until otherwise provided by Congress.

There was a high degree of uncertainty about federal authority in reference to state water laws. Prodded by Mead and the Wyoming delegation, the Reclamation Act in Section 8 said that,

> Nothing in this Act shall in any way interfere with the laws of any State or Territory relating to the control, appropriation, use, or distribution of water used in irrigation, or any vested right acquired thereunder

But in a rather contradictory statement, it also said,

> The Secretary of the Interior shall proceed in conformity with such laws, and nothing herein shall in any way affect any right of any State or of the Federal Government or of any landowner, appropriator, or user of water in, to, or from any interstate stream or the waters thereof

Any water the federal government did appropriate for these projects "shall be appurtenant to the land irrigated, and beneficial use shall be the basis, the measure, and the limit of the right." The words "appurtenant" and "beneficial use" eventually posed many legal problems. The U.S. Supreme Court's decision in *Kansas v. Colorado,* in 1907, narrowly defined federal authority over water, even on navigable streams. The decision bordered on the edge of making the Reclamation Act unconstitutional.[27]

The Act also declared that at least 51 percent of the funds collected in a state or territory from the sale of public lands must be spent there. This provision proved difficult to enforce because some arid states did not have the saleable land to support projects. The solution was to move funds and receipts from land-rich states to land-poor ones. Still, the law worked to promote multiple projects awarding at least one to each state and territory, if possible. The act also gave attention to the conditions of labor on government projects. In a reformist, pro-labor spirit, it declared, "That in all construction work eight hours shall constitute a day's work," and in another attempt to defend the rights of labor (white labor), the Act stipulated that "no Mongolian labor shall be employed thereon." This was in keeping with the anti-Chinese attitudes of the West's rank and file white labor force. Both conditions boosted the cost of projects, increased the indebtedness that settlers had to repay, and prompted charges of inefficiency and extravagance on the part of the Reclamation Service.

As many predicted, farmers were as slow to pay operation and maintenance charges for the delivery of water as for construction costs. The ten-year repayment schedule proved unrealistic in light of the heavy capital investments farmers had to make to bring new lands under cultivation. The funding crisis required additional loans to the Reclamation Fund from Congress and, in return, raised new doubts and criticisms about the program. The initial government underwriting of project construction made settlers indebted to the U.S. Treasury. The Reclamation Service soon discovered, as the General Land Office had a century earlier, that pioneer farmers were poor credit risks. Requests from both the Reclamation Service and water users that Congress delay or forgive debts provoked continued displeasure in Congress.

Congress was not the only critic. Settlers, themselves, complained about the Reclamation Service: its rules, its inefficiency, and the debilitating 160 acre limitation that prevented settlers from running larger and more profitable operations. Others argued that the limitation represented Reclamation's anti-monopolist idealism and commitment to small farming enterprises. Newlands proudly maintained that the provision prompted the breakup of "the existing land monopoly which has been so bothersome in California and the inter Mountain states." But the executive chairman of the American Irrigation Federation, G. L. Shumway, believed the 160 acre rule ran contrary to the spirit of the typical American, "the man of pluck, of energy, of resources, and if you please, of speculative tendencies." He went on to say that "the policy of theorizing engineers and experts" should not expect people to come into a new land and "toil for its development for a mere living." Controversy over the 160 acre limitation rarely abated. If the Reclamation Service conscientiously adhered to the provision, it ran afoul of ambitious, prosperous landowners who wanted to expand their operations; if it chose to look the other way in some cases, it invited charges that it favored larger corporate landowners in defiance of the law. No other provision of the Reclamation Act generated as much controversy or literature over the years.[28]

In any event, the Reclamation Act set a demanding course for those appointed to carry out its intents – the new Reclamation Service and its leadership. The 1902 law did not confer the power to centralize decision making and trump local authority, especially in attempts to claim adequate water for the new projects. Added to the vagaries of the law was the *Kansas v. Colorado* (1907) decision that stayed the hand of the federal government in regulating the waters of the West in favor of the states. More pointedly, the Reclamation Act prevented the federal government from interfering "with the laws of any State or Territory relating to the control, appropriation, use, or distribution of water...," and, "in carrying out the provisions of this act, [the federal government] shall proceed in conformity with such laws...." Neither the state governments nor state or federal courts permitted any significant expansion of federal authority over water. By the time the Reclamation Service came upon the scene, prior appropriation dominated western water law. Put simply, those who diverted water for use first had the highest claim upon the resource. Prior appropriation supplanted the principles of riparian water law dominant in eastern humid states. In the riparian tradition, those who owned the banks of a stream or water source commanded the use of the water under the condition that neither the volume nor the quality of the stream be significantly altered. Riparian law was already under revision in eastern states as industrialization demanded the use and diversion of rivers and streams for waterpower. Some western states, with a combination of arid and humid landscapes, adopted both the riparian and prior appropriation doctrines, i.e. the West Coast states and those adjacent to the Mississippi Valley.[29]

Consequently, the Reclamation Service often found itself in court asserting water claims against local and longer established interests within the states and territories. In its early years, the Reclamation Service insisted upon close cooperation between the state engineers and the reform of water laws before it would entertain irrigation projects within a state. Yet, the authority to institute new water laws and codes rested entirely with local governments. No federal water law or federal ownership of state waters emerged, as some had feared, from the Reclamation Act. What did occur was the entrance of the Reclamation Service into the complicated world of local western water law and rights to protect the vested interests of the new projects it sought to develop.

With established water law and most western water sources already over appropriated, the Reclamation Service was a Johnny-come-lately. The Reclamation Service could offer only to build dams and storage reservoirs to capture unclaimed runoff. This provided a larger useable water supply for the new projects. Reclamation also supported the doctrine of "beneficial use" announced explicitly in the Reclamation Act as "the basis, the measure, and the limit of the right" to water. It could, in some instances, undermine rights of appropriation, if the appropriation were being wasted or unused. The Reclamation Service's legal maneuvering did not always succeed. Proposed projects often stood at the mercy of local water arbiters and ownerships whose water rights predated the building of the projects. The Truckee-Carson Project in Nevada, later named the Newlands Project, stands out in this regard. The Reclamation Service built Lahontan Dam and Reservoir to store waters from the Carson and Truckee Rivers, but, at the same time, it believed that Lake Tahoe would serve as the backup reservoir in the dry months of late summer and fall. Tahoe residents thought otherwise. They asserted their water rights along the shores of the lake and others asserted their long-established diversion rights from the Truckee River. Irrigation advocates continued to argue that Lake Tahoe should be the primary storage facility for the exclusive irrigation water needs in the lower valleys. As John A. Widstoe, a professor of irrigation studies at Utah Agricultural College, wrote in 1928, "All other needs should become secondary to those of irrigation, in planning the utilization of Western lakes." The dilemma was that the property at Lake Tahoe was worth far more than the farmland around Fallon, Nevada, in the irrigation project.[30]

While the property owners at Lake Tahoe fended off the Reclamation Service, the Pyramid Lake Indian Reservation and tribal community at the other end of the Truckee River was powerless in the face of Newlands Project diversions from the river. Historically, the tribe drew much of its living from the fisheries of Pyramid Lake. But the diversion canal, completed in 1905 to carry water from the Truckee River to the Carson River in the Lahontan Valley near Fallon, reduced the flow of water into Pyramid Lake, thus changing natural conditions. Damage to Indian fisheries occurred despite the decision of the U.S. Supreme Court in the Winters Decision of 1908, which declared that Indian reservations possessed "reserved rights" to water for the purposes for which the reservation was established. And despite its use of Indian water, the Newlands Project still found itself in a never-ending battle for water throughout the twentieth century. Throughout the West, the Reclamation Service used the allotment of Indian lands under already existing legislation (Dawes Act, 1887 and Burke Act, 1902) to offer water to Indian lands in severalty in return for the reduction of Indian reservations that would also coincidentally open reservation lands for white settlement. The result was almost invariably a reallocation of lands that reduced the size of reservations.[31]

In addition to the struggle to secure adequate water for projects, the Reclamation Service faced problems of protecting project lands from speculators. When government surveyors appeared in the vicinity, sharp-eyed land claimants often rushed to file land claims with the local Land Office before the Secretary of the Interior could withdraw the site from entry. Once a reclamation project got underway, the private lands increased in value and could be sold to new farmers, who then planned to receive water under the terms of the project. This meant that some farmers carried a heavier debt for the purchase of lands than they had expected. Some have argued that it was unfortunate that the Reclamation Act did not authorize the Secretary of the Interior to withdraw all possible irrigable lands in a blanket declaration to discourage speculation and entry under the land laws of the United States. Such a mandate had been attempted with the USGS's Irrigation Survey in the late 1880s and early 1890s. The result was a political firestorm that ended the survey and eventually John Wesley Powell's tenure as the Director of the USGS. One successful precedent, however, occurred when Congress authorized this procedure in the creation of forest reserves in the 1891 Land

3.10. Theodore Roosevelt Dam and powerplant in 1909 during construction.

Revision Act. Under this act, the president wielded the power of proclamation to transform almost instantly vast mountainous acreages into forest reserves. This approach to forest conservation also produced protests, and by 1907, Congress withdrew the power of the president to proclaim forest reserves.[32]

Private landholders grasped immediately the possibilities of federal reclamation to drive up land values. Roosevelt Dam and the Salt River Project stand out as an example of how private land values zoomed upward when federal reclamation became the dominant force on the river. When future Secretary of Agriculture, Henry A. Wallace, visited the Salt River Valley in 1909, he noted, "The value of the land here is going to be measured by the reading of the gauge at Roosevelt Dam." The project drove up the value of both urban and rural land, with speculative land values leading the charge into the future. On some projects, individuals filed upon land but neither did they live upon it nor farm it. Instead, they merely held it for speculative purposes, delaying the overall development of a project.[33]

3.11. Walter Lubken labeled this 1911 photograph — "The Desert" before water is applied, directly under Arizona canal. This is the nature of our best orange land.

The reclamation program not only increased property values by enhancing the value of previously arid lands, but it also guaranteed a more reliable water supply to lands already under irrigation, as in the case of the Salt River Project. In addition, dams and reservoirs offered hydroelectric power as well as protection from the ravages of floods. Revenues from power proved an important source of funding for the Salt River Project, and they were a preview of the important role hydroelectricity profits would play in financing reclamation over the next half century. That government reclamation projects could draw immense benefits from hydroelectric development was a source of satisfaction to President Roosevelt. He, like Pinchot, believed that the waterpower of the West belonged to the people. While this arrangement was made by special agreement on the Salt River Project as early as 1904, Congress formally declared, in an act of 1906, that hydroelectric power developed in conjunction with a reclamation project should be used to finance and serve reclamation projects.[34]

3.12. Development of "orange land" on the Salt River Project resulted in another Lubken photograph: "Picking oranges in Salt River Valley near Mesa, Arizona." February 19, 1907.

3.13. Early days of construction at Theodore Roosevelt Dam and powerhouse.

But try as it might to repair Reclamation's program, Congress grew increasingly disappointed at its progress. In spite of rapid movement on the part of the Reclamation Service to launch projects, the immediate results were underwhelming. Newell wanted projects to be based upon their "intrinsic merits and feasibility" and not simply a "probability of an early refunding of the cost [that] will endanger the future of the work." In short, he was hesitant to launch projects where private irrigation was already in progress. He wanted to increase the opportunities for newcomers, not simply increase the wealth of those already on the land. Still, Newell realized he needed to strike partnerships with private irrigation to bring more immediate successes. The Salt River Valley, where irrigation was already underway before the arrival of the Reclamation Service, belied Newell's ideals of placing new people on the land. Notwithstanding that, the success of the Salt River Project reflected well on the Reclamation Service.[35]

In March 1903, the Secretary of the Interior authorized five projects: the Truckee-Carson (later renamed Newlands) Project, in Nevada; the Milk River Project, in Montana (pending agreement with Canada on distribution of water of the St. Mary and Milk Rivers); the Sweetwater (North Platte) Project, in Wyoming and Nebraska; the Gunnison (Uncompahgre) Project, in Colorado; and the Salt River Project, in Arizona. But the annual Reclamation reports made it clear that there were investigations and reconnaissance going on in every state and territory where reclamation law was applicable. That included: Washington, Oregon, California, Nevada, Idaho, Utah, Arizona, New Mexico, Colorado, Wyoming, Montana, North Dakota, South Dakota, Nebraska, Kansas, and Oklahoma, excluding

3.14. An early view of the USRS headquarters building on the Uncompahgre Project in Montrose, Colorado.

3.15. Two years, to the day, after President Theodore Roosevelt signed the Reclamation Act, Derby Diversion Dam was dedicated on the Truckee-Carson (Newlands) Project. Note the chartered train waiting for the party to return to Reno, Nevada, from the event. June 17, 1905.

Indian lands in that territory. Texas, as a non-public land state, was not covered under reclamation law until 1905-06.

The political leadership in both Congress and the executive branch encouraged the Reclamation Service to create numerous projects across the West by the 1920s. While the number of projects was impressive and in many cases represented engineering triumphs, there were many economic and social problems on the projects, despite the fact that the years from 1902 – 1919 were a "golden age" for American agriculture. By the 1920s, barely more than a million acres enjoyed government water — far short of earlier predictions. The Reclamation Service built impressive dams and reservoirs: Theodore Roosevelt Dam, Arrowrock Dam, and Shoshone (Buffalo Bill) Dam, but all was not well in rural America and the projects reflected it. Despite good times for agriculture, American cities and factories offered opportunities and rewards that became attractive alternatives to farming and the increasingly high investments required to make a farm, even on a government reclamation project. The Reclamation Service struggled against the tide

3.16. Senator Francis G. Newlands, on the far right, with Representative Franklin W. Mondell of Wyoming, and L. H. Taylor, project engineer for the state of Nevada, standing on the newly completed headworks of Derby Diversion Dam, just after speaking at the dedication of the dam. June 17, 1905.

3.17. Pathfinder Dam in Wyoming under construction in June of 1908.

on many fronts, including the repayment debt burden incurred by farmers for the construction of dams and canals – a burden that settlers were expected to pay back to the government. Yet, the idealism that created the Reclamation Service continued to drive it. In the process, it employed the tools of high-modern twentieth-century civilization: application of capital, scientific and engineering expertise, and an urban based labor force to build the necessary dams, reservoirs, and canals in an attempt to establish nineteenth-century homestead farming. The Reclamation Service was swimming against the tide even though many voices espoused the values of rural America and condemned the dehumanizing forces of urban industrialism.[36]

3.18. Construction equipment on the Milk River Project in Montana in 1906 and 1907.

Visions of Rural Life Reinforce the Enterprise of Reclamation

"Back-to-the-Land movements" (or as William Smythe phrased it, "placing the landless man on the manless land") were much in evidence. In periods of rapid transition, nostalgia flourishes. The good life on the land

3.19. Reclamation photographers captured what Reclamation hoped to achieve on its projects in these two images of the 1910s on the Okanogan Project in eastern Washington – homes and productive lands.

seemed to offer more certainties than life in dangerous cities. Still, urban life, with its greater freedom and opportunity, proved attractive. Most telling, migration from rural areas to the cities confirmed their allure. Some believed that America could have the best of both worlds. One among them was Liberty Hyde Bailey, Director of Cornell's Agricultural College. He advocated a "philosophy of country life in which he envisioned a new rural civilization based, above all else, on the concern of men for nature." He called it the "Rural Life Movement." While he embraced technological progress, he wanted to retain older values. "How to have both the new and the old was the critical problem," for Bailey. While many believed this was impossible, Bailey and his adherents carried their ideas to the national press and even to President Roosevelt.[37]

The Rural Life Movement gained the official attention of President Roosevelt with his appointment of a Country Life Commission in 1907. Another movement laden with nostalgia, but in some ways antagonistic to the Rural Life Movement, was the Back-to-the-Land Movement that began during the last years of Roosevelt's second term. It remained in evidence through the 1920s and received a temporary boost when the Great Depression called people back to the farm to escape urban unemployment. For President Roosevelt, the Reclamation Act of 1902 was a tremendous victory for the values that inspired both the Rural Life and the Back-to-the-Land Movements. The Reclamation Act, by promoting homemaking, made "for the stability of the institutions upon which the welfare of the whole country rests … actual homemakers … have settled on the land with their families," Roosevelt proclaimed. A report from his Country Life Commission in 1909 hailed the Reclamation Act for opening to settlement previously worthless land that "insures to settlers the ownership of both the land and waters." More to the point, the act ensured the continuation of the western land frontier now that water could be provided to its arid portion.[38]

3.20. Reclamation photographers also captured the failure of homesteads, as in this 1927 image taken on the North Platte Project.

The Rural Life Movement and the Back-to-the-Land Movement helped to validate government efforts to foster farming communities in the West. The two movements were not exactly the same, and neither advocated turning the clock backward to a pre-industrial era. Both saw value in rural living and farm life, even though the Rural Life Movement did not attempt to lure people away from the city to settle upon the land as did the Back-to-the-Land Movement. One study points out that the Back-to-the-Land Movement was in some respects "a bastard child of the country-life movement."[39] The latter received support and publicity from a presidential commission and published a handsome and slick periodical, *Country Life in America: A Magazine for the Home-maker in the Country.* It appeared monthly from 1900 to 1910, then on a semimonthly basis until 1912, when monthly publication resumed until 1942. Under the direction of one of its early editors, Liberty Hyde Bailey, Chairman of the Country Life Commission and principal author of its report, the magazine published such articles as "The Landward Movement"; "Could I Succeed on a Farm?"; "The Philosophy of the Soil"; "A Five-Acre Model Farm"; and "Cutting Loose from the City." The movement and magazine did not advocate a resettlement of population in the country. The goal was to improve rural life and thereby discourage the rapid flow of population to the city. The Rural Life Movement also supported an early version of suburbanization or the placement of city dwellers on rural acreages. From these suburban homesteads they could draw their living both from city occupations and the land itself. Rural life reformers wanted to make urbanites comfortable in a rural setting, bringing modern urban ideas of efficiency and order to the country.[40]

Bailey drew a sharp line between the Country Life Movement and the Back-to-the-Land Movement. In a 1911 publication he referred to "the present popular back-to-the-land agitation ... a city or town impulse, expressing the desire of townspeople to escape, or of cities to find relief, or real estate dealers to sell land." He believed not much should be expected of that movement as a vehicle to rejuvenate either rural or urban life. "But whatever the outward movement to the land may be," he said, "to effectualize [*sic*] rural society, for the people who now comprise this society, is one of the fundamental

3.21. Liberty Hyde Bailey as a young man, about 1880. Cornell University Library.

problems now before the people. The country-life and back-to-the-land movements are not only little related," he concluded, "but in many ways they are distinctly antagonistic." Overall, the object was to raise the quality of rural life for the purpose of keeping people on the land and diminishing the flight to the city. But country life could not be idealized by journalists and promoters to the point where city dwellers were unwisely tempted to try their hands at farming.

The warning became all too relevant for settlers on government reclamation projects who arrived with little experience and little ready cash to invest in their new venture. Others drawn to the land by the propaganda of the "dry farming" movement suffered similar misadventures when told to farm dry lands by using elaborate plowing and disking techniques. Urban migrants to the country could degrade rural life and risk ruining their own lives. But there was little doubt in the Country Life Commission's report of 1909 that rural values were fundamental to the health of the nation. Not only were farmers essential to the material wealth of the nation, "but in the supply of independent and strong citizenship, the agricultural people constitute the very foundation of our national efficiency." In prose echoing the English agrarian poet Oliver Goldsmith and, of course, William Jennings Bryan's Cross of Gold Speech of 1896, the Commission could not resist an unflattering remark about the parasitic nature of cities compared to the virtues of country life: "the city exploits the country, the country does not exploit the city."[41]

Building the Projects, Physically, Administratively, and Socially

Even at the beginning of the twenty-first century, a hundred years after passage of the Reclamation Act, the *New York Times*, perhaps America's most urban newspaper, editorially reminisced about the family farm: "Few institutions are more central – iconic, even – to America's self-image. The words themselves conjure up Norman Rockwell and a shared national heritage that extols self-reliance and the conquest of the frontier."[42] While the Reclamation Service found inspiration in this farm ideology, it faced immediate tasks: to build the structures — the dams, the reservoirs, the canals, the ditches — and, to level the land for the proper application and drainage of water. And Reclamation had little time to ponder rural sociology or the changing economics of farming.

In addition to Newell as chief engineer, other names fill the engineering rosters of the early Reclamation Service — Ignatius D. O'Donnell, Arthur Powell Davis, John L. (Jack) Savage, Morris Bien — and numerous project directors who were also engineers. Years prior to the Reclamation Act, engineers and surveyors worked for the USGS locating reservoir sites

3.22. Arrowrock Dam in 1915, at completion, with the construction camp still in place below the dam.

3.23. Buffalo Bill Dam Powerplant under construction in 1927 on the Shoshone Project in Wyoming.

and measuring stream flow.[43] Now the opportunity was at hand for these experienced people to move into the real work of the Reclamation Service. In the last two decades of the nineteenth century, engineering schools produced a wealth of talented and trained men who staffed the construction projects of both government and private firms. For most of the nineteenth century, engineers in the West gained their knowledge through experience, trial, and error. But with the growth of engineering training in technical schools and colleges, a new breed of engineer emerged. Their numbers far exceeded American engineers from the Army's U.S. Military Academy at West Point. That institution had served well to fill the ranks of the Corps of Engineers and develop a "cosmopolitan science ... the bunker of a bookish tradition that distanced army construction from improvised frontier technique," according to one historian of the Corps of Engineers.[44]

Outside the Army, many of the engineers in the West in the late nineteenth century were self-trained entrepreneurs, jacks-of-all-trades. But the younger generation had more formal training, as the number of graduates increased from established schools and from new land grant colleges brought into existence by the Morrill Act of 1862. In the 1870s, approximately two thousand men identified themselves as engineers; in 1880 seven thousand did so; by 1920 that number rose to 136,000. The number of engineering colleges increased from twenty-one in 1870 to 110 in 1896. Like other professions, such as medicine, law, and academics, engineers formed professional societies in the mid-, late-nineteenth century, e.g., the American Society of Civil Engineers (ASCE) was established — 1852. Historian Burton J. Bledstein identified the formation of these organizations as part of a broader movement toward the professionalization of American life. Professional organizations recognized educational training, standards of individual performance, and even commitment to the larger good of the society. Bledstein concluded that commitments to "professional standards" were an important aspect of middle class life.[45]

Because many aspects of capitalism were wasteful, private standards of professional commitment to the progressive values of science and engineering easily translated into "a crusading social movement" to reform capitalism. "Conservation provided them with the means of linking professional goals with a program of national reform." In this respect, science meant "analysis and planning in any engineering undertaking in reference to altering or improving physical structures and of nature itself." Engineering had great implications for a planned society. The new chief engineer of the

Reclamation Service, Frederick Newell, was among this new breed. His high hopes for reclamation communities grew out of his belief that if engineers brought together land and water, farmers, with the continuing guidance of the Reclamation Service, could build harmonious and prosperous farming communities.[46] As Newell wrote in the first chapter of *Irrigation in the United States* (1902): "In a wider sense, irrigation is taken to include the whole question of conservation and utilization of water in the development of the arid regions, and to embrace a discussion of features of social and political importance arising from the reclamation of the public lands." The chapter contains the clearest explanation of the idealistic goals of the Reclamation Service.[47]

Some engineers waxed even more enthusiastic about the possibilities for their talents, "We are the priests of material development, of the work which enables other men to enjoy the fruits of the great sources of power in Nature, and the power of mind over matter." The national figure of Herbert Hoover eventually epitomized the engineer and his social mission. Hoover's experience with mining engineering, with business, with the National Food Administration in World War I, and with Republican Party politics made him a prime example of an engineer who combined professional engineering values and the processes of scientific management into a program of social action. Furthermore, Hoover's assumption of a leadership role in negotiating the 1922 Colorado River Compact for the distribution of the waters of the Colorado River gave assurances of a successful outcome as he applied modern methods of problem solving to highly volatile political and economic questions. Little wonder that, after the interparty battles subsided over the Boulder versus Hoover name for the great dam on the Colorado, Congress officially affixed Hoover's name to the dam in 1947.[48]

But all was not engineering and social philosophy. It took bureaucratic organization to get the Reclamation Service up and running. Much of the organization came from the USGS. Meritocracies characterized modern governmental service bureaus, and the U.S. Reclamation Service, with designated grades of employment and pay, was no exception. Employment for beginning engineers depended upon success in a civil service exam that opened the way for them to be employed for a six months trial period as "engineer aides" at $60 to $75 a month. If they were kept on, they assumed a more permanent job as "assistant engineers" at salaries of $1,200 to $1,600 a year. Advancement depended upon "ability displayed, both in engineering operations and in matters regarding business ability and tact." Assistant engineers in charge of field parties often directed the work of engineering aides and temporarily employed teamsters, packers, and cooks. Middle engineers were the next level, "approaching middle age" with experience in irrigation work or construction, and they were paid upward from $1,800 a year. District engineers stood in the higher grades and oversaw reclamation in a particular drainage basin. Their district headquarters were generally located at the largest project under construction or in the nearest community. District offices kept financial records and sent "abstracts" to the Office of the Chief Engineer in Washington. The Reclamation Service regularly assembled boards of engineers to consider various undertakings and their problems. Reclamation engineers and consulting engineers comprised the boards. Usually, they consisted of three to five members and sometimes invited the assistance or advice of citizens and businessmen interested in a project at hand. Their chief function was to study proposals, designs, and consequences and make recommendations to the chief engineer. What could be termed "engineering by committee" often produced careful but conservative designs and solutions to problems.[49]

By the end of 1907, the Reclamation Service had authorized twenty-four projects – at least one in each of the original sixteen states and territories (except Oklahoma) mentioned in the Reclamation Act. Originally, Congress excluded Texas because it possessed no public lands, but added it to the list of Reclamation states in 1905-1906. The number of projects and their scattered geographical locations meant that the Reclamation Service found itself moving rapidly in several different directions. The results were not always consistent with rational, scientific planning. They were, however, consistent with the enthusiasm of the Roosevelt Administration and the political demands for projects in various localities. Most importantly, the availability of money from land sales in the western states made these years the heyday of reclamation.[50]

Project approval was as follows:

1903 Salt River Project, Arizona Territory
Milk River Project, Montana
Truckee-Carson (Newlands) Project, Nevada
Sweetwater (North Platte) Project, Nebraska
Gunnison (Uncompahgre) Project, Colorado

1904 Yuma Project, Arizona Territory
Minidoka Project, Idaho
Lower Yellowstone Project, Montana
Hondo Project, New Mexico Territory (abandoned in 1922)
Belle Fourche Project, South Dakota
Shoshone Project, Wyoming

1905 Boise Project, Idaho
Garden City Project, Kansas (abandoned in 1910)
Huntley Project, Montana
Klamath Project, Oregon
Umatilla Project, Oregon
Rio Grande Project, New Mexico Territory and in Texas by special permission of Congress
Strawberry Valley Project, Utah
Okanogan Project, Washington
Yakima Project, Washington

1906 Sun River Project, Montana
Williston Project, North Dakota (abandoned in 1926)
Carlsbad Project, New Mexico Territory

1907 Orland Project, California

1911 Grand Valley Project, Colorado

1917 King Hill Project, Idaho
Riverton Project, Wyoming

1926 Owyhee Project, Oregon
Vale Project, Oregon

The numbers were impressive. Initiating twenty-four projects in five years severely challenged the resources and talents of the Reclamation Service and prompted charges that the Reclamation Service had overreached itself. But what a reach it was, from North Dakota to Texas and from Nebraska to the Pacific Ocean. Building complex dams, delivering water, installing hydroelectric facilities, and organizing agricultural communities required immense skill and expertise. The Reclamation Service was confident that it possessed the skills to mix engineering with bureaucracy, land with local water laws, individual self-interest with community needs, and agriculture with the challenges of desert and mountain environments. Yet, in attempting to please all, the Reclamation Service disappointed many. This was especially true in the case of people drawn to the reclamation projects with expectations of quick success. That the Reclamation Service accomplished as much as it did in such a short period was a testimony to its leadership and supply of available American engineering talent in the early twentieth century. In addition, there was a general climate of approval prevailing in Congress and in the daily and periodical press.

Exuberance and even hubris contributed to the energy showed by the Reclamation Service in these first years. But pride often bred mistakes as well as impressive accomplishments. By 1910, a chorus of criticism emerged from Congress, the Administration, and the very settlers the program hoped to serve. The question of what went wrong has preoccupied both scholars and the popular press. Newell and his well-trained professionals tried to respond to the mandates of the Reclamation Act, political pressures from Congress, and the settlers themselves who wanted quick entry to lands irrigated by government water. Of course, settlers often entered the lands, as on the Truckee-Carson and Grand Valley Projects, even before water was available. In dam building, the Reclamation Service not only tried, but excelled. From 1903 into the 1920s, the Reclamation Service built some of the most impressive water diversion structures and the largest and highest dams in the Western Hemisphere and the world: Roosevelt in Arizona, Arrowrock in Idaho, Elephant Butte in New Mexico, Shoshone (Buffalo Bill) and Pathfinder in Wyoming, Belle Fourche in South Dakota, and the Gunnison Tunnel in Colorado. Because of this, even before Hoover Dam, the

3.24. A modern view of Elephant Butte Dam and powerplant on the Rio Grande in New Mexico.

3.25. "Waiting for Water" on the Grand Valley Project in western Colorado. August 2, 1913.

Reclamation Service had established a world-wide reputation for design and construction.

Dam building eventually became the jewel in Reclamation's crown, but doubts about the slow settlement of the reclamation projects haunted the Reclamation Service from its inception. The Reclamation Act was not simply the application of improvements such as those contained in the typical rivers and harbors bill. Reclamation was more than improving rivers. And while in many ways it was an extension of past legislation that encouraged the creation of family farms, it was a bold step in the direction of government activism. As a new government bureau, the Reclamation Service, suddenly became responsible for opening new lands by way of water delivery and building of communities. The law dictated a repayment schedule for the building of the water delivery systems and imposed acreage limitations to keep the small farm character of a community. This effort reached beyond the formidable challenges that the civil engineer faced in building dams and manipulating rivers. The effort involved social engineering in ways never before assumed by American government at any level, let alone the national government.[51]

Still, it was not a plunge into state socialism. Later critics, who lambasted the expansion of the federal government under the New Deal of the 1930s, sometimes pointed to reclamation as another unneeded and unwanted program. But reclamation was the creature of an earlier era that sought to preserve American values rather than a program to meet a dire economic crisis. The Conservation and Progressive Movements combined during an era of relative prosperity in the first decade of the twentieth century to create an imagined idyllic life of the nineteenth-century countryside in the arid lands of the West. To do so, Congress called upon the Reclamation Service to employ engineering skills in water manipulation on the public lands to promote western development. The new projects would, in turn, encourage a social and land distribution policy attractive to the settlement of families upon the land. Congress had first intended this in the original Homestead Act of 1862, and now the Reclamation Act was to ensure the same for the arid lands. But trouble often arises from the best of intentions.

Of all the service bureaus created in the Progressive Era, including the Forest Service, the National Park Service, and the service bureaus within the Department of Agriculture, the Reclamation Service faced the most complicated and varied demands upon its talents and resources. The Forest Service dealt with trees, land, and forage, not with peopling the land or constructing huge dams that required tremendous organization and the best engineering minds in the country. Also, the National Park Service did not face the task of transforming lands but rather simply defending them, catering to tourists, protecting scenic beauty, and excluding Native Americans from their traditional resource use within the parks. Probably the only other agency with a comparable task was the Panama Canal Commission that Congress authorized only a few months after the passage of the Reclamation Act. Clearly, both the Reclamation Service, with its mandate to build new communities in the arid lands, and the Panama Canal Commission, charged to dig and construct a water route through disease filled swamps to connect two oceans, required the resources that only the federal government possessed. And once government proved its success with these tasks, it could embark upon others. Reclamation was an entering wedge, in Newlands's mind, for an ever larger role for the federal government in American life.

3.26. Steam shovels #230 and #222 are just about to meet in the Culebra Cut on the Panama Canal. May 20, 1913. Steam shovels were used extensively in Reclamation construction projects. After completion of the canal, Reclamation received some of these steam shovels for work on its projects. Official Photograph of the Panama Canal Commission – Panama Canal Graphics Section. Provided courtesy of William P. McLaughlin.

In 1915, the celebratory Panama-Pacific International Exposition in San Francisco marked the completion and opening of the Panama Canal. No similar celebration occurred for the Reclamation Service. Although it completed comparable feats in its series of high dams, its work in building irrigated communities contrasted rather than compared to the trans-isthmus canal achievement. The Reclamation Service was a distinctly civilian operation, conducting its work under the authority of a new civil service bureaucracy. The new bureaucracies that were supposed to embody the growth of the modern, service-oriented state in the Progressive Era, however, often reflected the weaknesses of American government rather than its strength. Given the structural divisions in American government — divided sovereignty in a federal system and division of powers into legislative, executive, and judicial functions within the federal government itself — a broad-gauged program to irrigate western lands that required expertise in many fields besides engineering was a gamble, to say the least.[52]

In these early years, the tasks facing the Reclamation Service were as complicated as they were numerous. Much work already had been done by the USGS in scouting out and surveying reservoir and dam sites, but this very activity excited the appetites of speculators. They tied up lands requiring the Reclamation Service to buy them out, often by invoking eminent domain powers in expensive court proceedings. Most challenging, and at the same time rewarding, the Reclamation Service showed its engineering capabilities in the building of major dams. It quickly, some would say eagerly, moved beyond constructing simple diversion dams such as the Derby Diversion Dam, completed in 1905, on the Truckee River for the Newlands (Truckee-Carson) Project.

3.27. Walter J. Lubken photographed the Derby Diversion Dam three days after its dedication. June 20, 1905.

Not only did construction of waterworks demand attention, but so did the water rights for the new projects. Placing farm families on the land and persuading them to stay on it was the most difficult task of all. Prospective farmers often had neither the capital nor the skills to settle upon a project and make a go of irrigation agriculture. Without successful and happy settlements, the Reclamation Service faced serious problems. On the other hand, could a bureau so new and inexperienced be expected to achieve so much in so little time? The arid lands question presented a daunting task to the Reclamation Service, Chief Engineer Newell, and the cadre of engineers he gathered for this grand experiment with government, land, water, and agriculture. In addition, the Reclamation Service found itself encumbered by a host of clerical rules as it tried to operate within the Department of the Interior — a department that had recently had its share of scandal associated with land deals in California, Oregon, and Washington.[53]

Yet, political scientists who study the evolution of bureaucracy note that the Reclamation Service was one of the most unencumbered agencies in the government at its outset. The Secretary of the Interior (which really meant the Reclamation Service) had the power to designate projects and the power to spend freely from the Reclamation Fund without consulting Congress. In addition, the fund was "revolving" in a "self contained funding scheme," which meant that it renewed itself through the continued sale of public lands in the West and the expected repayment for water by project settlers. This was an almost unprecedented grant of authority and discretionary power by Congress to a new governmental bureau. Within a decade, however, the Reclamation Service lost the confidence of Congress and much of the discretionary powers extended to it.[54]

In Defense of the Early Years

The commitment by Congress, the Executive Branch, and the federal bureaucracy to the challenges of reclamation — dam building, reservoir creation, delivering water, the production of hydroelectric power, community building — was too great to permit the entire program to fall prey to its critics. This was true, in part, because the Reclamation Service early recognized the power of publicity and getting its message out to the public as well as Congress. That message emphasized great works, the conquest of the desert wilderness, and the advance of civilization in the West.

3.28. J. H. Quinton, supervising engineer on the Uncompahgre Project, and C. J. Blanchard at the entrance to the west portal of the Gunnison Tunnel near Montrose, Colorado.

C. J. Blanchard, who ostensibly was employed as a statistician, was instrumental in launching the publication of *Reclamation Record* in 1908, the Reclamation magazine that became *New Reclamation Era* in 1924, and in 1932 evolved into *Reclamation Era.* (It ceased publication in 1983.) As the voice of the Reclamation Service, the publication invariably delivered an upbeat message about the projects, engineering feats of the bureau, and agricultural advice to homesteaders. It was similar to other farm publications with regular columns on animal husbandry (i.e. "Pig Points") and crop and garden suggestions. As early as 1911, Blanchard experimented with the moving picture as a means to depict and advertise the impressive works of the Reclamation Service. Film crews recorded the presence of ex-President Roosevelt at the ceremonies to mark the completion of Roosevelt Dam near Phoenix, Arizona, in 1911. Such films received wide distribution, advertising the good work of Reclamation building essential dams in remote areas of the West.[55]

Not to be overshadowed by these technological achievements were Reclamation's redemptive social and economic goals in underwriting homemaking in the West as expressed in Newell's words: "the reclamation of places now waste and desolate and the creation there of fruitful farms, each tilled by its owner." All was predicated upon a faith that, in the words of one scholar, "small farmers would inherit the earth," and echoed Smythe's often quoted *Genesis* 2:10 that agriculture had its origins in irrigation and even a biblical blessing, "And a river went out of Eden to water the garden." Other supporters quoted *Isaiah* 35:1, "The wilderness and the solitary place shall be glad for them, and the desert shall rejoice, and blossom as the rose."[56]

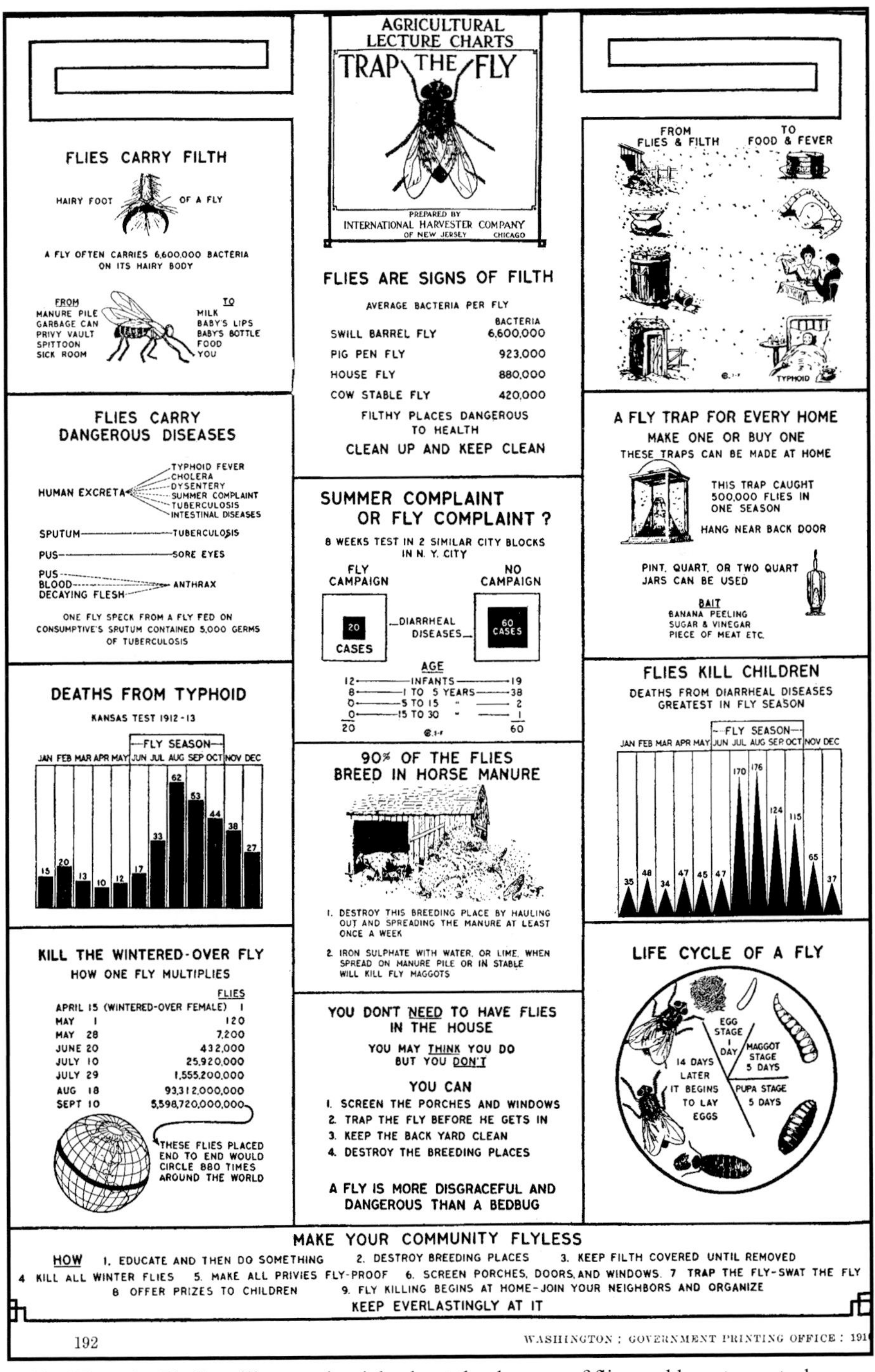

3.29. "Trap the Fly," an illustrated article about the dangers of flies and how to control them. *Reclamation Record,* April 1916. An example of educational material provided to Reclamation settlers.

3.30. This float of farm produce at the Pioneer Day parade in Ogden, Utah, on July 24, 1934, alluded to an often quoted biblical phrase cited by irrigation supporters. It can be found at Isaiah 35:1: "The wilderness and the solitary place shall be glad for them, and the desert shall rejoice, and blossom as the rose."

Putting Idealism to Work on the Land (Including Indian Land)

Behind the starry-eyed idealism in much of the literature of reclamation stood the harsh realities and hard work of bringing new lands under the ditch. Newell and Assistant Chief Engineer Arthur Powell Davis faced the task of gathering a wide range of engineering talent to keep pace with the numerous projects undertaken. The speed of the entire enterprise might have inspired caution in those less optimistic and committed. Congress also showed a heightened enthusiasm for reclamation in continued legislative activity that added provisions to expand and improve the original Reclamation Act, but not necessarily to fetter the administration. The Reclamation Service enjoyed a long leash that came with the novelty and youthfulness of a newly created government bureau charged with "breaking new ground." Gifford Pinchot used this phrase for the title of his autobiography, *Breaking New Ground* (1947), that tells of his pivotal role in the creation of a new service bureau, the U.S. Forest Service. The words also serve to describe the uncharted ground facing the fledgling Reclamation Service. [57]

Not content to confine reclamation to the public land states and territories authorized in the Reclamation Act, Congress quickly extended it to Texas and Indian reservations. Reclamation entered Texas in 1905 with the Rio Grande Project's dam in New Mexico Territory that could also serve land in Texas. The following year, in 1906, Congress included all of Texas within the purview of the Reclamation Act, although it was not a public land state. While this might suggest that Congress was not overly concerned about revenues from

3.31 Chief Engineer Arthur Powell Davis, Director Frederick Haynes Newell, Division Engineer Hiram N. Savage (Northern Division), and Division Engineer Louis C. Hill (Southern Division) at Arrowrock Dam on August 16, 1911, near the beginning of construction.

3.32. Reclamation's Elephant Butte construction camp in 1909. Note the variety of housing.

land sales to sustain the Reclamation Service, there was not another project in Texas until the early 1940s. Elephant Butte Dam, after several setbacks, was finally completed in 1916. It provided the necessary water reserves to serve land in New Mexico, Texas, and Mexico.[58]

At the beginning of the twentieth century, American Indian reservations reeled under the land reductions instigated by the Dawes Allotment or Severalty Act of 1887. Congress offered the Reclamation Act on a selective basis to Indian reservations under the fourth section of the Allotment Act, which permitted the conversion of reservation lands into private lands for individual Indian farmers. In doing so, Congress permitted what amounted to further incursions upon reservation lands by using reclamation to reduce lands under communal (reservation) ownership. An act of 1904, entitled "Reclamation of Indian Lands in Yuma, Colorado River and Pyramid Lake Indian Reservations," directed the Secretary of the Interior to open for entry, "any lands on said reservations which may be irrigable … as though the same were a part of the public domain." Indians were eligible to receive five- to ten- acre irrigable homesteads, but "The remainder of the lands irrigable in said reservations shall be disposed of to settlers under the provisions of the reclamation act." The price of this land was based upon its value prior to the construction of irrigation facilities, which meant that white settlers obtained bargain prices for lands that quickly increased in value as irrigation became available. Money from sale of lands was to be credited to the Bureau of Indian Affairs (BIA) and expended for the benefit of the reservation "from time to time" by the Secretary of the Interior. The land payments, however, most often went to retiring the debt for the construction of irrigation projects built to benefit white farmers. Initially, individual Indians did not have to sign repayment contracts. The 1904 act set the pattern for subsequent Indian

3.33. The first load of rock to be placed in the foundation of Elephant Butte Dam, 1912. Note the stenciling on the flatcar "U.S.R.S." This and Reclamation's Yuma Valley and Boise and Arrowrock Railroads connected with major railroad lines and interchanged cars with them. Though Reclamation construction rail lines often did not connect to interstate carriers, they were present at many early day projects to haul materials.

reclamation acts. In 1908, after the passage of several acts for specific reservations, Congress passed a general act authorizing the Secretary of the Interior to make agreements with Indian tribes "for the reclamation of lands allotted to them under the general allotment act."[59]

3.34. Congressional interest in Reclamation's projects, in spite of all the issues, remained high. Here a congressional party inspects Elephant Butte Dam during construction in 1915.

The Reclamation Service was happy to reclaim Indian lands. As early as 1884, Congress created a general irrigation fund to be expended at the discretion of the Secretary of the Interior for irrigation on reservations. Out of this grew the Indian Irrigation Service that existed at the time of the passage of the Reclamation Act. The work of the Indian Irrigation Service had not been impressive and Congress willingly permitted the new Reclamation Service a role in Indian irrigation. By 1913, Congress revived the Indian Reclamation Service that permitted the Bureau of Indian Affairs (BIA) to develop reservation water resources. In 1914, Congress shifted repayment responsibilities from the tribes to individual Indian settlers, placing what amounted to a discouraging burden on the development of Indian agriculture. In 1920, it set the repayment period on Indian-owned allotments at twenty years. The reclamation of Indian lands, while offering irrigation to farmlands, flew in the face of many Native American cultural traditions. This was part of the plan. Irrigation agriculture could serve "to

3.35. Indian children on a wasteway structure on the Two Medicine Main Canal. Blackfeet Project, Montana. About 1910.

civilize" Native Americans, breakdown their primitive communal traditions, and turn them into property holders who could undertake market agriculture and compete with and emulate their white neighbors who were settling upon former reservation lands. One thing was certain, however: Indian reclamation opened the way for non-Indians to acquire former reservation lands because the small acreage offered to individual Indians left many acres

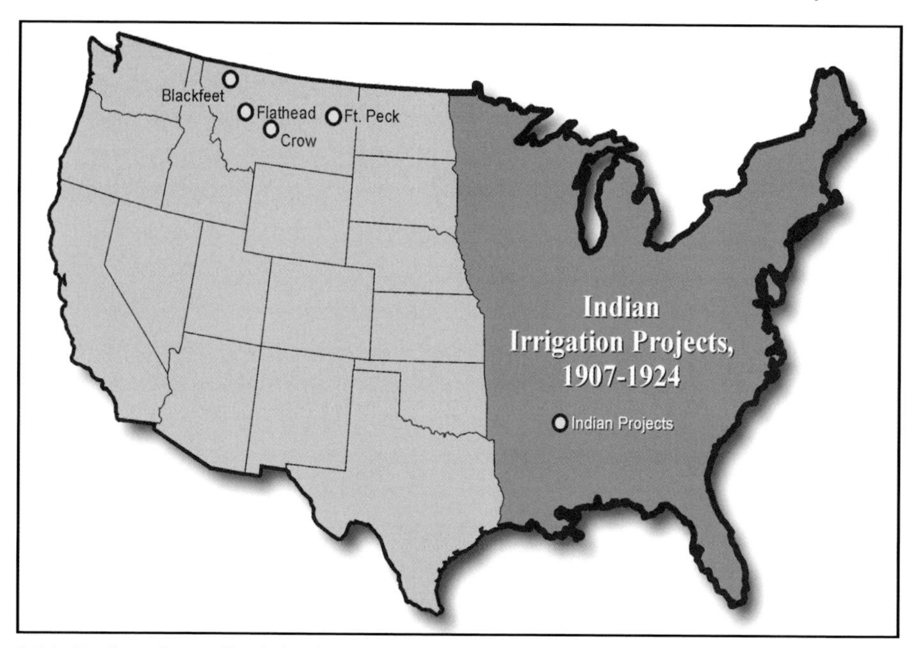

3.36. Reclamation Indian irrigation projects.

on a reservation unoccupied and open to non-Indian sale and settlement. Reclamation seemed cut to order for this task. For Native Americans it resulted in thousands of acres lost from Indian or tribal reservation ownership without opening much opportunity.[60]

The experience of the Pyramid Lake Paiute Reservation in Nevada is a case in point. In 1905, the Derby Diversion Dam on the Truckee sent water away from the Paiute reservation to the Truckee-Carson (Newlands) Project. Not only was water removed from the Truckee River which drained into Pyramid Lake and sustained its fishery, but it was transferred by canal to the Carson River or Lahontan Basin to serve the irrigation project that also encompassed another reservation — the Fallon Indian Reservation of the Paiute-Shoshone Tribe. Some Indian land was potentially good for irrigation, which quickly subjected the small reservation to allotment and overall reduction in the size of the Fallon reservation. In the Stillwater section of this reservation, Indians were offered five-acre irrigation allotments. The reduction, of course, freed up land for non-Indian settlers. Reclamation's record in dealing with the reclamation of Indian lands and "Indian water" largely reveals Congress's failure to protect Indian property. Agreements between Bureau of Indian Affairs and the Reclamation Service to share the benefits

3.37. Reclamation's Indian projects provided construction jobs. Here Indian teamsters are working Fresno scrapers on construction of the Two Medicine Main Canal on the Blackfeet Project about 1910.

of government reclamation under Indian allotment arrangements became a much criticized aspect of western water history.[61]

Professor Pisani characterizes the relationship as one of "uneasy allies." Political scientist Daniel McCool notes the rise of "iron triangles" in western states to thwart and limit Indian water rights that had received unexpected recognition in the U.S. Supreme Court's Winters Decision of 1908.[62] The Winters Doctrine, arising from the decision, held that the reservations were entitled to enough water to serve the purposes for which they were created. This meant that Indian reservations had open-ended "vested rights" to water, threatening established as well as future western water users. In response, western constituencies, according to McCool, often resorted to the politics of the "iron triangle." The three legs of the triangle included: (1) local organizations, sometimes powerful water users' associations; (2) Congress; and (3) government bureaus. Local constituencies and organizations pressured Congress to adopt programs favorable to them and not Indians, whose voices were less than audible in the political arena. Congress obliged its constituencies by passing legislation antagonistic to Indian water claims, permitting white settlers to share water and lands reserved for Indian use. When Reclamation undertook projects to bring government water to Indian reservations, there often was a *quid pro quo*. After the demand for Indian reclamation had been satisfied, Reclamation invited non-Indians to apply for and take ownership of surplus lands within the project. This made the Bureau of Reclamation a participant in implementation of the Dawes Allotment Act. While Reclamation ardently defended its efforts to improve

3.38. The isolated location of Theodore Roosevelt Dam demanded creative supply measures. This USRS sawmill in the Sierra Ancha Mountains provided construction lumber and was photographed by Walter Lubken on July 14, 1904.

Indian water resources, particularly for irrigated lands, by the late twentieth century it admitted that some projects resulted in loss of Indian land. [63]

3.39. Reclamation built the "Apache Trail" from Mesa, Arizona, to Theodore Roosevelt Dam as a supply haul road. Twelve horses haul 5.5 tons of cement back down the Apache Trail from the USRS cement plant at Roosevelt for use in construction on the Granite Reef Diversion Dam. March 27, 1907.

Where reclamation projects formerly included both public and private lands, now Indian lands were also opened for the work of the Reclamation Service. In return, settlers need only offer a promissory ten-year note, either individually or through their local water users' association, to pay back at no interest the costs of constructing water delivery works on the project – dams, canals, laterals, etc. Endeavors on Indian lands further dispersed the operations of the Reclamation Service, which were already under criticism for their wide geographical distribution. Partly this occurred because the Reclamation Act mandated that the program spend fifty-one percent of revenues from land sales in the states and territories where collected (a provision removed in 1910).[64] Members of Congress wanted projects for their districts, and the President wanted as many projects in the West as feasible. The Reclamation Service did not shrink from the requests. By 1905, Secretary of the Interior Hitchcock had fully committed the annual income of the Reclamation Fund, an average of 2.5 million dollars each year. From 1903 to 1906, the Secretary of the Interior authorized twenty-four reclamation projects. Work did not proceed so quickly that it exhausted the money on hand, but, significantly, in 1906, the new Secretary of the Interior, James R. Garfield, separated project authorization and project funding.

Projects could be authorized, but nothing would occur until funding became available. Only four additional projects received funding between 1907 and 1918.

This is to say that, by 1907, when Newell officially became Director, the Reclamation Service bumped up against strict financial limits. Not only did the number of projects strain the Reclamation Fund, but the Reclamation Service consciously chose to overbuild structures both to insure safety and to convey the impression of durability and permanence in the eyes of the public. An ongoing inflation, provisions in the Reclamation Act that excluded Chinese (Mongolian) labor (previously used so advantageously by railroad companies), and work shifts limited to eight hours made for high labor costs. While it contracted work out to private companies who still had to adhere to these requirements, the Service also directly employed workers. In remote work places, it provided services that included stores, hospitals, doctors, and even recreational facilities.[65] Scarcity of labor in remote areas also increased labor costs and costs for transporting materials were high. Project costs often outran estimates,

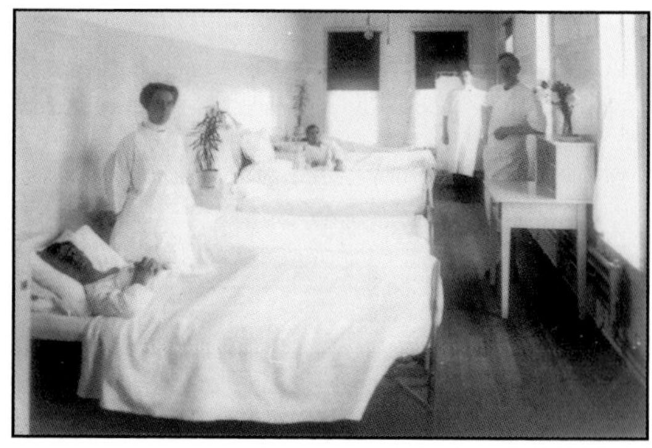

3.40. The hospital at Arrowrock Dam. April 18, 1912.

3.41. Mess hall kitchen at Arrowrock Dam. February 28, 1912.

3.42 USRS engineers' mess at Arrowrock Dam. February 28, 1912.

and that prompted charges of inefficiency, waste, and even fraud from critics and water user associations under repayment contracts with the Reclamation Service. The complaints, however, did not immediately bother the Reclamation Service as it continued to fly high in these first heady years.

Questions of the Law

A myriad of legal problems confronted the Reclamation Service as it began work. The fortunes of the Reclamation Service became entangled in a series of legal questions not addressed in the Reclamation Act of 1902. Most often, they involved the question of where federal authority ended and local (state, municipal, and county) authority began. In many instances, ambiguities in the law required expert legal counsel and the creation, almost immediately after passage of the Reclamation Act, of a legal division to handle problems. This division litigated water rights, examined titles, initiated condemnation procedures, prepared public notices and agreements with water user associations, filed suits, and provided advice to state and federal officials and water users. The chief legal officer of the Reclamation Service was Morris Bien, in Washington, D.C., and there were seven western offices headed by District Counsels from the late 1900s to the mid-1910s.

Bien took an expansive view of the authority and prerogatives of the Reclamation Service in western lands and waters. An 1879 engineering graduate of the University of California, Berkeley, Bien joined the USGS survey work under Major Powell in the Rocky Mountain region. After Powell's departure, he involved himself with legal matters of rights-of-way on public lands with the General Land Office and took a law degree at Columbian University (later George Washington University), in Washington, D.C. After passage of the Reclamation Act, Newell brought him from the General Land Office to the Reclamation Service to assume the position of a Supervising Engineer, although his work was primarily legal. From his office in Washington, D.C., Bien often administered the new bureau while Newell was out of town. In 1903, he officially assumed the office of supervising engineer in charge of the Lands and Legal Division that placed him in charge of the legal affairs of the Reclamation Service until 1924. One of his major jobs, which also involved a good deal of travel to localities in the West, was the transfer of water rights held by earlier settlers and irrigation companies to the Reclamation Service's new projects. This meant drawing up articles of incorporation for water users' associations and securing agree-

3.43. Morris Bien, the head of Reclamation's early legal efforts. *Reclamation Record,* June 1920.

3.44. The Lake Tahoe Dam controls the outlet into the Truckee River.

ments from them to repay the costs for the construction of irrigation works to the government. In his memoirs, he recounts one of his major failures: the attempt to gain control of Lake Tahoe for the new Truckee-Carson Project. The effort failed, in part, because the Reclamation Service was unwilling to pay the demanded price of $100,000 for the necessary water rights. Regretfully, he said, "We would have been saved a world of trouble and expense if we had bought it then."[66]

Most distressing to the Reclamation Service was the chaotic condition of western water law. It had been developed by state courts and legislatures in the light of the needs of various interest groups and resource users within the states. Prior appropriation and riparian doctrines intermingled in some states, while, in the predominantly arid states, prior appropriation dominated. In most states "beneficial use" limited prior appropriation. Bien assumed the task of trying to convince western states, and oftentimes specially appointed state water commissions, to adopt model water codes fashioned by him in the interest of the Reclamation Service. The goal was greater uniformity of water laws from state to state.[67]

This would ease the work of the Reclamation Service when it built projects that not only made additional water supplies available from new dams and reservoirs but also drew upon, redirected, or purchased long-standing water rights. Friendly state water codes could open the way for the Reclamation Service to go about its tasks under the provisions of the Reclamation Act, which required Reclamation to conduct its operations within the bounds of state water laws. On the other hand, the Reclamation Act appeared to suggest that, based on its original sovereignty and ownership, Congress possessed a reserved water right to improve federal public lands. It also implied federal control over interstate rivers. If all of these factors could be brought into play, negotiations over water rights and the demands of the Reclamation Service could easily defeat, for example, the riparian water rights of property owners along the shore of Lake Tahoe. These property owners resisted drawing down the Lake for the benefit of farmers in the Nevada desert. If the Reclamation Service possessed such sweeping authority, it might even embark on large scale river basin developments. Before it became apparent that many states would balk at accepting his water codes, Bien followed this legal course with a good deal of confidence. He thought the federal government should assume supreme authority in interstate water questions and over any new water the Reclamation Service would develop or impound.[68] Some state officials feared the growing federal presence in the activities of Reclamation and resisted. The struggle for uniform water codes never succeeded — much to the dismay of the Reclamation Service and its chief legal officer.

The reform campaign, however, reflected the confidence and power with which the Reclamation Service approached local governmental bodies in those early years. With the approval of the Reclamation leadership in Washington in 1903, Newlands persuaded the Nevada legislature to pass an administrative water reform act. It created Nevada's Office of State Engineer whose first duty was to register all water rights and issue all future rights. This was an important reform for a state that had long wrangled over water rights. One provision of the Act required the Secretary of the Interior to nominate Nevada's State Engineer, subject to approval of the Governor. In addition, the state legislation provided for the appointment of district water commissioners, also nominated by the Department of the Interior. After registration and approval of water rights in the state by the State Engineer's Office, the commissioners were to supervise the use of water on each stream according to a list of water rights "and to serve the government and its grantees their water according to their rights," according to Newlands.

The law, Newlands believed, would ensure that trained and experienced men, "above all, impartial men — administer the control of the stream." Newlands suggested that a compliant state legislature would ensure that Nevada would receive one of the earliest Reclamation projects. Nevada's new Senator asserted "that there are many things in this act, and particularly in the spirit of this act, that can be emulated by our sister states and territories." Newlands noted that President Roosevelt, on a recent visit to the state, suggested that the water legislation was "a model" for other states."[69] Yet, less than a decade later, in 1911, the Nevada legislature rescinded the power it had ceded over administration of its water to the Secretary of the Interior — falling in line with a general western reaction against the growing power of the federal government in water and land matters.[70]

3.45. Justice David J. Brewer wrote the *Kansas v Colorado* decision in 1907.

Meanwhile, the U.S. Supreme Court, in the case *Kansas v. Colorado* (1907), rejected the Reclamation Service's arguments that the government had authority to regulate interstate streams.[71] Justice David J. Brewer noted that, since the federal government had not established a broad authority over the streams, the states had filled the vacuum and their laws and court decisions could not be suddenly swept away by the federal government. The case also involved what a later historian called the defeat of the "dead end of exclusive riparian rights."[72] The U.S. Justice Department entered the case on behalf of the Reclamation Service to undermine, if not to abrogate, the riparian doctrine in the western states. The Reclamation Service held strongly to the view that the defeat of the doctrine would serve to strengthen and expand the water claims of western reclamation projects. A favorable decision would also work to undermine the doctrine in the eastern states and open the way for national control of interstate streams by the USGS. Greater national authority over the streams raised the prospect of drainage and river basin reclamation projects in the eastern United States for the Reclamation Service. In other words, *Kansas v. Colorado* halted the plans of the Reclamation Service to become a national bureau with projects and works in the river basins of the East as well as the West. While the decision curbed the ambitions of the Reclamation Service, Bien took even a darker view. He believed it cast doubt upon the constitutionality of the Reclamation Act itself.[73] Perhaps, fortunately for the continued life of the Reclamation Service, the Supreme Court never addressed this issue.

Still, the West was no small field for ambition. The twenty-five projects launched between 1903 and 1912, plus two more by 1918, and two in the 1920s, with the huge undertaking of the Hoover Dam project in 1928, took all the energy, diplomacy, and finance Reclamation could muster. The Reclamation Service accepted the challenge of thirty projects before 1930. The challenges that this new bureaucracy faced could have hardly been greater as it moved forward under the imperfect directions and authority of the Reclamation Act. In the process, it oversaw the building of dams, reservoirs, irrigation works, water delivery systems, powerplants, and towns. At the same time, it assumed the tasks of operating and maintaining reservoirs, dams, and the controversial administrative decisions regarding project construction costs that directly affected repayment schedules. The Reclamation Service took pride in bringing engineering know-how together with government administration and financing to achieve the development of the West's water resources and the beginning of its hydroelectric production.

Of Dams and Hydroelectricity

Dam building meant not only reservoirs, canals, and ditches, but also hydroelectric power. The era of western reclamation came upon the scene just as the age of electricity dawned in America and became timely partners as they grew together in the twentieth century. While the 1902 law had not considered hydroelectric power, Congress passed the Town Site and Power Development Act of April 16, 1906, to enable the Reclamation Service to withdraw 160 acres of land for town site lots and to utilize the production and sale of electricity for the building of reclamation projects. The Salt River Project immediately benefitted from this act. Electricity generated at Roosevelt Dam not only ran the onsite cement plant and provided conveniences to dam workers, but it also provided power to Phoenix and project participants some sixty miles away. On the Salt River Project electricity not only provided domestic conveniences, it also enabled farmers to pump ground water to supplement the water supply from the Roosevelt Dam reservoir. Most important, Congress permitted the proceeds from electrical power sales and leases to be used to pay off the project's construction costs.[74]

The hydroelectric power career of the Reclamation Service began in 1908 on the Strawberry Valley Project in Utah. That powerplant is now operated by the water users, and the oldest plants now run by Reclamation went into service in 1909 on the Salt River and Minidoka (Idaho) Projects. As the Reclamation Service increased the pace of dam building, there arose

3.46. The grounds and powerhouse of the Upper Spanish Fork Powerplant, Strawberry Valley Project, in September of 1910.

3.47. The interior of the Strawberry Valley Project powerhouse in early 1906. This powerhouse is now operated by the water users.

an accompanying array of formidable powerhouses from which water turbines, dynamos or generators, and transformers poured power into transmission lines that ran to project settlements and to cities thereby changing the landscape of the West. Electricity became the "paying partner of irrigation."[75] As official policy, however, the production of power remained secondary to water storage, delivery for irrigation, and flood control. Most of the early dams were far removed from urban power markets. But, ultimately, the mysterious new power of the age reached everywhere — into farm, factory, home, and urban commerce. The dams, the penstocks, the powerhouses, and transmission lines announced that the age and the structures of the "technological sublime" were at hand in the West to compete with the natural wonders of mountains, lakes, and forests. American memory long associated that mysterious force with the kite experiments of Benjamin Franklin and the inventions of Thomas Edison. Henry Adams now saw it embodied in the dynamos, "symbol of infinity," and manifested from remote mountain canyons to distant cities.[76]

But with the entrance of the Reclamation Service into the business of power production and distribution, there also began a long-standing controversy. Private power companies feared the competition from what would be known as "public power" — dams and powerplants built by not only the federal government but also states and municipalities. The public power implications of reclamation legislation must have come as a surprise to the private power industry, which was attempting to secure power sites and provide services to cities under conditions of a "beneficial monopoly." The public vs. private power question became one of the most incessant questions of the Progressive Era. It would be fought out at the state and local levels into the post-World War II period.

The Reclamation Service avoided much of the early struggle by limiting the power it generated to the needs of its own projects. According to the body of reclamation law, hydropower production must first serve the project's irrigation goals, then the general needs of the project. Only if there was surplus power could it be distributed for sale to points outside the projects. Still, the example of the Reclamation Service providing electricity to its own communities and even returning the profits to them was not lost upon Progressives at the municipal and national level who came to advocate municipal socialism for basic utility services and that meant electrical power. If the services were natural monopolies, a government bureau, rather than a private corporation, might just as well assume ownership and operation and put the interests of the people before the interests of stockholders. The advocacy often took the form of revoking company franchises and buying them out. It was a struggle against private corporations who saw profitable opportunities in the development of new power companies. Private enterprise was in no mood to yield electrical power production and distribution to municipal governments, state governments, river basin development projects under Reclamation, or any other federal bureau. In the next decades, private power interests mounted an enormous campaign against public power, pointing out inefficiencies in government enterprises and arguing that public power was a hoax upon the very people it was supposed to serve. Over the next fifty years, the issue of the proper relationship between business and government resulted in bitter struggles that brought forth charges and countercharges about the evils of socialism versus the excesses of capitalists exploiting a helpless public.

Immediate Issues and Criticisms

Meanwhile, Reclamation Service leaders had more immediate controversies in their own backyard. Water user associations and settlers on the projects balked at the schedules for repayment of construction costs. Ongoing costs of operation and maintenance also became a standing issue. The language of the Reclamation Act did not clearly address who should assume these charges, but Morris Bien saw them as a "great drain" upon the reclamation fund and a threat to the future of the program. In 1905, he decided that operation and maintenance costs should be assumed by the projects rather than by the Reclamation Service. Opportunity arose with the Minidoka Project in Idaho. In the first public notice of the charges to be assessed on the project, Bien noted that an annual charge would be made per acre for Operations and Maintenance (O&M) expenses. Still, he knew that these charges should be specified in an act of Congress. When he drew up a bill for a reclamation project on the Wapato Indian Reservation in eastern Washington, he inserted a provision "for the repayment of the charge for operation and maintenance as required in the Reclamation Act." He hoped it would set a precedent.

3.48. Morris Bien maneuvered legislation and practice so water users were responsible for paying all current operations and maintenance charges. 1915.

From 1905 to 1911, seven other similar pieces of legislation relating to reclamation projects included the provision. Bien said that not even Director Newell realized how he had maneuvered the legislation so that the projects would assume the O&M costs. Water user associations soon challenged the costs and, in 1911, brought the matter into U.S. District Court in Washington as a complaint against the Sunnyside Division of the Yakima Project.[77] It ruled in favor of the Reclamation Service. A U.S. Circuit Court reversed the decision, but finally the U.S. Supreme Court ruled in 1913 to confirm the right of the government to collect the charges. Bien considered this a great victory and estimated that it saved the government about two million every year or thirty million dollars through 1930. Imposition of the additional charges provided yet another point of contention between settlers and the Reclamation Service.[78]

A. E. Chandler, California water law expert, hydrographer for the Reclamation Service, and former state engineer of Nevada, complained about an article in the *Reclamation Record* of March, 1909. It praised water users' associations for their educational efforts that prepared them for self-government and the point in the future when they could "assume larger and larger control of local affairs." Chandler saw such organizations as "a force for evil rather than for good." Officially, however, the Reclamation Service regarded the associations as essential for developing the organizational skills and leaders who could take over "the intricate and difficult matters of water distribution." Often, they were the local corporate organizations that assumed responsibility for debt incurred in the building of the projects and acted as a taxing body upon water users and landowners to collect annual repayments and manage water distribution upon a project. State legislatures usually recognized the associations as irrigation districts with powers to tax and collect payments to retire construction debts. This relieved the Reclamation Service of many onerous duties. The associations also prepared project settlers for assuming complete management of the projects when the Reclamation Service relinquished its control. Personnel in Bien's office, or Bien himself, often drafted water user association agreements. With a view to local autonomy, the Reclamation Service took steps to turn over operation and maintenance work to the associations once the projects were going concerns. The arrangement did not work out entirely to the satisfaction of the Reclamation Service, confirming some of Chandler's fears about these associations. By 1909, bitter criticism of the Reclamation Service on the part of water associations and others came to the attention of Congress.[79]

The enormous challenges of bringing reclamation to so many diverse projects predictably provoked a host of critics on all sides. Abiding by the 160 acre limitation clause in the Reclamation Act was a main source of friction. This became all too evident when the projects partnered with or took over an already existing private irrigation company or served participating private landholders. Decisions that enabled family members to take up ownership of excess lands (over the 160 acre limitation) helped to make the restriction more acceptable. The most common complaints, however, grew out of repayment schedules which farmers and water associations were hard pressed to meet. Many settlers labored under these additional loans necessary to bring new lands under cultivation as well as their own inexperience as irrigation farmers. The combination created individual discontent and discouragement to the point that many fled the projects.

Others believed the Reclamation Service had made serious mistakes in laying out the new projects by failing to conduct soil studies to test not simply for fertility and alkali, but for poor drainage. "Seepage," in those days, meant that some soils held water, resulting in high salt content that rendered them infertile. Providing drainage became costly and many critics believed that the Reclamation Service should either have planned better or assumed the cost of drainage itself. In response to complaints about the hard times on the projects, the leaders of the Reclamation Service, notably Director Newell, believed that many of the early settlers did not possess the mettle and determination needed to succeed. Discontent on the projects also compounded a fundamental problem facing the Reclamation Service: it was running out of money. The revolving reclamation fund was not being replenished. Expenses far exceeded returns as the Reclamation Service engaged in the building of numerous projects in the first seven years of the reclamation program.[80]

From 1900 to 1910, government reclamation costs increased by six times, reaching an average of fifty-five dollars per acre, while private irrigation costs increased only by half as much. [81] By 1909, the financial crisis in the Reclamation program was of such proportions that both the House and the Senate launched investigations. Congressmen closely questioned the leadership of the Reclamation Service, asking for reports on each project underway. The Reclamation Service responded with well-honed and illustrated testimony. Hiram N. Savage, supervising engineer for projects in the northern Great Plains, bolstered his testimony with a stereopticon emphasizing views of the condition of lands before and after irrigation made possible by the construction of Reclamation's dams and canals. Congressmen were impressed, but they did not hold back their questions about costs: what were the operation and maintenance costs and what did it cost to bring water to the land? Savage offered the figure of seventy-five cents an acre for maintenance and operation. Newell quoted a low figure of thirty to forty dollars per acre for construction charges per acre — an underestimate on most projects, as it turned out. For testimony, the Reclamation Service brought to Congress C. J. Blanchard who, although officially a statistician, increasingly served as chief publicist for the organization. He saw to it that articles on the Reclamation Service appeared in a wide range of periodicals. Most important, as the use of the stereopticon before Congress indicated, Blanchard saw the importance of photographic images in acclaiming the work of Reclamation. He was quick to employ moving pictures to record the opening of reclamation projects and the completion of dams. Blanchard's work revealed how important the Service thought it was to create a favorable image before Congress.[82]

But all did not go well. In these 1909 House Hearings, Director Newell admitted to the already well-known deteriorating financial position of the Reclamation Service. It had spent far more than it took in from either land sales or in repayments from project settlers. Nebraska Congressman Gilbert M. Hitchcock ventured that the Reclamation Service already had spent the originally authorized $50,000,000 million in the Reclamation Act, but Newell reluctantly corrected him. The figure was $52,000,000. Newell hastened to point to successes beyond the simple figures that showed deficits. In Arizona, on the Salt River Project, there were thriving farms and a large reservoir and dam under construction with every possibility of a successful payback. Newell lauded Arizona as a country that is at "present perhaps the best opportunity for showing the benefits of irrigation, as the whole civilization there is founded upon the water supply." Beyond this bright spot, Newell and others emphasized that the high cost of labor on the projects in remote places in western states played a major role in cost overruns. This occurred in spite of efforts to use "cooperative works" as on the Huntley Project in Montana. In these undertakings, settlers involved themselves in building the projects to earn "cooperative certificates," or what amounted to government scrip to be used to meet repayment assessments. Newell also related that Apache Indians were being used to good advantage on the Salt River Project. At first, they were not paid as much as white workers, but as their work improved they received equal pay.[83]

3.49. Threshing the first crop of grain on the Huntley Project, near Billings, Montana, in November of 1908.

The Reclamation Service's revenue had been exhausted, yet the various projects were not near completion, and many settlers were unable or unwilling to meet their repayment schedules. Congress was in no mood to scrap the program, and western representatives had a vested interest in the continuation and prosperity of federal reclamation efforts. In his official annual reports,

3.50. Sugar beets on the Stewart Ranch, Huntley Project, 1914.

Newell was optimistic about the prospects for Reclamation, and westerners could cite increases in real estate values, a faster pace of business activities, and a 51 percent increase in the population of the sixteen western states from 1900 to 1910. Certainly, reclamation played some role in this impressive growth. Congressman George Norris of Nebraska thought that the financing of reclamation should not be tied to western land sales, but should come directly from the Treasury (as many backers of federal reclamation believed would eventually occur). Little public land remained within Nebraska, and the state faced a dim future in terms of reclamation projects if that was the only source of financing. [84]

The 1909 Senate and House Hearings signaled a changing climate of opinion. The undercurrent of suspicion toward reclamation, present in the debates over the Reclamation Act, now became emboldened after Roosevelt's retirement from the Presidency and the election of his successor, William Howard Taft. The Taft Administration's appointees, as leading Progressives soon discovered, were not entirely friendly to the larger mission the federal government assumed under Roosevelt. No matter how cogently progressive conservation proponents argued that conservation served the self-interests of business, many in business circles saw the movement as a threat. Richard Ballinger, Taft's new Secretary of the Interior, epitomized these suspicions.

3.51. President William Howard Taft speaking at the opening of the Gunnison Tunnel on September 23, 1909. Uncompahgre Project, Colorado.

He joined the critics of Newell's administration of the Reclamation Service, and he particularly did not like the policies of Chief Forester Pinchot, the star of the Progressive Conservation Movement. The tension and differences exploded in the first year of the new Administration in a fight known as the Ballinger-Pinchot Controversy, which led to Pinchot's dismissal. Ballinger's name became a lightning rod drawing the criticism of Progressive conservation forces and disapproval amongst the leadership of new conservation agencies created in the Roosevelt Administrations – Newell, Pinchot, and even A. P. Davis. According to one historian of conservation, "It was Ballinger's fate to undertake a course of action ... which threw the conservation program into confusion and threatened to reverse it altogether." [85]

Davis recalled that Secretary of the Interior Ballinger came into office especially critical of Roosevelt's and the previous Secretary Garfield's conservation policies. The withdrawal of potential waterpower sites on the public lands (especially, in National Forests on Pinchot's recommendations) and over three million acres of land for future reclamation projects infuriated Ballinger, and he set about restoring them to entry under the public land laws. Pinchot's and Roosevelt's aggressive expansion of the national forest system was a source of contention, as were the operations of the Reclamation Service that had enjoyed the confidence of the previous Administration. In fact, Davis said that Ballinger and his appointed legal advisor, Oscar Lawler, an attorney from California, "exhibited venomous antagonism … for the Forest Service and the Reclamation Service." In the fall of 1909, Davis had just returned from Panama and work in Puerto Rico, where he designed irrigation systems to serve large sugar plantations, when Secretary Ballinger summoned him to his office. The Secretary's "unfriendly attitude" toward the Reclamation Service was immediately apparent. He was particularly concerned with the "publicity" put out by the Reclamation Service in newspapers and magazines.[86]

C. J. Blanchard's promotional activities, in particular, nettled Ballinger. The information from Blanchard's office, according to Ballinger, was neither "modest nor truthful." Blanchard's success as a gifted publicist for the Reclamation Service drew fire from many sources and Ballinger was now in a position to halt the propaganda. Any forthcoming articles should be cleared with him before publication, he ordered. A series of articles by Blanchard in *National Geographic* between 1906 and 1910 depicted the Reclamation Service as the perfect government service bureau performing the noble work "of reclaiming for home-builders an empire which in its present state is uninhabited and worthless." The Reclamation Service provided settlers with an opportunity to achieve a livelihood on the land, increased the property values of the West, and expanded markets for eastern manufactured goods. Blanchard also praised the engineering works of the Reclamation Service. He described the Shoshone (Buffalo Bill) Dam on the Shoshone Project in Wyoming, as the highest masonry dam in the world at 310 feet, and the Pathfinder Dam, on the North Platte Project, as holding back a reservoir of 1,025,000 acre feet of water. Some of the more innovative engineering feats of the Reclamation Service also received his attention. On the Uncompahgre Project in Colorado, Blanchard publicized the construction of the Gunnison Tunnel that brought water from the Gunnison River beneath the mountains that separated it from the Uncompahgre River. The articles, appearing not only in *National Geographic* but also in other publications, featured richly illustrative photographs throughout the texts that left little doubt about the great works the Reclamation Service was performing in the arid West. Blanchard also noted the urban amenities available to the various projects. Rural isolation had been banished in the government projects. Rural delivery of mail, daily newspapers, telephone service, traveling libraries, centralized schools, and even trolley lines into the towns, Blanchard insisted, brought "the desert farmer within the stimulating currents of the world's thoughts."[87]

These all-too-rosy depictions made Blanchard a prime target for the enemies of the Reclamation Service and helped bring a fury of criticism down upon Director Newell. According to Chief Engineer Davis, Secretary Ballinger came into office refusing "to do business with the Director" and Ballinger suggested that Davis should replace Newell as Director in a pending reorganization. He also ordered Davis to produce a list of all the withdrawals for hydroelectric purposes by the previous Secretary because he deemed these unconstitutional and intended to restore them. When the Secretary of Agriculture, James Wilson, submitted additional withdrawals recommended by Chief Forester Pinchot for the necessary approval by the Interior Department, Ballinger rejected them. The refusal provoked an open dispute between two members of the cabinet that the President had to take public moves to reconcile. But that did not end it. Soon followed the highly charged Ballinger-Pinchot Controversy that first led to the dismissal of Pinchot as chief forester, then to an outcry in the Progressive press over Taft's reactionary and anti-conservation policies, and ultimately to Ballinger as a liability for the Taft Administration. [88]

By the end of 1910, Taft gratefully accepted the resignation of Ballinger, but only after a joint House-Senate investigation of the entire affair. The lengthy hearings produced thirteen volumes of testimony, including a summary report. In the public eye, the conservation press vilified Secretary Ballinger and the Taft Administration for breaking with the progressive conservation policies embraced by the Roosevelt Administration, especially on reclamation and forestry. During tough questioning, Newell admitted that only two projects had been completed in the "over twenty…, which are turning back a revenue." The committee's Republican majority insisted that too many projects had been started, and "It would have been better if a less number of projects had been in process of construction at the same time, as more funds, more energy, and more speed could have been obtained in such case." In response, Newell explained that there were always additions being made to the irrigation projects, and, "Probably the word completion is about as applicable to an irrigation system as to a city. Whenever the city will be completed our irrigation system will be completed." The majority was also critical of the "Cooperative Certificates" that the Reclamation Service issued in return for settlers' work on building the irrigation projects. The certificates could be applied toward repayment of their construction cost obligations. Congressmen believed these stretched the meaning and intent of the provisions in the Reclamation Act. Secretary Ballinger had justifiably suspended their issuance in 1909. The Democratic minority report took the opposite tack: it praised the Cooperative Certificates as an excellent administrative choice to help farmers on the projects meet their obligations and said that, given the experimental character of the program and the demands made upon it to build so many projects, "It is not a matter of surprise that some mistakes were made, but rather that there were not more." [89]

With Ballinger's departure, the Reclamation Service escaped for the moment a significant reorganization, if not a halt in its entire enterprise. It might have been to Ballinger's advantage to concentrate his attack on the Reclamation Service rather than Pinchot and the powerful progressive press that backed him. Although Taft supported Ballinger against Pinchot, the dismissal of the Chief Forester became largely a Pyrrhic victory, but one that diverted attention away from the Reclamation Service.

In Congress, Representative Sereno Payne of New York directed a withering attack against the Reclamation Service for its mistakes. The Congressman pointed out that projects had been undertaken hastily, cost overruns had exhausted the Reclamation Fund, which Director Newell liked to refer to as a "trust fund," and settlers lived in poverty on projects that had not yet received water. By 1910, the reclamation program was clearly in trouble, but not according to the Reclamation Service's *Annual Reports*, overseen by Newell. If there were difficulties, the program was still popular enough among western Congressman. Congress refused to do away with the program. Instead, it bailed out the Reclamation Fund. In 1910, Congress appropriated $20,000,000 directly from the Treasury as a "loan" to enable the work of the Reclamation Service to continue. Reclamation had survived. [90]

Congress's rescue of Reclamation imposed some restrictions and even humiliation. It authorized the issuance of $20,000,000 in bonds repayable from the Reclamation Fund. The money could be expended upon existing projects and their planned extensions, but only after an investigation by a panel of five engineers from the U.S. Army Corps of Engineers, appointed by the President. The engineers, drawn from the Reclamation Service's rival organization in the War Department, were to investigate all projects underway as to engineering and financial feasibility. Moreover, the panel was to report upon the desirability of further investment of funds into particular projects. The removal of these decisions from the Reclamation Service and its engineering staff carried with it a clear insult to the judgment and trustworthiness of its leadership. While it did not find any serious engineering problems, the Army panel recommended no further expenditures either from the loan or from the general Reclamation Fund "except for necessary maintenance and operation" for the following works: Orland, California; Garden City, Kansas; Kittitas, Wapato, and Benton on the Yakima Project, Washington; and the Carlsbad and Hondo Projects in New Mexico. [91]

Although willing to embarrass the Reclamation Service, the Taft Administration, was in no mood to undercut or destroy it in the wake of the Ballinger-Pinchot Controversy. Such a foray would only further damage any conservation credentials that the Administration retained. The new Secretary of the Interior, Walter L. Fisher, avoided confrontation with Director Newell and backed a move by Congress to sell government water outside of the projects to private landowners, but with the 160 acre limitation requirement attached. Wyoming's Senator Warren attached his name to the 1911 Act that permitted the government to deliver water, under the Reclamation Act, to private corporations and irrigation districts instead of simply permitting a private landowner to contract with the government or a project. The act removed all doubts about the authority of the Secretary of the Interior to sell project water to private irrigation companies, provided that the water did not serve landholdings in excess of 160 acres. By 1912, application forms were in circulation for "Water-Right Application for Lands in Private Ownership and Lands other than Homesteads under the Reclamation Act." Needless to say, the availability of water for private lands added to their value and to the value of property near government projects. In the same year, Congress optimistically recognized that some project settlers might be on the verge of completing their payments to the government when it made provisions for all settlers on projects to obtain patent to their lands and "a water right for irrigation," if all payments for land and water delivery construction costs were paid. [92]

While the Reclamation Service survived attacks in Congress and criticism from officials in the new Taft Administration, its troubles were not over. The political situation in 1912, during the national presidential campaign, became highly charged as the popularity of the Taft Administration plunged in the face of Progressive criticisms that it had not lived up to the promises of Theodore Roosevelt's Progressivism, especially in the realm of conservation and the protection of natural resources from the predators of great wealth. The resulting administrative paralysis granted the Reclamation Service and its leadership a reprieve from administrative criticism in Washington, especially after the resignation of Ballinger. But what would the election of 1912 bring? The announcement by Theodore Roosevelt that he would again seek the Presidency to restore Progressivism to the White House electrified his supporters. Yet, his decision to mount a third party campaign opened the door to victory for the Democratic Party candidate, Woodrow Wilson. Wilson also identified himself as a Progressive, but how would the new Democratic administration treat the federal reclamation program? Aside from Newlands and his ilk, many western Democrats had been thoroughly critical of Roosevelt's western conservation policies, and there had not been a Democrat in the White House since the days of Grover Cleveland (1893-1897).[93]

Endnotes:

1 Francis G. Newlands to Leslie M. Shaw, Secretary of the Treasury, January, 1906, Francis G. Newlands Papers, Box 13, Sterling Library, Yale University, New Haven, Connecticut.
2 "Description of Reclamation Service Exhibit, Panama-Pacific International Exposition, San Francisco, Cal.," *Reclamation Record*, 6 (June 1915), 257-8; Mark Fiege, *Irrigated Eden: The Making of an Agricultural Landscape in the American West* (Seattle: University of Washington Press, 1999), 17.
3 Rauchway, *Murdering McKinley*, 71-4; Paul W. Glad, *McKinley, Bryan, and the People* (Philadelphia: Lippencott, 1964).

[4] Pisani, *To Reclaim a Divided West*, 279.
[5] Pisani, *To Reclaim a Divided West*, 280-85.
[6] "A National Forest Preserve," *The Nation*, 37 (September 6, 1883), 201.
[7] Theodore Roosevelt, *An Autobiography* (New York: The Macmillan Company, 1914, originally published Scribner, 1913), 409; Gifford Pinchot, *Breaking New Ground* (New York: Harcourt, Brace, Jovanovich, Inc., 1947); see Char Miller, *Gifford Pinchot and the Making of Modern Environmentalism* (Washington, D.C.: Island Press, 2001) for an excellent assessment of Pinchot's impact upon Roosevelt's conservation policies and McGee's influence on Pinchot, 154-55; Flack, *Desideratum in Washington*, 97; 154-5; Donald J. Pisani, "A Conservation Myth: The Troubled Childhood of the Multiple-Use Idea," *Agricultural History* 76 (Spring 2002), 154-71.
[8] United States Presidents, *A Compilation of the Messages and Papers of the Presidents, 1789-1918*, volume 13, 1896-1901 (Washington, D.C.: Government Printing Office), 6656-7; Dorothy Lampen, *Economic and Social Aspects of Federal Reclamation* (Baltimore, Maryland: The Johns Hopkins Press, 1930), 37-8.
[9] Pisani, *To Reclaim a Divided West*, 312; Theodore Roosevelt, *State Papers as Governor and President, 1899-1909* (New York: Charles Scribner's Sons, 1926), 106, 108.
[10] *Compilation of the Messages and Papers of the Presidents*, 6659.
[11] Pisani, *To Reclaim a Divided West*, 312-3; Michael G. McCarthy, *Hour of Trial: The Conservation Conflict in Colorado and the West, 1891-1907* (Norman: University of Oklahoma Press, 1977) notes opposition in Colorado to expanded federal authority over resources in the Progressive Era, 46-74; Lilley and Gould, "The Western Irrigation Movement," 71.
[12] Roosevelt, *An Autobiography*, 409.
[13] Frederick Haynes Newell, *Irrigation in the United States* (New York: Thomas Y. Crowell & Company, 1902), 3.
[14] Richard White, *"It's Your Misfortune and None of My Own:" A History of the American West* (Norman: University of Oklahoma Press, 1991), 399-400.
[15] Orsi, *Sunset Limited*, 373-5.
[16] Robert W. Righter, *The Battle Over Hetch-Hetchy: America's Most Controversial Dam and the Birth of Modern Environmentalism* (New York: Oxford University Press, 2005), 118, 197.
[17] Ellis L. Yochelson, *Charles Doolittle Walcott, Paleontologist* (Kent, Ohio: Kent State University Press, 1998), 398-400; Robinson, *Water for the West*, 19; Charles Doolittle Walcott, "Vast Extent of Arid Lands," *National Magazine*, 15 (1902), 571-2; Mary C. Rabbitt, *Minerals, Lands, and Geology for the Common Defence and General Welfare, 1879-1904*, volume II (Washington, D.C.: Government Printing Office. 1980), 327.
[18] Yochelson, *Walcott*, 403.
[19] "Newell Diary," 58, typescript, Frederick H. Newell Papers American Heritage Center, University of Wyoming, Laramie, Wyoming, hereafter cited as Newell Papers; Newell, *Irrigation in the United States*, 1-2; for comments on Newell's "naïve faith in the seemingly unconstrained bounty of technology," see Donald C. Jackson, "Engineering in the Progressive Era: A New Look at Frederick Haynes Newell and the U.S. Reclamation Service," *Technology and Culture* 34, No. 3 (July 1993), 574.
[20] U.S. Army Corps of Engineers, Hiram M. Chittenden, "Examination of Reservoir Sites in Wyoming and Colorado," (Washington, D.C.: U.S. Army Corps of Engineers, 1898), 2815-2922 (especially 2862-4); Gordon B. Dodds, *Hiram Martin Chittenden: His Public Career* (Lexington: University of Kentucky Press, 1973), 24-41; Chittenden, "Government Construction of Reservoirs in Arid Regions," 256-7.
[21] Kluger, *Turning on Water*, 27.
[22] Kluger, *Turning on Water*, 28-9.

[23] Lee and Pisani use the word "conjoined" in their article "Reclamation," in Howard R. Lamar, *The New Encyclopedia of the American West* (New Haven, Connecticut: Yale University Press, 1998), 949.
[24] Kluger, *Turning on Water*, 37; Stevens, *Hoover Dam*, 246.
[25] Kluger, *Turning on Water*, 58, 72; Pisani, *Water and American Government*, 44-5.
[26] Jackson. *Building the Ultimate Dam*, 11-2; McCullough, *The Johnstown Flood*; George Bancroft, *History of the United States of America: From the Discovery of the Continent*, volume 6 (New York: Appleton, 1882, originally Boston: Little, Brown and Company, 1854), 5-6 as quoted in Ernest A. Breisach, *American Progressive History: An Experiment in Modernization* (Chicago, Illinois: University of Chicago, 1993), 7. For the relationship of dams to economic progress in this period, see James D. Schuyler, *Reservoirs for Irrigation, Water Power, and Domestic Water Supply* (New York: John Wiley & Sons, 1909).
[27] U.S. Department of the Interior, Bureau of Reclamation, *Federal Reclamation and Related Laws Annotated*, volume I, Richard K. Pelz, editor (Washington, D.C.: Government Printing Office, 1972), 76; Pisani, *Water and American Government*, 41.
[28] Francis G. Newlands, "National Irrigation Works," *Forestry and Irrigation*, 8 (February 1902), 65; G. L. Shumway, "Give Settlers a Free Hand," *The Irrigation Age*, 22 (March 1907), 145.
[29] Theodore Steinberg, *Nature Incorporated: Industrialization and the Waters of New England* (New York: Cambridge University Press, 1991); Donald J. Pisani, "Beyond the Hundredth Meridian: Nationalizing the History of Water in the United States," *Environmental History* 5 (October 2000), 466-82; Chandler, *Elements in Western Water Law*; Harry N. Scheiber and Charles W. McCurdy, "Eminent-Domain Law and Western Agriculture, 1849-1900," *Agricultural History* 49 (January 1975), 126-7.
[30] Douglas H. Strong, *Tahoe: An Environment History* (Lincoln: University of Nebraska Press, 1984); John A. Widtsoe, *Success on Irrigation Projects* (New York: John Wiley & Sons, Inc., 1928), 118.
[31] Alvin Josephy, Jr., "Here in Nevada: A Terrible Crime," *American Heritage Magazine* 21, 4 (June 1970), 153-73; Daniel McCool, *Command of the Waters: Iron Triangles, Federal Water Development, and Indian Water* (Tucson: University of Arizona Press, 1994); A. J. Liebling, "Lake of the Cui-Ui Eaters," a series of four articles originally published in the *The New Yorker*, 30 (January 1, 8, 15, 22, 1955) republished in Elmer R. Rusco, editor, *A Reporter at Large: Dateline—Pyramid Lake, Nevada* (Reno: University of Nevada Press, 2000); James Hulse, "Indian Farm or Inland Fishery? The Pyramid Lake Reservation," *Halcyon* 15 (1993), 131-48; John Shurts, *Indian Reserved Water Rights: The Winters Doctrine in Its Social and Legal Context, 1880s-1930s* (Norman: University of Oklahoma Press, 2000); Robert B. Campbell, "Newlands, Old Lands," *Pacific Historical Review,* 71 (May 2002), 222-8.
[32] Harold K. Steen, *The U.S. Forest Service: A History* (Seattle: University of Washington, 1976), 26-37, 99-100; McCarthy, *Hour of Trial*, 52.
[33] Paul Wallace Gates, *History of Public Land Law Development* (Washington, D.C.: Zenger Publishing Co., Inc., 1968), 661-3.
[34] Henry F. Pringle, *Theodore Roosevelt: A Biography* (New York: Harcourt, Brace and Company, 1931), 430-1.
[35] United States Geological Survey, *First Annual Report of the Reclamation Service*, 17.
[36] R. Douglas Hurt, *American Agriculture: A Brief History* (Ames: Iowa State University Press, 1994), 221.
[37] Barbara Allen, *Homesteading the High Desert* (Salt Lake City: University of Utah Press, 1987) says an "urbanized citizenry raised the specter of a nation bereft of its traditional conservative values." One response was the Back-to-the-Land Movement, "a synthesis of the city and the country," 124; William L. Bowers, *The Country Life Movement in America, 1900-1920* (Port Washington, New York: Kennikat Press, 1974), 46-8.

[38] Roosevelt, *An Autobiography*, 398; Stanford J. Layton, *To No Privileged Class: The Rationalization of Homesteading and Rural Life in the Early Twentieth-Century American West* (Provo, Utah: Brigham Young University, 1987), 23-4.
[39] Layton, *To No Privileged Class*, 37.
[40] Layton, *To No Privileged Class*, 13.
[41] Hardy W. Campbell, *Campbell's 1907 Soil Culture Manua*l (Lincoln: Nebraska State Printer, 1907); Mary W. M. Hargreaves, *Dry Farming on the Northern Great Plains: Years of Readjustment, 1920-1990* (Lawrence: University Press of Kansas, 1993); L. H. Bailey, *The Country-Life Movement in the United States* (New York: The Macmillan Company, 1911), 2; Layton, *To No Privileged Class*, 13, 10.
[42] "Farmland Bubble," (editorial) *New York Times*, December 26, 2003.
[43] Steponas Kolupaila, "Early History of Hydrometry in the United States," *Journal of the Hydraulics Division, Proceedings of the American Society of Civil Engineers* (January, 1960), 1-51; U.S. Department of the Interior, U.S. Geological Survey, "Report of Progress of Stream Measurements for the Calendar Year 1897," F. H. Newell in the *Annual Report of the U.S. Geological Survey*, volume IV, Hydrography (Washington, D.C.: Government Printing Office, 1899), 162-72.
[44] Todd Shallat, *Structures in the Stream: Water, Science, and the Rise of the U.S. Army Corps of Engineers* (Austin: University of Texas Press, 1994), 79.
[45] Pisani, *Water and American Government*, 24; Burton J. Bledstein, *The Culture of Professionalism: The Middle Class and the Development of Higher Education in America* (New York: Norton, 1976).
[46] Pisani, *Water and American Government*, 25; Edwin T. Layton, Jr., *The Revolt of the Engineers: Social Responsibility and the American Engineering Profession* (Cleveland, Ohio: The Press of Case Western Reserve University, 1971), 109; Jackson, "Engineering in the Progressive Era," 550.
[47] Newell, *Irrigation in the United States*, 2.
[48] As quoted in Pisani, *Water and American Government,* 24; Layton, *The Revolt of the Engineers*, 73, 179; Kendrick A. Clements, *Hoover, Conservation, and Consumerism: Engineering the Good Life* (Lawrence: University Press of Kansas, 2000), 85; Merritt, *Engineering*, 110-35.
[49] U.S. Geological Survey, *Third Annual Report of the Reclamation Service, 1903-04*, 39-41.
[50] Gates, *History of Public Land Law Development*, 658; Lampen, *Economic and Social Aspects*, 53-4; Ellis L. Armstrong, editor, *History of Public Works in the United States, 1776-1976* (Chicago: American Public Works Association, 1976), 316.
[51] Donald J. Pisani, "Reclamation and Social Engineering in the Progressive Era," *Agricultural History* 57 (January 1983), 46-63.
[52] Pisani, *Water and American Government*, 291-2.
[53] Gates, *History of Public Land Law Development*, 585-91; Elmor R. Richardson, *The Politics of Conservation: Crusades and Controversies, 1897-1913* (Berkeley: University of California Press, 1962), 19.
[54] Daniel P. Carpenter, *The Forging of Bureaucratic Autonomy: Reputations, Networks, and Policy Innovation in Executive Agencies, 1862-1928* (Princeton, New Jersey: Princeton University Press, 2001), 330.
[55] Carolyn Gentry, editor, *The Roosevelt Dam*, Roosevelt Memorial Association Film Library. Located at the Library of Congress, American Memory, http:\\memory.loc.gov\ammem\collections\troosevelt_film\
[56] Newell, *Irrigation in the United States*, vii, 1; White, "*It's Your Misfortune*," 403.
[57] Miller, *Gifford Pinchot and the Making of Modern Environmentalism*, 361-2.
[58] Pelz, *Federal Reclamation and Related Laws Annotated*, volume I, 98-9, 119; U.S. Department of the Interior, Water and Power Resources Service, *Project Data* (Denver, Colorado: Government Printing Office, 1981), 1052.

[59] Pelz, *Federal Reclamation and Related Laws Annotated*, volume I, 90-1; Brookings Institution, *U.S. Reclamation Service*, 28; Leonard A. Carlson, *Indians, Bureaucrats, and Land: The Dawes Act and the Decline of Indian Farming* (Westport, Connecticut: Greenwood Press, 1981); Leonard A. Carlson, "Federal Policy and Indian Land: Economic Interests and the Sale of Indian Allotments," *Agricultural History* 57 (January 1983), 33-45; Pisani, *Water and American Government*, 170-1, 178-9.

[60] Francis Paul Prucha, *The Great Father: The United States Government and American Indians*, volumes 1 & 2 (Lincoln: University of Nebraska Press, 1984), 891; Pisani, *Water and American Government*, 160; Donald J. Pisani, "Irrigation, Water Rights, and the Betrayal of Indian Allotment," *Environmental Review* 10 (Fall 1986), 157-76; R. Douglas Hurt, *Indian Agriculture in America: Prehistory to the Present* (Lawrence: University Press of Kansas, 1987).

[61] Josephy, "Here in Nevada"; McCool, *Command of the Waters*, 181-2; Martha C. Knack and Omer C. Stewart, *As Long as the River Shall Run* (Berkeley: University of California Press, 1984); John M. Townley, "Reclamation and the Red Man: Relationships on the Truckee-Carson Project, Nevada," *The Indian Historian* 11, No. 1 (1978), 21-8; William D. Rowley, "The Newlands Project: Crime or National Commitment?," *Nevada Public Affairs Review* (No. 1, 1992), 39-43.

[62] *Winters v. United States*, 207 U.S. 564, 28 Sup. Ct. 207 (1908), affirming two appellate decisions by the Circuit Court of Appeals, Ninth Circuit, 143 F. 740 and 148 F. 684 (1906).

[63] Pisani, *Water and American Government*, 154-5; McCool, *Command of the Waters*, 5, 173-4. For a concise discussion of the origins of the Winters Doctrine, see Daniel McCool, *Native Waters: Contemporary Indian Water Settlements and the Second Treaty Era* (Tucson: University of Arizona Press, 2002), 9-16; Pisani, "The Dilemmas of Indian Water Policy, 1887-1928", Char Miller, editor, *Fluid Arguments: Five Centuries of Western Water Conflict*, (Tucson: University of Arizona Press, 2001), 78-90.

[64] Pisani, *Water and American Government*, 113; Brookings Institution, *Reclamation Service*, 29-30.

[65] Warne, *Bureau of Reclamation*, 88; Brookings Institution, *Reclamation Service*, 56-57.

[66] Morris Bien, "Autobiography," typescript, Morris Bien Papers, American Heritage Center, University of Wyoming, Laramie, 8, hereafter cited as Bien, "Autobiography."

[67] Pisani, *Water and American Government*, 38-40.

[68] For an informative discussion of riparian rights see John Norton Pomeroy, *A Treatise on the Law of Water Rights* (St. Paul, Minnesota: West Publishing Company, 1893) and the more succinct entries by Pomeroy, "Riparian Rights," and "Water Courses," *Johnson's New Universal Cyclopedia: A Scientific and Popular Treasury of Useful Knowledge,* 4 volumes, Frederick A. P. Bernard and Arnold Guyot, editors-in-chief (New York: A.J. Johnson & Co., 1881).

[69] Francis G. Newlands to J. H. Dennis, August 4, 1903, Newlands MSS (Yale) as quoted in Rowley, *Reclaiming the Arid West*, 108-9.

[70] Bien, "Autobiography," 17; Rowley, *Reclaiming the Arid West*, 108; Pisani, *Water and American Government*, 38.

[71] *Kansas v. Colorado*, 42 Sup. Ct. 552 (1922), 259 U.S. 419, 463, 66 L. Ed. 999.

[72] Starr in *Material Dreams* writes of California escaping "the dead end of exclusive riparian rights" by understanding Australia's early experience with irrigation, 8, 18.

[73] Pisani, *Water and American Government*, 40-1. For a thorough discussion of *Kansas v. Colorado,* see James Earl Sherow, *Watering the Valley: Development along the High Plains Arkansas River, 1870-1950* (Lawrence: University Press of Kansas, 1990), 103-19; Michael J. Brodhead, *David J. Brewer: The Life of a Supreme Court Justice, 1837-1910* (Carbondale: Southern Illinois University Press, 1994), 162-4.

[74] Smith, *Magnificent Experiment*, 82; *Project Data*, 643; P.M. Fogg, "A History of the Minidoka Project, Idaho to 1912 Inclusive," typescript (Boise: Idaho State Library).

[75] Toni Rae Linenberger, *Dams, Dynamos, and Development: The Bureau of Reclamation's Power Program and Electrification of the West* (Denver: Bureau of Reclamation, 2002), 23.
[76] David E. Nye, *American Technological Sublime* (Cambridge, Massachusetts: The MIT Press, 1994), 142; Adams, *The Education*, 380.
[77] The "Sunnyside Project" was known, later, for a time as the Sunnyside Unit and today is known as the Sunnyside Division of the Yakima Project. See Pfaff, *Harvests of Plenty*.
[78] Bien, "Autobiography," 27-8.
[79] A. E. Chandler to Morris Bien, March 22, 1909, RG 115, Entry 3, Box 88, Bureau of Reclamation Records, National Archives and Records Administration (NARA), Denver; *Reclamation Record*, 1 (March 1909), 33.
[80] John M. Townley, "Soil Saturation Problems on the Truckee-Carson Project," *Agricultural History,* 52 (April 1978), 280-91.
[81] Pisani, *Water and American Government*, 53; Raymond P. Teele, *Irrigation in the United States* (New York: D. Appleton and Co., 1915), 235.
[82] U.S. Congress, House, Committee on Irrigation of Arid Lands, *Hearings before the Committee on Irrigation of Arid Lands of the House of Representatives*, February 1909, 7, 14, 15, 68-81.
[83] U.S. Congress, House, *Hearings of Committee on Irrigation of Arid Lands*, 16, 40, 42, 45.
[84] Gates, *History of Public Land Law Development*, 663-4.
[85] James L. Penick, Jr., *Progressive Politics and Conservation: The Ballinger-Pinchot Affair* (Chicago: University of Chicago Press, 1968), 24-5.
[86] Arthur Powell Davis typescript memoir, A. P. Davis Collection, Box 13, file "Ballinger Controversy 1909-1910," 41-3, American Heritage Center, University of Wyoming, Laramie, Wyoming. Hereafter cited as Davis Papers.
[87] Examples of C. J. Blanchard's writing for the *National Geographic* received publication in book form in 1908 and 1909: *Home-making by Government: Account of 11 Immense Irrigation Projects to be Opened in 1908* (Washington, D.C.: Press of Judd & Detweiler, Inc., 1908); *Call of the West: Homes are Being Made for Millions of People in Arid West* (Washington, D.C.: Press of Judd & Detweiler, Inc., 1909).
[88] Davis Papers, "Ballinger Controversy," 41-5; A. P. Davis to Richard Ballinger, Secretary of the Interior, October 2, 1909, Davis Papers, Box 13, file "Ballinger Controversy," 36-8, 43.
[89] U.S. Congress, Senate, *Investigation of the Department of the Interior and the Bureau of Forestry*, volume 1, 61st Congress 3rd Session, Senate Document 719, 2, 84-5, 90, 186-7; Strom, *Profiting from the Plains*, 95.
[90] Gates, *History of Public Land Law Development,* 663; F. H. Newell to John M. Stahl, January 2, 1909, RG 115, Entry 3, Box 296, NARA, Denver.
[91] *Fund for Reclamation of Arid Lands,* Message from the President of the United States Transmitting a Report of the Board of Army Engineers in Relation to the Reclamation Fund, 61st Congress, 3rd Session, HR, Document 1262, January 6, 1911 (Washington, D.C.: Government Printing Office, 1911), 10-1.
[92] Patents and Water-Right Certificates, Act of August 9, 1912, (37 Stat. 265), Pelz, *Federal Reclamation and Related Laws Annotated,* volume I, 166, 177-83; Department of the Interior Form, 7-273, RG 115, Entry 3, Box 103, NARA, Denver, Colorado.
[93] Miller, *Gifford Pinchot*, 236-8.

3.52. In 1914, the Bassett Ranch near Clint, Texas, on the Rio Grande Project, prepared its fields to plant cauliflower.

3.53. The Fort Shaw office of the USRS on the Sun River Project, Montana. March 27, 1907.

CHAPTER 4

LIMITS

Introduction

Many western Democrats opposed the conservation policies of Roosevelt's Republican Administration and of Chief Forester Gifford Pinchot, but like Democrat Newlands, of Nevada, they favored national reclamation as "the highest form of conservation." When President Taft's Secretary of the Interior Richard Ballinger attempted to rein in the influence of Pinchot and, in the view of many, turn resources over to exploitation by business interests, he infuriated the progressive wing of the Republican Party. To some Republicans, the 1910 Ballinger-Pinchot fight represented a betrayal of the Conservation Movement that Roosevelt had entrusted to the Taft Administration. Still, Ballinger's views made more sense to western Democrats than Pinchot's regulations on timber cutting and grazing within the national forests. The Reclamation Service, however, with its dedication to the development of resources, appealed to both Democrats and Republicans. In fact, more Democrats supported the Reclamation Act of 1902 than Republicans. Reclamation was a brand of conservation all of the West could rally around. Enthusiasm prompted great expectations. But, as often is the case, when hopes are high, disappointments are great.

Sober Reflections and Times

Paradoxically, prospects of increased public and congressional criticism loomed when the Democratic Party won the presidency and a majority in Congress. As the Wilson Administration set up shop in Washington, Director Frederick Newell braced the Reclamation Service for even closer, more critical supervision during the closing months of 1912. Friends of reclamation increasingly realized that success meant more than building dams, storing water, and making it available to farmers. Magnificent structures were rising in the West, but economic, social, and agricultural barriers to irrigation projects seemed to be everywhere. Project farmers professed an inability to repay construction costs, in spite of "Leniency Acts" to defer payments. The Reclamation Service had launched too many projects too soon, and private speculation in project lands, lack of markets, inexperience with irrigation agriculture, and low morale on the projects slowed progress and raised the specter of a limited future for the Reclamation Service.[1]

In a conciliatory tone, Newell admitted past mistakes. He declared in the 1912 *Eleventh Annual Report of the Reclamation Service* that it was now appropriate to present a review of "operations and results" accomplished by the Reclamation Service over the past decade, "and note the lessons taught thereby." The message asserted that the Reclamation Service was capable of learning from its mistakes. Reclamation advocates once saw national reclamation as a continuation of the generous public land policies pursued by the United States in the nineteenth century. Now, with some inconsistency, Newell said that reclamation was a break with the past, "largely pioneer in character and carried on at localities remote from lines of transportation and centers of population." To emphasize the challenges faced by the Reclamation Service, he explained that "conditions could not be anticipated nor safely predicted." From Newell's point of view, reclamation was more than a business proposition. It embodied the highest ideals of Progressive idealism: "By making use of the waste places of the arid region through the utilization of the waste waters, rendering the lands available as productive homes for citizens whose energies might otherwise have been wasted through lack of opportunity to secure a home."

The Director admitted that the early Reclamation Service had operated under, as he put it, "fallacies." He listed them: (1) underestimation of project costs, based on earlier irrigation projects undertaken by private companies — the government projects served less accessible land and hence, were more expensive; (2) the assumption that bringing water to the land would be enough for farms to prosper when the land often required leveling, heavy cultivation with fertilizers, and soil testing to determine deficiencies; (3) the problem that many soil types and situations drained poorly when subjected to irrigation water, resulting in "swamping" and the build up of salts or alkali in the soils that could only be remedied by the building of expensive drains; (4) the belief that, if crops were produced, they would automatically return profits with no consideration given to available markets or the types of crops or varieties most profitable under the conditions of soil, climate, and available transportation facilities for a specific project; and, finally, (5) the assumption that competent farmers would settle upon the lands. Finding good farmers became the most difficult part of the problem.

According to Newell, the failure of this human element was a greater challenge than the engineering problems. Oftentimes the first wave of settlers appeared on the projects with far more of a speculative temper than farming experience. They either sold out to speculators or abandoned their claims. The result, Newell wrote, had been that the land was opened to agriculture very slowly. Much land fell into the hands of speculators and richer farmers who held the land solely for increased values.[2] While Newell saw these as lessons learned, "the grand irony," as one source observed, was that the Reclamation Service was in no position to learn from its mistakes. Too many projects already undertaken had exhausted the Reclamation Fund. There was little prospect of starting new ones that might profit from the lessons of the past decade.[3]

A New Secretary of the Interior, Project Issues, Newell's Eclipse

A more pressing issue for the leadership of the Reclamation Service was the watchful, critical eye of the new Secretary of the Interior, Franklin K. Lane, appointed by recently elected Democratic President Woodrow Wilson. Lane was an ambitious Democratic Party politician from California, strong supporter of reclamation, water users' associations, and a friend to Nevada's reclamation-minded Senator Newlands. Lane had been District Attorney for the city of San Francisco. As such, he had favored the city's efforts to dam the Tuolumne River in Yosemite National Park's Hetch Hetchy Valley. Lane was anything, if not, outspoken and aggressive. If the Reclamation Service thought it had endured troubled times under Secretary Ballinger, Director Newell soon perceived that Secretary Lane was a far graver and more determined meddler into the bureau's affairs. Lane came to office determined to reorganize and reform the Reclamation Service, starting with the removal of Director Newell or at least a dilution of his power.

To this end, the Secretary called representatives of all the project water users' associations to Washington. They first met on May 2, 1913, at which meeting Newell and Chief Engineer A. P. Davis fielded questions and took criticisms. The atmosphere in the meeting quickly turned "acrimonious." Davis related that Senator Newlands became concerned that the Reclamation Service was not mounting a vigorous defense at the conference. To which Davis replied that Newell thought, "We had better let things pass over with little comment, and perhaps we would have opportunity to insert something in the record [later]." Newell gave the senator the impression that Secretary Lane instructed that neither Newell nor the chief engineer should participate in the conference in a manner that might discourage criticisms. When Newlands asked Secretary Lane about any efforts to quiet Newell's and Davis's responses, he received assurances that this was not the case. The Senator's inquiry pointed to a rift between the Secretary and the Director, which disturbed him "very much," according to Davis. Newlands intimated to Davis that the Secretary was acutely aware of the "unpopularity" of Newell, making the Senator press Davis for an explanation of Newell's shortcomings. Davis could only reply that Newell had insisted that water users not be permitted to escape their debt to the government.[4]

The Service saw most project critics as ne'er-do-well agitators, troublemakers, and chronic complainers. In a confidential memo of November 12, 1912, Newell directed project engineers and managers to compile a list or "Who's Who among Chronic Complainants" on their projects. He explained that these people often wrote or telegraphed Washington. A short statement about each of these people should be identified with the letters "K.K.K." in the upper right hand corner for filing convenience. The letters stood for "Known Knockers and Kickers." The report should contain the name, address, and location of the farm owner and comment on acres under cultivation, the nature of the crops, and the degree to which the farmer was successful or unsuccessful. The report should also disclose whether the "Knocker" abided by the rules of the project. Newell believed that many complaints came from people who did not even own land in the projects, or who had been appropriating water illegally.[5]

Over the next year, a flurry of reports flooded into Washington from the projects marked K.K.K. Almost a month after Newell's confidential directive, he sent a scolding letter to the project manager in Sunnyside, Washington, ridiculing the manager's response that there were no "chronic complainants" on the Sunnyside Division in the Yakima Project. Newell said that he personally knew of some who fell into that category and "who are liable at any moment to make trouble … to explode at any moment." It was wise to have their names with a memorandum on each. Newell concluded: "As you are probably aware the next few months are possibly among the most critical for the continued existence of the Reclamation Service and of the principal men connected with it, so that it is vital to the continuation of the work to take every reasonable precaution along these lines."[6]

Project Manager D.W. Cole of the Truckee-Carson Project was quick to comply. Cole's reports dated January 3, 1913, were labeled with the K.K.K. notation in the upper right hand corner of the page and samples included:

J. S. GRAY, Fallon, Nevada

Homestead Entryman. Director of local water User's Association. Irrigable acres 71 with 56 acres under cultivation. Is a hard worker and successful farmer. Was formerly a railroad switchman and yardman. Is of somewhat bumptious proclivities. A socialist, rather radical talker at the public meetings, and imagines himself to be fair and broad-minded while in fact is rather narrow. Has been critical of project management with particular hostility to the irrigation manager but has not refused to observe project regulations. On the whole is a desirable citizen.

J. A. GALBRAITH, Fernley, Nevada

Homestead Entryman. 40 acres irrigable. 38 acres irrigated. Less successful than majority of neighbors. Claims excessive cost in preparation of land, though land was easily leveled. Formerly S.P. Station Agent at Fernley. Generally troublesome, and defiant in attitude toward Service. Has stolen water on several occasions. Careless irrigator, using excess of water, duty 1912 8.72 acre ft. per acre; 6.17 acre ft. per acre average for his district. Nevertheless claims discrimination against him. Has written Project Engineer threatening legal action to secure satisfaction in alleged conspiracy to deprive him of water.

F. H. SEARS, Fallon, Nevada

Homestead entry -80- acres. 78 acres irrigated. Good crop returns. Inexperienced irrigator. Not especially skillful, but should be classed as successful farmer. Formerly in law practice in Chicago. Aspires to political career. Prominent at all public meetings of farmers in which he seeks to take leading part. Respects rules of service and has not been a chronic kicker so far as local management is concerned, though he has on one or two occasions criticized our methods. A "self-made" man with comparatively little schooling, but fairly intelligent and well informed. Prominent promoter of Water Users Association.

MATT JOHNSTON, Fallon, Nevada

Homestead Entryman. Irrigable 33 acres, 23 acres irrigated. Admits inexperience in farming. Has always rented his land though maintaining residence. Volumes have already been written about Matt in connection with his refusal to pay charges pending drainage of his land.

J. B. YOUNG, Fallon, Nevada.

Private Land Owner. Irrigable 112 acres. Irrigated 60 acres. Fairly successful as farmer. Active politically, having been on one or two occasions an unsuccessful candidate for member of state legislature. Generally mistrusted and disliked in the community. "Loud-mouthed" and boastful. Rather inclined to disregard service rules and regulations. Makes frequent complaints against service and is constantly criticizing project management.

Meanwhile, the *Reclamation Record* acknowledged the unhelpful attitude of "complainers" on the projects but noted that men of every character made their way to the projects. It reported that, "They Have 'Em in Australia, Too," when it ran an article of that title from the *Irrigation Record*, Leeton, Australia, that said, "It is hard to tell the 'kicker' from the well-known gentlemen who is always 'agin' the Government."[7]

The confidential request for names and lists of complainers on the projects reflects the deteriorating relationships between the leadership of the Reclamation Service and project settlers. In Congress, however, good will for the Reclamation Service ran strong among western representatives. Senator George W. Norris of Nebraska voiced his continued enthusiasm for reclamation when he declared that he supported any measures "that will tend to give the Reclamation Service enlarged powers for the accomplishment of good." But the Secretary of the Interior pursued a more critical path, as was apparent in the water users' conference in May 1913, when he listened sympathetically to complaints against the Reclamation Service and Director Newell. Even Senator Newlands lost confidence in Newell. In reply to further questions from the Senator about Newell's shortcomings, Davis further replied that probably Newell's initial failing was an inability to say "No." In the early years of Reclamation, Newell yielded to local political pressures and permitted the Reclamation Service to take on too much. Delays on projects caused by lack of funds exposed settlers to hardships as they waited for water. Newlands also wanted to know if Newell lacked tact. Davis cautiously responded that Newell worked well within the organization, but he was not tactful in dealing with the Secretaries of the Interior, "nor with the Water Users, due to his desire to enforce his own ideas." This reflected Newell's rigidity on the issue of repayment schedules that should be kept. It also suggested that Newell commanded great loyalty within the organization, as indicated by Davis's steadfast refusal to accept appointment as Director either from Ballinger or Secretary Lane. [8]

At the end of May, Davis was on a field trip in New Mexico and read in an Albuquerque newspaper that the Secretary of the Interior had placed the Reclamation Service under the control of a five member commission, one of whom was Davis. By the end of the year, Newell still wore the title of Director, but he was in charge only of the "Scientific, Statistical, and Historical Division." Chief Engineer Davis was in charge of the "Engineering and Technical Division." William A. Ryan headed the "Fiscal and Accounting Division." Will R. King was in charge of "The Legal Division." And Ignatius D. O'Donnell headed the "Division of Operation and Maintenance," also called "Supervisor of Irrigation," based in Billings, Montana.

4.1. Secretary of the Interior Franklin K. Lane quickly had a picture of "The Reclamation Commission" inserted in the Reclamation Record, January 1914. William A. Ryan, I. D. O'Donnell, Arthur P. Davis, Will R. King, Frederick H. Newell, and Secretary of the Interior Lane.

While Davis wanted to believe that this was a constructive move on the part of Secretary Lane, the circumstances under which he learned of the shakeup in leadership indicated that it was a highhanded, arbitrary move on the part of the Secretary. Since each of the members now reported directly to the Secretary, Newell's position as Director, in the opinion of Secretary Lane, was largely in name only.[9]

Problems of repayment continued to dog the Reclamation Service. Many project settlers urged that the repayment schedule be extended. As a member of the Senate Irrigation Committee, Senator Newlands said the committee had an open mind on the question of the extension, but was averse to anything that approached "repudiation" of debts. Repudiation would threaten the Reclamation Fund and the projects not yet completed as well as curtail the initiation of new projects. A representative from the Salt River Valley Water Users Association wanted to know if the Senator would apply the word "repudiation" to resistance on the Salt River Project to pay higher fees to bring water to additional lands. While the Salt River Project was in good shape on the repayment issue, it advocated an extension of time for repayment of construction charges and resisted the spiraling construction costs of the Reclamation Service on new additions that might show up in either construction or operation and maintenance charges. Repayment delays alarmed Newlands for he was aware that the Reclamation Revolving Fund was already failing to revolve.[10]

Complaints from the Salt River Project were particularly alarming. Farmers there seemed to have the best of all possible worlds. Their association enjoyed access to cheap electricity and revenues from its sale to the city of Phoenix and surrounding communities. The power revenue helped pay operation and maintenance charges as well as original construction costs. According to William S. Cone, an official on the Salt River Project, the development of electrical power served three purposes: (1) to bring in revenue to help pay the operating expenses of the project, (2) to help develop the resources of the region by furnishing cheap power to local industries, and (3) to help make farm life more comfortable by putting modern electric appliances within the reach of all settlers. Yet, despite these advantages, complaints occurred.[11]

The search for new sources of funding became almost an obsession with the Reclamation Service. Since public land sales were the source of funding, Newell also looked at returns from timber sales and grazing charges in the national forests. The tie between forest lands and irrigation had always been emphasized in the progressive conservation literature. Forests were indispensable to the water supply of irrigation projects. Forest protection meant water supplies for irrigators and forest protection also included the regulation of grazing on forest lands to protect normal stream flows.[12]

In 1906, the Forest Service levied fees for grazing stock animals within national forests. Very soon these fees brought in several millions of dollars a year, exceeding revenues from timber sales until 1910, and periodically for a decade thereafter. Since grazing in the national forests often supplemented the income of reclamation project settlers and abetted the combination of agriculture and grazing prevalent on the projects, Newell reasoned that tapping income from national forest grazing fees for the Reclamation Service was a possibility. This, of course, did not please officials in the Forest Service. Chief Forester Gifford Pinchot jealously guarded income from grazing fees for forestry operations. In 1908, when Congress required payments from grazing fees be deposited to the general treasury and not credited directly to the Forest Service, Pinchot became increasingly wary of the designs of other bureaus upon revenues generated by the Forest Service. As late as 1914, Newell still pressed for grazing proceeds. He understood that a grazing bill was under consideration by Congress with the proceeds to go to the states. Considering the close relationships between forest lands and irrigation, Newell asked why the proceeds of grazing fees should not "go to the building of Reclamation works?"[13]

Complaints about the high costs of government construction of hydraulic works irked the Reclamation Service. An article in *Engineering News* of August 21, 1913, by John E. Field, State Engineer of Colorado, on "The Cost of Reclamation Service and Other Irrigation Projects in Colorado," asserted that "the original cost of private enterprises is about one-half that of Government enterprises." Chief Engineer Davis hastened to reply that Mr. Field, "would have found a higher average cost per acre for private than for public," if he, "had made due allowance for the better character of work, the more complete construction of lateral systems and drainage works and had confined his statement to the facts concerning projects built within the last ten years, [the period during which] Reclamation projects have been built."[14]

Repayment Problems Continue

At a September 1913 conference of Supervising Engineers at Lake Tahoe, California, the repayment question dominated discussions. When Secretary Franklin K. Lane arrived on the last day of the conference, he went immediately to the issue of extensions. The supervising engineers had recommended extensions, but interest payments would be required. The most generous recommendation suggested only seven years beyond the original ten year repayment period and a graduated schedule postponing higher payments to later years, thus leaving farmers with less burdensome payments in the earlier years of their enterprise. Over the summer, Lane had traveled to reclamation projects, Indian reservations, and national parks assessing the water resources and the work underway on the various projects. He observed progress but also "disheartening" failures in the planning and administration of projects, especially on the northern plains in Montana. As a result, Lane grew even more sympathetic to the plight of the project settlers and water users' associations and sympathized with their complaints against the Reclamation Service.[15]

Stung by the criticisms and charges against the Reclamation Service on his summer tour, Secretary Lane emphasized that project managers were crucial in dealings with farmers. "If the manager is indifferent or snobbish or not sympathetic and without apparent human interest," Secretary Lane said, farmers quickly became alienated. "The Service is," he said, "judged largely by the man in local charge and the standard of project managers must be raised and maintained to the highest degree." Demands were rampant for new projects, but Lane knew that there was little money available. The Reclamation Service already had enough to do. Lane said that the work in hand was "larger and more important than that on the Panama Canal." He was sensitive to the prevailing spirit on the projects: "If the people on the farms are continually dissatisfied then it will not be possible to obtain more money to continue the work." Project settlers needed "a hopeful spirit" that would inspire Congress with the good works of the Reclamation Service and perhaps elicit new appropriations. In his work, *Water and American Government*, historian Pisani notes that Lane "was the first Secretary of the Interior to recognize that the problems of federal reclamation were psychological as well as economic."[16]

Ironically, Secretary Lane echoed some of Newell's sentiments, for Newell repeatedly said that "the human element is by far the great problem.

I find that throughout the arid West, the same condition exists, namely, the need of good farmers to utilize the lands which have been provided with water at large expense." Where Lane and Newell differed was in their assessment of the source of the problem. Newell saw it in the deficient character of project farmers and Lane saw it in policy shortcomings of the Reclamation Service. Whatever the source of the problem, the Secretary viewed Newell as a political liability because he had become the prime target for critics. The Director had to be removed if the Reclamation Service was to move ahead. When Elwood Mead, the longtime expert on irrigation agriculture, returned from his employment as an irrigation expert in Australia in early 1914, Secretary Lane offered him the directorship of the Reclamation Service. Mead decided not to become entangled with the problems of the Reclamation Service and left again for Australia, briefly, before returning permanently to the United States in 1915 to assume a position at the University of California in Berkeley.[17]

During Secretary Lane's first year in office criticism of the Reclamation Service by water users and hearings in Congress that were sympathetic to their plight were important precursors to the passage of the Reclamation Extension Act in August 1914. This legislation doubled the repayment schedule to twenty years and put into effect a graduated increase in the percentage to be repaid. During the first four years only 2 percent of the cost came due, then 4 percent for two years, and finally 6 percent for the subsequent fourteen years. The schedule was consistent with the supervising engineers' recommendations at the Lake Tahoe meeting in September 1913, but the legislation failed to include the engineers' recommendations that interest charges be attached to the repayment schedule. Other provisions of the 1914 law opened the way for water users' associations or irrigation districts to assume "the care, operation and maintenance of all or any part of the project works," prior to the repayment of construction costs. [18]

While Congress was generous to the water users in this legislation, it restricted the future powers of the Secretary of the Interior and the Reclamation Service. This was particularly true in the choice of projects. Thereafter, a project must receive congressional approval, which also meant approval of appropriations from the Reclamation Fund, which had previously been exclusively under the control of the Department of Interior. Even decisions to expand an already-established project now rested with Congress.[19] This rescinded the policy favored by early Progressives who believed that most of these decisions should be made under the guidance of or by experts without the interference of politics and congressional logrolling. Ideally, the Progressive Conservation Movement believed that trained professionals of the Reclamation Service should choose qualified projects on the basis of their feasibility without the interference of self-interested members of Congress. In the early years, however, the Reclamation Service had self-interests too. It showed little sign of making decisions based on science or efficiency. Projects were often built for political reasons, in deference to individual members of Congress. The Reclamation Service's independence came to an abrupt end with the 1914 legislation. While choosing projects now fell into the hands of Congress, there was little choosing to do: the Reclamation Fund was so depleted that few, if any, new projects were possible.

By the end of 1914, it was clear that the Reclamation Service was under new management. When Secretary Lane notified Davis that he would assume the title of "director and chief engineer," he effectively removed Newell from the governing commission, telling Davis that Newell should report to him "as one of the consulting engineers to whom you may refer anything that you desire." Significantly, Sydney B. Williamson, a former engineer with the Panama Canal Commission, became chief of construction to inject some of the efficiency and energy of the Canal's administration into the lagging fortunes of the Reclamation Service. Judge Will R. King, Ryan, and O'Donnell remained in their positions. Under the circumstances, of course, Newell could not remain with the Reclamation Service. He formally submitted his resignation in May 1915 and accepted a position as Head of the Civil Engineering Department at the University of Illinois.[20]

4.2. Sydney B. Williamson served as Chief Engineer of the U.S. Reclamation Service from 1915 to 1916.

Meanwhile, Lane established a "Board of Review" to investigate conditions on the projects, particularly construction charges and operation and maintenance costs. Of the three appointees, one represented the Reclamation Service; another, the water users; and the third a broader public interest. Their recommendations and findings then went to a Central Board of Cost Review made up of Elwood Mead, back in the United States after his employment in Australia from 1907 to 1915; General William L. Marshall, consulting engineer to the Secretary; and Ignatius D. O'Donnell. The local review boards offered water users participation, at least in an advisory capacity, in the decisions made pertaining to costs for repayments, the perennial issue of costs for operation and maintenance, and extra costs for "betterments" such as drainage works relating to seepage.[21]

Extension Acts and Faith in Agriculture

The reforms of 1914 conceded a great deal to the demands of project settlers and their waters users' associations. Newell later complained that they gave away too much and rewarded improvident settlers and greedy speculators. Newell asserted that the extension of the repayment schedule to twenty years without the imposition of interest fees on the money loaned was contrary to good business sense and an insult to the character of industrious project farmers. Mead made similar observations in a 1915 report to Secretary Lane in which he argued that settlers should pay interest on money loaned and go through a screening process before they could take up land on the projects. Those lands should be classified and valued according to their relative agricultural value, and that steps should be taken to halt speculation. The 1914 legislation, while playing to the interests of the water users, did not solve the financial problems of the Reclamation Service, or, for that matter, the financial problems of project settlers as they faced the multiple problems of starting successful irrigated farms. Critics within the Reclamation Service and in Congress suspected that the repayment extension in the Reclamation Extension Act of 1914 simply compounded problems and merely postponed the ultimate reckoning with debt.[22]

For some on the projects, the reforms of 1914 fell short of what they expected. Officials in the Reclamation Service tried to meet the continuing criticism by talking up the generosity of the government in extending the payment schedule to twenty years, graduating payments, and relinquishing control and management of the completed projects to either water users' associations or irrigation districts at an early date. A letter in the *Reclamation Record* by H. G. Tyson, Jr., of Caldwell, Idaho, argued that reclamation settlers should not have to repay the costs of reclamation projects given the fact that appropriations for river and harbor improvements had no such condition attached. This viewpoint reflected an insightful understanding of the legislative history of the Reclamation Act and the relationship of its passage to river and harbor improvements. In 1900, the debates over national aid to western irrigation turned on the issue that Congress provided aid in plenty to improve harbor and waterways in the Midwest and East but gave no similar aid to the interior West. This view sketched a long-held conviction that reclamation was an "entitlement" that the federal government should underwrite.

Nevertheless, there were other settlers who accepted the largesse and good faith of Congress and Secretary Lane in the passage of the Reclamation Extension Act of 1914. The President of the Klamath Water Users' Association stressed the need for settlers to establish cooperatives to market their products directly to consumers and cut out the "middleman." Cooperation, he said, would enable the settler to take advantage of the favorable terms provided by the 1914 act. He was confident enough to declare, "The opportunity to make good is now open and the settlers are relied upon to do their part. With willing minds and grateful hearts we should rise above our former despondency and work as never before. Make good or die in the effort should be our motto."[23]

From the Okanogan Project in eastern Washington came the advice that water users' associations provided a ready-made organization from which farmers could form cooperatives for marketing, bulk purchases, and resale to farmers. Many saw association cooperatives as the key to solving the farm problem. Reclamation projects with their water users' association presented ready-made laboratories and an organizational structure to test the advantages of the farm cooperatives. Water users' associations were legal and permanent organizations in which all were perpetual members on the project. Such organizations required no solicitation or extra dues and did away with a multiplicity of meetings, which would be required if there were separate cooperatives. As explained in a letter from a farmer to the *Reclamation Record* in 1914, "The regular association has a set of officers, the secretary at least on pay, or part pay, and a small increase to him might in many cases suffice to do a great amount of business much cheaper than if several new companies were formed." By late 1916, Omak Fruit Growers, Inc. emerged on the Okanogan Project as a for-profit corporation that simplified the marketing and distribution while eliminating profit reducing competition among the farmers. In early 1914, C. J. Blanchard, ever the optimistic publicist for reclamation, reported a growing number of cooperatives from packing houses on the Uncompahgre Project to a creamery on the Klamath Project. Farm wives showed a great deal of interest in meetings organized by the state agricultural colleges to deliver lectures and answer questions about cooperative associations.[24]

In 1914, Congress not only extended the repayment period, it also took steps to extend scientific and technical aid to agriculture. It authorized funds for six agricultural experiment stations to be located on selected projects to explore crop disease and soil problems, and to identify optimal crops to plant under the project's particular climatic and soil conditions. Since 1887, Congress supported agricultural experiment stations at the land grant colleges, and now it extended this service to some of the reclamation projects. In addition, the Smith-Lever Act, or Agricultural Extension Act of 1914, provided federal aid for a system of agricultural extension agents associated with the land grant colleges and counties to promote agricultural knowledge in local farm communities and with individual farmers.[25]

IRRIGATED
HOMESTEAD
LANDS

Now Open to Entry under the Truckee-Carson Project in Churchill County

75 Choice 40- and 80-acre Farms lying west of Fallon open to entry
September 22, 1914

TERMS AND CONDITIONS

THE LAND is FREE

Water Rights furnished by the U. S. Reclamation Service at $60 per acre, payable in 16 installments in 20 years, **Without Interest.** First installment of $3.00 per acre, payable at time of filing. Next payment due 5 years later

Residence on the land 7 months a year for three years necessary to secure title

Cultivation of 1-4 of irrigable area in 3 years, and 1-2 in 5 years is required

Water Supply under the Great Lahontan Reservoir is permanent and assured

Lands in Private Ownership, with or without water rights, may be purchased **Now** at attractive prices. **As yet** there has been no inflation of land values

CHURCHILL COUNTY

is one of the best sections in the Entire West for dairying, stock raising, truck gardening, sugar beet culture and general farming. Fallon has a half million dollar beet sugar factory which will operate next season, under extremely favorable price conditions. Fallon has the most up-to-date creamery plant in Nevada, and high-grade cows can be bought on the easy payment plan.

The Opportunity of a lifetime for the homeseeker exists RIGHT NOW in Churchill County. For further information Communicate with

Project Manager U. S. Reclamation Service

or Sec. Churchill County Chamber of Commerce, Fallon, Nev.

4.3. An early promotional poster for the Truckee-Carson (Newlands) Project.

Congress, in its 1914 agricultural legislation, demonstrated a faith in farming. Still, for its survival in an industrializing economy, agriculture needed to become savvy and modern. The legislation of 1914 (the Smith-Lever Act that supported a system of county agricultural agents and acts to aid the reclamation projects) expressed Congress's desire first to modernize American agriculture and second to secure the future of its irrigation and water development program in the West. This was a not-so-tacit acknowledgement that the Reclamation Service had neglected the everyday agricultural challenges on its projects: soil types, insect and disease threats to crops, plant and animal breeding, and economic questions relating to economies of scale, possible markets, and distance from markets.

The Reclamation Service's home in the Department of Interior, and Interior's long rivalry with the Department of Agriculture, denied settlers

4.4. Even as farmers learned the economic realities of project life and repayment on Reclamation projects, Reclamation, settlers, and promoters emphasized the abundance and variety of crops on the projects. Yakima County Fair, Yakima Project, 1907.

the agricultural knowledge they needed to prosper. The 1914 legislation promised to make farmers successful participants in the emerging industrialized economy of the twentieth century. Success on the land offered the sure alternative to the pitfalls of urban life that bred, "Tuberculosis, Anarchy and Socialism … the Triplets Born of the Tenements." The correspondence of Reclamation Service officials, agricultural college experts, and water users' associations indicated that only experienced and educated farmers had any prospect for success.[26]

Problems of Land and Water

The challenges of making a good living on reclamation projects were endless. The efforts differed from midwestern farms, which enjoyed good soils and adequate, if not always predictable, rainfall. Farmers on reclamation projects took up farms under the Homestead Act (not in excess of 160 acres). They were given the land but assumed the obligation of paying their share of carrying water to that land, according to the number of acres settled upon. But the best land was in private hands by the time federal reclamation came upon the scene. There were not enough public lands on which to build projects, and those available were often located in undesirable places on the northern Great Plains with poor soils and short growing seasons. In the Southwest, District Counsel for the Reclamation Service, P. W. Dent, observed that public lands were often crucial to a project's growth and well-being. He said, "Experience on the Rio Grande project has led to the conviction … that those projects are fortunate which are designed to irrigate only public land or at least where government land predominates." The presence of private lands under old irrigation systems complicated projects when the courts, as they did in New Mexico, upheld customary water rights. This delayed the construction and completion of the Rio Grande Project.[27]

The scarcity of good public land suitable for irrigation required the Reclamation Service to build water projects to serve a combination of marginal public lands and better lands already in private ownership. While established farmers had already acquired the prime lands, speculators moved early to purchase other lands in the vicinity upon the announcement of a federal reclamation project. Often, government farmers had to buy land from private landowners at market (or speculative prices) and then face additional indebtedness for the payback of water charges. These heavy obligations upon new settlers required enough capital at the outset to make a go of it as well as extraordinary determination and self denial, or failing that departure from the land. The latter choice befell too many settlers. Some reclamation officials, Newell among them, came to believe this was a part of the natural process of weeding out incompetent farmers. This group often sold its claims and improvements at a loss to a second wave of better disciplined settlers who tried to profit from this transfer of capital improvements, but oftentimes they too sold out to a third-wave settler who finally succeeded. The term "relinquishments" described the value of the improvement that the earlier entrants made upon the land before they relinquished the holdings, often for little or no compensation, to successors.

The process, however, raised a great hue and cry as many went broke and complained about the false promises of government reclamation. The second generation of settlers often acquired land which carried outstanding construction charges and past operation and maintenance (O&M) charges. These lessened the value of the relinquishments. A point of enduring friction persisted in the Reclamation Act's limitation on the sale of water to no more than 160 acres for any individual holding. Section 12 of the Reclamation Extension Act of 1914, however, tried to address the problem of the high land prices demanded on private holdings in excess of 160 acres in new projects. The Extension Act said that the Secretary of the Interior may subdivide land in excess of 160 acres to be sold at a price designated by the government. Land not subdivided and sold under these terms would be excluded from project water. The original holder, of course, was entitled to water for 160 acres, if there was an agreement to subdivide and dispose of the excess lands. But the provision afforded little relief because it could not be applied to projects already underway, and, when it was applied, the holders sold their land to another purchaser "at such price as he is able to extort." The process introduced another middleman at the expense of the settler. Even the exclusion of the land from a project did not prevent the landowner from holding it at an ever rising price to sell to a smaller holder who would be tempted to purchase it with the probability that the government would eventually extend water rights to the plot in a project expansion.[28]

Congress was at a loss to foil the speculative tendencies of Americans when it came to land, and this was another ingredient in the rising

4.5. Over the years, supply methods for basic construction materials used on Reclamation projects have evolved radically. This 1918 image shows the way sand and gravel were obtained for construction on the Yakima Project along the Tieton Main Canal.

tide of criticisms from the projects. Speculators often exaggerated the productivity of the land and duped inexperienced settlers who then became disgruntled and resentful against the Reclamation Service. One settler on the Tieton Project in Washington State noted, "This project has been boosted outrageously by speculators, commercial clubs, etc., and honestly sometimes because the boosters were not farmers and most of those who were disappointed were not farmers and naturally believed what they were assured they could do."[29]

While land speculation created cost burdens in the development of farms, the ongoing questions of water delivery costs and contracts for repayment were ever present on the projects. Layers of organization between the Reclamation Service and settlers added to misinterpretation of its rules for water use and charges for water and other services on the projects. While the Reclamation Service set the charges for repayment, they varied according to the amount of land receiving water. Individuals made their payments not directly to the Service but to water users' associations or irrigation districts with which the Reclamation Service had made the agreements for repayments and O&M charges. Assigning responsibilities to a local organization or a government entity created a set of problems and, more conspicuously, forums for the expression of criticisms, especially in the case of the water users' associations. District Counsel Bernard E. Stoutemeyer favored water districts over private water associations. He believed that districts eased the relationship between the water users and the federal government because: (1) "The irrigation district brings into the project all of the lands in a solid body, and … keeps down to a minimum the cost per acre for building and operating the project," (2) the signed contract between the district and the government "gives the government greater security in proceeding with the work," (3) "the irrigation district has greater efficiency in the collection of charges on account of the taxing power, and … can collect from uncultivated speculative holdings," (4) people are used to paying taxes on time, (5) the tax does not prevent the farmer from securing a second mortgage or borrowing on his property, and (6) the general public is more comfortable with the federal government making contracts with "various subdivisions" of government.[30]

Whether the Reclamation Service dealt with the settler or through an intermediary organization, criticisms persisted. In spite of the reforms, aid, and extension of payments provided for in the legislation of 1914, these fundamental problems continued. From the congressional arena came complaints that Reclamation appeared incapable of either paying for itself or even paying the operation and maintenance charges. Yet, the projects benefited towns and cities whose businesses had been largely increased by the construction of the irrigation projects, "resulting often in doubling or trebling land values in those cities in a very short time." Congress, the defenders of

THE FARMER'S VIEW.

OVER-IRRIGATION IS A MENACE BECAUSE

---

1. Smaller crop yields are obtained for each unit of water used.

2. More plant food is taken up by the plant for each pound of crop.

3. The quality of the crops is greatly reduced.

4. Straw is produced at the expense of grain.

5. Plant food is washed out of the soil.

6. Lower-lying lands become water-logged.

7. Other dry lands are cheated of irrigation water.

8. The extension of the irrigated area is hindered.

9. A wholesome community spirit is lowered wherever water is wastefully used.

Utah Agricultural College;
Utah Conservation Commission.

THE ENGINEER'S VIEW.

(Cartoons published by courtesy of the Idaho Daily Statesman, Boise, Idaho.

4.6. The *Reclamation Record* tried to educate farmers about good farming practice, as shown by this page from the September 1914 issue.

the Reclamation Service argued, should be content to see the multiplication of property values and the growth in western businesses supplying the new projects. [31]

Agricultural experts from the land grant colleges often complained that farmers gave scant consideration to the careful application of water to their fields and crops. Few farmers knew what was meant by the term "duty of water" or what amount of water could serve their crops most efficiently. Too much water prevented proper plant growth and in many cases injured plants. Farmers should take necessary steps to improve the location and grade of their ditches and crop land. Inadequate attention to these details could increase the duty of water two-, three-, and even four-fold. Water in the reservoirs was like money in the bank, but, if that water was squandered upon its application, it was tantamount to throwing money away.[32]

Currents of Change against the Background of War

Efforts to reform the Reclamation Service and make for more settler-friendly projects went forward at a time when American agriculture suddenly enjoyed wider markets. Increased demand for food staples came from Europe, which, after August 1914, was locked in what would be called the Great War. Rising prices, caused by increased demand for American farm products, stifled for a time criticisms, particularly from midwestern farm states. Some in the Department of Agriculture saw little advantage from inflated wartime prices. "If all the things farmers have to buy rise in price on the average as much as all the things they have to sell there will be no gain to them as a class," wrote one official at the outset of the war. [33]

Still, the war in Europe seemed remote to officials of the Interior Department and Reclamation Service. Reorganization continued to haunt the Reclamation Service as the unwieldy five-member commission at its head struggled to meet demands of a critical Congress and a sharp-eyed Secretary of the Interior. Secretary Lane continued to prod the Reclamation Service to be more responsive to the complaints and needs of project settlers represented in the water users' associations and irrigation districts. In December 1914, Arthur P. Davis assumed the position of "Director and Chief Engineer," and Sydney B. Williamson took the office of "Chief of Construction." Williamson brought to the Reclamation Service much experience. He had supervised federal dam and lock construction on the Tennessee River in the

1890s and went to work for the Panama Canal Commission Service in 1907. Immediately after his appointment to the Reclamation Service, he sought to promote more direct communication of project managers with an office located in the West, either in Salt Lake City or Denver. Secretary Lane chose Denver.[34]

By November of 1915, a directive from the Secretary of the Interior, which accepted most of Williamson's recommendations, announced a major reorganization. (1) Headquarters was to remain in Washington, D.C., but leadership was now to be vested in the director/chief engineer, as chairman, plus the chief counsel and the comptroller. (2) In addition to the office in the nation's capital, the office of chief of construction was to reside in Denver, Colorado. The chief of construction received appointment by the Secretary of the Interior on the recommendation of the director and chief engineer. All matters relating to construction, purchasing, or disbursement were to be managed through Denver. Also, departments of purchasing and disbursing were to be in Denver, which became a major executive office for the Reclamation Service in 1915. (3) A project manager or engineer on each project controlled all employees in the construction and operation of a project and "will be held strictly responsible for the economical and efficient administration of the project offices."

There were to be four distinct divisions created in the Reclamation Service that reached from top to bottom: (1) Executive and Engineering. This division encompassed the Director and Chief Engineer as the executive officers and controlled all employees engaged in investigating, constructing, operating, and maintaining projects. The Director issued instructions to the executive office in Denver to carry out policies and undertake assigned work. The Director supervised all employees in the Washington office except the Legal Division and the comptroller. The chief of construction, in Denver, represented the director in the field and supervised employees in construction and operation and maintenance on the projects. Project managers or engineers reported directly to the chief of construction on a monthly basis. All communications directed to Washington were to pass through the chief of construction's office in Denver. (2) Legal Division. This division was to be led by the chief counsel, who would correspond directly with the district counsels regarding legal affairs, but would communicate with the executive department through the chief engineer. (3) Fiscal Division. The Comptroller headed the inspection division charged with conducting the inspection of all fiscal practices and accounts. Any irregularities found in audits were to be reported to the director and chief engineer. (4) Supervisor of Irrigation. This office was to be maintained in Billings, Montana. Its function was to advise and counsel water users on the best practices for irrigating and cultivating lands, the development of markets, and all questions affecting the welfare of settlers and water users.[35]

The work of the Division of Irrigation tried to open the Reclamation Service to the Department of Agriculture and cooperation with USDA experts now assigned to the projects. To this end, both Secretary Lane and Director Davis renewed an invitation to Elwood Mead to join the Reclamation Service upon his return from Australia to Berkeley and the University of California. Mead finally accepted an appointment as chairman of a Central Board of Cost Review to consider the finances and repayment problems of the reclamation projects. He wished the office to be located at Berkeley so that he could simultaneously perform his faculty duties at the university as Professor of Rural Institutions. His position also enabled him to explore the possibility of introducing the Australian plan of irrigated communities to California. By 1917, the California legislature approved a settlement at Durham in the middle of the Sacramento Valley, and, in 1919, another at Delhi in Merced County. Meanwhile, in response to an invitation to serve as a consultant to the Reclamation Service, Mead wrote Davis, expressing a willingness to help out. With the recent departure of Newell from the Reclamation Service, Mead believed he could operate with a clean slate, voicing recommendations that in previous years might have been rebuffed.[36]

The economic difficulties faced by federal reclamation, according to Mead, were not only the challenges of farming reclaimed land, but also inflated private land prices, higher than expected construction charges, and, finally, the need for farm credit. Farm organizations had long backed a federally supported farm loan program. A National Conference on Marketing and Rural Credits convened in Chicago in 1915 to voice support of federal farm loans. A. P. Davis, however, believed that Congress would be reluctant to provide credit to farmers on the reclamation projects because the reclamation laws were already "a sort of rural credit by advancing money without interest for the construction of irrigation works and allowing a long time in which it is to be paid back." He concluded there would probably be considerable opposition in Congress to any more aid to the western farmer. Still, "Judge" Will R. King, chief counsel to the Reclamation Service, noted that the United States had invested more than a hundred million dollars in the reclamation enterprise and that Congress was not likely to see that investment wasted for lack of a government-backed rural credits program.[37]

In 1916, Congress made a decisive move in the long campaign on the part of farm advocates to provide low cost agricultural credit. The Federal Farm Loan Act established twelve regional Farm Loan Banks. Their function was to grant loans to farm cooperative associations from which farmers could borrow. Reclamation projects, whose water users' associations were already virtual cooperatives, seemed to be ideal organizations to take advantage of the legislation. But there was a major barrier in the legislation. The farm loans required lands as collateral, and in reclamation projects the federal government held a lien against much of the land until the construction charges were paid. Only then could the settlers receive full title and use their property as security for the farm loans. Congress had already tried to meet this problem with the Patents and Water-Right Certificates Law of August 9, 1912. After three years of residence and proof of cultivation, a settler could obtain a patent and a water-right certificate. Still, the government lien lurked in the background, and the loans would not be a possibility for three years into the enterprise. [38]

4.7. Will R. King, a member of Secretary Lane's "Reclamation Commission."

It was obvious that those who possessed private lands on the projects were in a better position to take advantage of the farm credit program, and the new credit program favored more well-to-do farmers who owned their lands and could borrow readily against them. As many predicted, Congress refused to tailor a farm credit bill to the special needs of project settlers in spite of suggestions that it establish water users' banks on each project based upon the value of water delivered in the project. The annual report of the Reclamation Service for 1917 complained that the Federal Farm Loan Law excluded most project landholders. If government farmers could not borrow money, the government projects would not be able to keep pace with other land where there was no barrier to the loan act. Moreover, private lending institutions stood all too ready to increase interest rates.[39]

A year after the 1914 Reclamation Extension Act eased the repayment schedule, Secretary Lane still received complaints that total repayments were too high. Could it be that the complaints were justified and that project costs should be revised downward? In addition to Mead's Central Cost Review Board, local cost review boards were appointed for the projects. The local boards sent their recommendations to the Central Board which then made final recommendations to the Secretary of the Interior. The Central Board, chaired by Mead, contained two other members; General William L. Marshall, consulting engineer to the Secretary of the Interior; and Ignatius D. O'Donnell, Supervisor of Irrigation for the Reclamation Service. But the effort to revise costs in favor of the settlers on most projects failed. Construction costs usually exceeded original estimates, not through incompetence or waste, because labor and materials simply cost more than expected. Or so Mead's board concluded. One early history of the Reclamation Service noted: "This body found that high charges for water were not responsible for the evils of the project, but 'inflated land prices, high freight charges, high interest rates, alien landlordism, and normal and not actual compliance with the regulation fixing the size of farm units that closely verges on fraud,' were the true cause of the difficulties." The review boards achieved little for settlers who found more relief from their economic circumstances, however temporary, in the higher prices brought by the war.[40]

4.8. General William L. Marshall, consulting engineer to the Secretary of the Interior in 1914.

Mead believed that better farming meant successful farming, including the application of science to agriculture and the adoption of efficient cost effective methods in agriculture to reduce waste such as economical use of water and market analyses to determine the best crops to raise. While he believed in the initiative of the individual farmer, his experiences in Australia also had developed a faith in governmental paternalism. Farm families facing the challenges of irrigation agriculture could not be expected to achieve successful farming in this new commercial age without help from government in the form of demonstration farms and other programs. Although authorized by Congress, demonstration farms had been discontinued on the projects. Model agricultural communities would be ideal, but they would require more money than Congress seemed willing to invest. Most important, farmers needed a system of rural credits backed by government and a system of aid to the projects that avoided putting all of the gains in the hands of speculators and those renting out their lands rather than into the pockets of the people actually on the land.[41]

Secretary Lane wanted the reorganization of 1915 to be carried out without creating "any spirit of antagonism to our plan or organization." Unfortunately, "a stiff formalism" had prevailed in the past, he claimed, that was "evidently offensive to some of the project managers." Lane recognized the need for "system" in the organization, but he wanted to encourage "the spirit of cooperation, the working together, and a recognition that conclusions are arrived at only after consultation." Still, friends of the Reclamation Service expressed concern that the organization was demoralized in spite of the reforms — or, for that matter, because of them. From his new university position in Illinois, former Director Newell asked Commissioner Davis for comments on his new booklet, "Engineering as a Career." In the same letter Newell noted that, "matters have quieted down to a point where you can really show better" — by which he meant that Davis could operate more easily and efficiently in an atmosphere of "mutual confidence and respect." In referring to the Reclamation Service prior to his departure, he remarked, "There is probably no element … so destructive to administrative efficiency as suspicion which feeds upon itself and develops certainties out of the vaguest rumors." Finally, he added a wistful postscript: "I long to see the arid lands again!"[42]

Within the Reclamation Service, however, the reorganization slowed the completion of projects underway. Even one longtime supporter of the Reclamation Service, Congressman Frank Mondell of Wyoming, complained about the North Platte Project. According to Davis, at a meeting in the presence of senators and others, the Congressman "was very vehement in his denunciation." When Davis explained that construction delays were caused by the reorganization in Denver, no one was mollified. The Congressman implied that the engineers were purposely delaying work to prolong their jobs. Davis directed the Chief of Construction in Denver to "get after this actively and also give attention to proper activity at other points, especially the Flathead." He noted that Frank Crowe (having assumed duties there after the sudden death of project superintendent, Ernest F. Tabor) was new at the Flathead Indian Irrigation Project in western Montana and might not yet be fully in command of the situation. Davis wanted the Denver Office to push work on the Flathead and he asked, "Please see that this work is properly pushed, as well as that at other points."[43]

The turmoil within the Reclamation Service quickly drew the concern of outside observers. Water engineer and consultant, J. B. Lippincott, formerly involved in Los Angeles' appropriation of Owens Valley water and a one time employee of the Reclamation Service, saw a decline in the organization's morale. In late 1915, he wrote Senator Newlands that, "I have been distressed during the past two or three years to see the Reclamation Service become demoralized and broken up. The *esprit de corps* is gone." He had been a part of the movement "to create public sentiment that organized the Reclamation Service." Great difficulties had been overcome by "the old Reclamation Service," which he recognized, "as the finest body of engineering and constructive talent. The work of the Reclamation Service has been well done and the integrity of the personnel of the organization cannot be successfully attacked," he believed. But there were enemies burrowing from within, according to Lippincott. One was Commissioner Ryan, who now worked in the San Francisco office. "He welcomes adverse criticism and accusation," and encourages despondent farmers to voice their discontent. He feared that many inexperienced people were put into positions of command in the Reclamation Service, which was likely to cause more distress. He told Senator Newlands that A. P. Davis should be retained as the real head of the Reclamation Service. Davis should be "the real executive and that those opposing him should be removed."[44]

Publicly, the Reclamation Service remained optimistic. One opportunity presented itself at the Panama-Pacific Exposition in San Francisco in 1915. The exposition celebrated the completion of the Panama Canal under the engineering expertise of the Corps of Engineers. Reclamation backers had long compared the two undertakings: linking the Atlantic and Pacific Oceans and watering the deserts. Both represented the power of government to create new environments. As visitors entered the exhibit, they passed beneath a trellised bower to face large windows that looked out "upon a beautiful mountain-rimmed desert valley that has just been reclaimed by the United States Government," according to one account.[45]

Although Secretary Lane was not about to replace Director Davis because it would cause too much disruption in an organization that was already suffering from morale problems, he continued to snipe at waste, inefficiency, or inappropriate activities. In the latter category, he found C. J. Blanchard's photography and filmmaking questionable. Like Secretary Ballinger before him, Secretary Lane found Blanchard, ostensibly a statistician, overstepping his duties by performing work also for the Forest Service. From Montana, Lane telegraphed Davis that he had discovered in local newspapers that Blanchard was traveling about the country doing motion pictures for the Forest Service. He did concede that if Davis was of the opinion he should be engaged visiting Reclamation projects taking pictures, he had no objections, "but I do not understand his assignment to any other work" when he is employed by the Reclamation Service. In reply, Davis explained that he had curtailed Blanchard's work as publicist for the Reclamation Service "on account of the criticism that might arise there from." In response, Blanchard became "quite active in securing subscriptions and making other arrangements with railroads, water users' associations, fair associations, chambers of commerce" to do this kind of work because he enjoyed it so much. The Forest Service made arrangements to pay for his services and travel, Davis further explained, so that the Reclamation Service was compensated. He also worked for the Reclamation Service while on these trips because "the work of the Reclamation Service is so closely connected geographically with that of the Forest Service that an extra expedition for doing the Forest Service work would merely double the cost of the whole, and refusal on our part to cooperate would work to the great detriment of both the Forest Service and the Reclamation Service." The arrangement was beneficial to both agencies, explained Davis.[46]

The War and Reclamation

War in Europe boosted agricultural prices for the 1915 harvest season. It seemed likely that the demand for food in Europe would rescue reclamation farmers and solve their repayment problems. Congressional expenditures for war preparedness and then the war itself, upon American entrance into the conflict in April of 1917, meant an end to congressional funding of domestic programs. Congress did authorize two additional reclamation projects in 1917 but refused to pay for constructing them in the face of mounting wartime deficits. On the brighter side, prosperity on the projects meant repayments for construction costs might revive the "revolving fund."

The burgeoning wartime economy offered "a reprieve" to the economic difficulties of the reclamation program and lifted the prospects of farmers on the projects.

Notice to Reader.—When you finish reading this magazine place a 1-cent stamp on this notice, hand same to any postal employee, a it will be placed in the hands of our soldiers and sailors at the front. No wrapping—no address.—A. S. BURLESON, *Postmaster General.*

Reclamation Record

ISSUED MONTHLY BY THE RECLAMATION SERVICE
DEPARTMENT OF THE INTERIOR, WASHINGTON, D. C.

Better Business : Better Farming : Better Living

THE BIGGER THINGS ARE EASIER TO DO THAN THE SMALLER THINGS—AND THERE'S LESS COMPETITION.

Volume 9, No. 1 PRICE {NOTHING FOR OUR WATER USERS. FIFTY CENTS A YEAR FOR OTHERS. January, 191

NEW YEAR'S GREETING TO THE RECLAMATION FARMERS.

The water and the thirsty land have been united. On more than a million acres of reclaimed land you have established your homes and have subdued the desert to profitable agriculture with an annual harvest valued at $50,000,000. During the past year, in response to the President's appeal to the farmers for increased food production, you added 200,000 acres to the cultivated area of your farms, and are now preparing for greater efforts in 1918. Your contributions to the Liberty Loan have been generous. Patriotically and loyally you have given your sons to the cause of democracy.

The Nation has reason to be profoundly grateful for the abundant evidence on every hand of enduring love and service for the cause of liberty. May the New Year bring you a full measure of health and prosperity.

U.S. Reclamation Service

4.9. In support of World War I, Reclamation's "seal" was adapted to a patriotic theme for the January 1918 front cover of the *Reclamation Record.*

The same period, however, witnessed the departure of Elwood Mead from the Reclamation Service to serve as head of California's Commission on Land Colonization and Rural Credits, where he could pay more attention to the Durham and Delhi colonies in California. Mead's absence robbed the Service of the most well-informed and talented person to advise it on its problems, but the economic indicators suggested that all would be well in future years. With these bright prospects now on the horizon, some in the Reclamation Service realized it labored under the legacy of past mistakes. While they concluded that many mistakes had been made, most would have agreed with F. G. Hough, project farmer, that, "Mostly they have been mistakes of the head, and not of the heart." The idealism of the Service remained intact as cultivated land within the projects increased from 1912 through 1918. Cultivated land increased by 241 percent from 923,000 to 2,229,000 acres, and irrigated land increased by 77 percent, from 694,142 acres to 1,225,480 acres. Still, it was clear that the cost of bringing those lands under cultivation had soared along with crop prices.[47]

Davis had essentially replaced Newell as the Director of the Reclamation Service in December 1914, but he resented the five-member advisory board, of which he was a member. He, and even Mead, advocated the independence of the director from the advisory board (or what was called, "the Commission"). Davis soon told Secretary Lane that, "The Commission form of administrative organization

4.10. A patriotic cartoon from *Reclamation Record* in September of 1918.

of this Service has proved not to be efficient," and called the commission system an exercise in "plural authority." On several occasions, especially on the Uncompahgre Project in Colorado, promises regarding the construction of drainage works were made both orally and in writing by one commissioner. Those promises committed the Service to spend a million dollars, even though no single commissioner had the power to authorize appropriations. Davis constantly complained of the "influence of plural authority" as "injurious to the internal organization of the Service." The Legal Division, under "divided authority," sometimes escaped the administrative control of the Director and embarrassed the Service.[48]

Squabbles over the administrative structure of the Reclamation Service paled in the face of American entrance into the European war. Among the prominent wartime slogans, such as "War to Save Democracy" and "War to End all Wars," was also the slogan that pertained to American farmers: "Food Will Win the War." From his position as head of the National Food Administration, Herbert Hoover asked farmers to increase their production, and as agricultural prices climbed to record peaks, farmers responded by raising more food and fiber and cultivating even marginal lands. This meant that irrigable lands, once dismissed as unnecessary for American food production, now assumed an importance previously unimagined by the opponents of reclamation. The United States entered the war in 1917, just as George Wharton James finished his history of the Reclamation Service: *Reclaiming the Arid West: The Story of the United States Reclamation Service.* James portrayed the Reclamation Service as an "army of peace" dedicated "to the high democratic ideal that originated it."[49]

But what role could the Reclamation Service play in the war across the Atlantic besides encouraging food production and expanding cultivation on the projects? Many of its engineers departed to the soggy fields of France to help build massive earthworks and battlements. The Department of the Interior designated a wing of the American Hospital at Neuilly, France, for the care of wounded or sick employees of the Department at the front. Reclamation employees sent contributions to the hospital in a campaign coordinated by Mrs. Franklin Lane. Meanwhile, project managers developed programs to bring uncultivated lands into production for the war effort. They classified lands as privately held uncultivated lands, unsettled lands open for entry, land not yet open for entry, and state lands. The unentered public lands, of course, were government lands subject to the Reclamation Service. These lands could be prepared for irrigation as rapidly as possible and rented out to competent farmers. If, in the process, they were made attractive to new settlers, this would contribute to the long term development of the project. There were also plans afoot to bring "idle" Indian lands into cultivation. When the Department of Agriculture voiced opposition to the Interior Department actively developing farms, Congress refused to appropriate money for the scheme. [50]

In terms of new construction, the Reclamation Service's Chief of Construction, Frank Weymouth, thought that little in the way of new construction work should be done during the war because of high costs and labor shortages. Congress did open one new stream of income for the Reclamation Service in 1917: revenue from royalties and rentals of potassium deposits on government land. Potassium nitrate was an important ingredient in the manufacture of gun powder or explosives. Congress also declared that, during the national emergency, the Secretary of the Interior could suspend residency requirements (not always strictly enforced) for settlers on private and other lands receiving water from reclamation works. This permitted the Secretary to encourage production by allowing water to be allocated to lands not farmed by their owners, who lived elsewhere. Under new rules, owners could rent out their lands to neighboring farmers and even bring tenants onto the land. There was an impressive 13 percent increase in irrigated lands on the projects during the war. The Reclamation Service's annual reports list irrigated acreage for 1916 as 1,010,000 acres and for 1918 as 1,141,516 acres. The really impressive gain occurred in the value of the crops raised on lands watered exclusively by government water and on Warren Act lands which could mix both private and government water. The 1916 crop value was close to $40,000,000, but in 1918 the estimate was well over $100,000,000. This resulted from a combination of increased production and dramatic price increases.[51]

4.11. Frank E. Weymouth, 1915.

Under the pressures of war, Secretary Lane yielded to the recommendations inside and outside the Reclamation Service to abolish the commission that he had created in 1914. Both Davis and Elwood Mead enthusiastically greeted the decision. Mead was now in Washington, D.C., serving in Herbert Hoover's Food Administration in the Department of Agriculture. His job was to promote "harmonious and efficient cooperation" with the Reclamation Service. He spent most of his time visiting and advising the twenty-four reclamation projects throughout the West. Mead hoped to put into practice a wider social and economic program to aid and develop agricultural communities on the projects, but the opportunity did not arise. The short duration of the war (eighteen months) afforded little time for ambitious new schemes, and there was little money for innovation on government reclamation projects.[52]

During the war, however, government construction of highways and railroads on some projects helped with the construction of large works. Most importantly, the construction of large dams and canals required careful supervision of the materials used, particularly cement. During these early years, the Reclamation Service maintained two cement testing laboratories, one in Denver and one in San Francisco, to check and prescribe cement tests and standards. On July 1, 1917, the Bureau of Standards, in the Department of Commerce, took over these duties, relieving the Reclamation Service of the cost of maintaining its own laboratories.[53]

In spite of Secretary Lane's attitude that further extensions could be made, Davis and others in the leadership of the Reclamation Service continued to take a hard line against any further liberalization of the repayment obligations for the projects. In 1918, the Secretary proposed to grant a special extension to the Yuma Project, which Davis believed would be "a violation of law" followed by publicity detrimental to the cause of Reclamation. He asked Gifford Pinchot to call upon Secretary Lane to "save the cause of Reclamation from dire disaster." Davis and other leaders in the Reclamation Service feared that not even the prosperity of the war years would spur project farmers to repay their debt to the government. And repayment would alleviate the financial circumstances of the Reclamation Service.[54]

D. W. Cole, now Project Manager of the Boise Project, took a similar line in a letter to a water user who complained about repayment contracts. He argued that interest free payments, previous extensions on payments, and

Government irrigation systems that are superior in construction and operation to private systems offered settlers far greater opportunities to succeed. He chided water users on the Boise Project for not creating irrigation districts to deal with the Government "in a business like fashion." If this had occurred, there would be no need for the Government to press for payments individually. As he explained the situation to the water users, the government gave

Reclamation Record

Better Business : Better Farming : Better Living

SAVING MORE IS BETTER THAN RAISING MORE.

Volume 6, No. 3 — March, 1915

THE COST OF THE RECLAMATION RECORD.

A careful estimate of the actual cost of editing, printing, and distributing the Reclamation Record for the entire year 1914 shows that it amounted to less than 8 mills, or four-fifths of a cent, per irrigated acre.

Up to the present time we have not received a single adverse comment about the Record. We have, however, received numerous congratulatory letters from water users dilating on the great value of the Record to them. Our own opinion is that it is well worth while. Commissioner O'Donnell's "better farming" articles alone have, we believe, resulted in a large increase in the average value of crops per acre, and there are still more valuable articles to come.

We wish, however, to get the opinions of all the men and women on the projects. Sit down NOW and write us what you think of the Record.

REVISION OF PROJECT COSTS.

Secretary Lane's Letter.

January 20, 1915.

The Director and Chief Engineer of the Reclamation Service.

My Dear Mr. Davis: It is my desire that a revaluation shall be made of the project works of all projects or units of projects as to which the construction charges have been heretofore announced by public notice. As previously announced to the water users' associations, it has been my intention to revise the estimated cost of each of these projects or units of projects upon which construction charges heretofore announced were based.

It is my desire that a board of revision for each of the projects shall be appointed, to be composed of three members, one to be selected by the water users of the project, one (an engineer or project manager) to be appointed by the Reclamation Service, and the third member to be selected by these two in the following manner: In each of the main divisions of the Reclamation Service presided over by a supervising engineer, all of the persons selected on each of the projects to represent the water users and the Reclamation Service shall unite in nominating three persons from among whom I shall select a third member to serve on each of these boards, the same person serving on each board.

The person so selected shall not be a water user on any project, shall not be either directly or indirectly interested in a project, and shall not be a person who is now or has at any time been in the employ of the Reclamation Service.

To illustrate the process of selection: In the southern division there are the Salt River, Yuma, Rio Grande, Carlsbad, Hondo, and Strawberry Valley projects. For each of these projects, whether construction charges have heretofore been announced or not, there shall be appointed by the water users and by the Reclamation Service, their representatives as above described. These 12 persons shall gather at some central point and nominate three persons

81080°—15——1 97

4.12. Possibly for political reasons, Secretary of the Interior Franklin K. Lane chose to announce a reevaluation of reimbursable project costs very publicly. Reclamation published his letter to Director A. P. Davis on the cover of the March 1915 *Reclamation Record,* which was then distributed to most water user managers and settlers as well as Reclamation employees.

settlers land that had practically no value until an $80 per acre irrigation loan was placed on top of it, thereby giving the settler the chance to make it worth $200 per acre. These arguments reveal a Reclamation Service that clearly believed settlers profited in many ways from government reclamation and should cheerfully meet their obligations to repay the cost of delivering water to their lands. Resistance revealed a moral failure on the part of project farmers that frustrated the upper echelons of the Reclamation Service.[55]

The number of employees of the Reclamation Service declined during the war. In June 1916, the work force totaled 5,410 persons. The Annual Report divided employees into the following categories: educational, 507; noneducational, 1,154; and laborers, 3,749. Contractors working for the Reclamation Service employed an additional 672 workers. In June 1919, 3,819 people worked for the Reclamation Service: educational, 555; noneducational, 951; and laborers, 2,313. Companies under contract to the Reclamation Service employed 112. The largest decline was in the category of laborers and other employees of private contractors. The figures suggest that the physical works of reclamation were not expanded during the war.

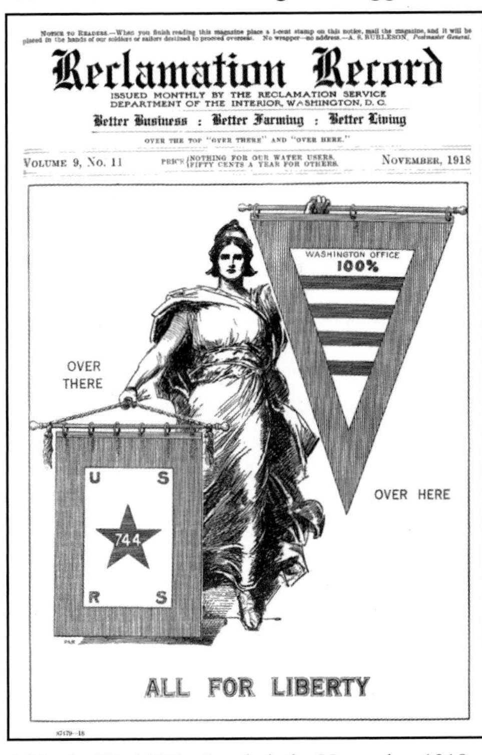

4.13. As World War I ended, the November 1918 cover of *Reclamation Record* celebrated the number of Reclamation employees in Europe and at home working for the war effort.

But workers who stayed with the Reclamation Service during the wartime crisis continued to enjoy worker's benefits that were among the most progressive of the day. Employees early enjoyed accident compensation and a medical care plan. In the work camps associated with Reclamation projects, sanitation and housing provisions had to be provided in remote locations. The Reclamation Service often operated mess halls, where workers paid for their meals, mercantile stores, and recreational facilities where Y.M.C.A. workers organized activities. The latter's recreational activities were employed in an attempt to keep alcoholic liquors out of the work camps. Generally, the messes and the mercantile stores showed a profit to the Reclamation Service. Such operations emulated the pattern and workings of private industrial company towns organized around lumbering, mining, or railroad maintenance facilities in many locations in the West. On almost every project, local physicians were under contract for services from the hospital fund accumulated through worker payroll deductions of $1 to $1.50 a month. By June 30, 1917, the hospital fund showed a net profit of $26,836.33. Civil service physicians were directly employed on the more remote projects, such as the Milk River Project in Montana and Jackson Lake reservoir in northwest Wyoming. During the construction of Arrowrock and Elephant Butte Dams, hospitals were built and equipped for workers but were dismantled after completion of the dams and the dismissal of the work force. In cooperation with the War

4.14. The kitchen for the mess hall at Avalon Dam on the Carlsbad Project in New Mexico, about 1913. In remote construction sites, such conveniences were important to workers.

Department, the Reclamation Service continued to inoculate its field force against typhoid. The 1919 annual report spoke of inoculations against pneumonia and the distribution of influenza vaccine upon request.[56]

While the war deferred the building of new reclamation works, the leadership of the Reclamation Service continued to plan for the future. Davis and Mead talked of Davis's long projected plan for the development of the Colorado River system and a big dam on that fabled river of the West. Secretary Lane roused once again the voice of long-time arid lands advocate, William E. Smythe, by commissioning him to study the role of reclamation in the postwar period of economic readjustment. The Secretary recognized that reclamation projects might find opportunity in any comprehensive government program to bring the economy back to a peacetime footing. What better person to call upon than Smythe, the indefatigable "irrigation crank" and promoter, to encourage returning soldiers to take farms on the projects. He did not disappoint.

As Smythe looked into his crystal ball, he saw industrial unrest sweeping across Europe after the war and threatening the United States if it did not take measures to anchor its population upon the land. Veterans should be offered a farm on reclamation projects as part of a soldier-resettlement program. Congress would do well to formulate such legislation immediately. Smythe's suggestions were in line with Lane's belief that a new "era of democratic planning" was at hand. Idealism aside, soldier resettlement could go a long way toward easing the economic adjustments in the post-war period. Lane and others believed that the reclamation projects held part of the answer to restoring a peacetime economy.[57]

In an article in *Reclamation Record*, C.J. Blanchard warned that if soldiers and sailors were not invited back to the land, "the farms of this country will lose permanently thousands of sons and agriculture will decline correspondingly." At the end of the war, the Department of the Interior received almost 200,000 inquiries about settlement opportunities on western reclamation projects. No wonder officials expected Congress to respond generously when they testified that, "The boys are coming home with the idea that this Government is going to do something for them." In testimony before Congress, Secretary Lane not only favored a soldier resettlement bill for reclamation projects in the West, but broadened the proposal to include the South and the East. For years, the Reclamation Service had looked to the South to apply its reclamation skills to idle swamplands. Now, Secretary Lane saw the chance to include the South in the reclamation agenda under the banner of "soldier resettlement." The result would be a Reclamation Service that was truly national in scope.[58]

Furthermore, Secretary Lane emphasized that the new settlements should involve community planning and improvement of social and economic conditions. The 1914 Extension Act authorized the Secretary to set aside land in the projects for community centers. Acting on their own initiative, project settlers willingly appropriated money to provide modern and well-equipped school buildings with qualified teachers. The annual reports of the Reclamation Service began to emphasize that "life on many of the projects has been made attractive, and the country has lost its isolation and loneliness." Oftentimes women were "the key to the success" on

4.15. In 1927, Reclamation was formulating plans for the South when this image of the Blodgett Naval Stores Company turpentine still was taken in Mississippi.

the new projects, and their needs for community had been neglected in the past. Secretary Lane saw too many instances of farmers' wives driven to insane asylums because of isolated living conditions and lack of community. "I would have settlement at the center of the farm area with good schools, motion picture theater, and other conveniences," he recommended. By 1915, the annual reports praised women for forming community clubs:

> More than 200 women's organizations have been reported, a large percentage of them being affiliated with State and National federations. That they are already an important

4.16. The Spring Lake School on the Klamath Project was a modern structure in August of 1916.

4.17. A 1914 home economics class in the "all electric" high school in Rupert, Idaho. The novelty of using electricity in the high school was made possible because the Minidoka Project began producing hydroelectricity in 1909.

factor in the upbuilding of the West is well recognized and they are working side by side in effective cooperation with

If you have not yet planned your garden a great pleasure surely awaits you. The seed houses are putting out some very attractive catalogues which they are glad to send free upon request, and the wise gardener will not delay securing these. It is rumored that there is a great scarcity of seed this year and unless you obtain your supply early you may have to entirely revise your garden map. While you are sending for circulars don't forget that the Department of Agriculture has bulletins galore on every phase of "gardenology," and that they will be glad to furnish you free advice of experienced men on the culture of vegetables and flowers. There are two particularly valuable bulletins. One gives directions for making a hotbed which soon pays for itself even in a small garden, although several of the early vegetables and flowers can be started in boxes in the living room. The other bulletin contains suggestions for laying out your garden plot to the best advantage, for with proper planning and rotation in planting you may have an abundance of green vegetables throughout the entire season.

4.18. Mrs. Louella Littlepage's column, "Project Women and Their Interests," in the *Reclamation Record* of March 1917 tried to pep up interest in the coming gardening season.

boards of trade, chambers of commerce, and other organizations for better farms, better health, better schools, better communities, and better homes.

Secretary Lane believed the settlement must be tied to the farm and vice versa. He pointed to the successful development of cooperatives, aided by the Department of Agriculture, in the citrus fruit industry of southern California. Those cooperatives produced high quality fruit at fair prices. Finally, new farmers needed education and a government loan program for success. And success it would bring if community planning, government aid in education, and farm credits could be forthcoming. In the South, results would be spectacular on newly reclaimed lands because the abundant rainfall of the region would ensure good crops, and the costs and problems of irrigation works would be far less.[59]

Congress was in no mood for bold moves. Yet, the practice of rewarding soldiers with western lands or land bounties reached back to the American Revolution. Now that only marginal lands remained in the West, reclamation projects took on new importance. To make western arid land settlements attractive, a program of agricultural loans, agricultural education, and other assistance must occur. Both the Department of Agriculture and the Reclamation Service hoped in vain for a favorable response from Congress. At one point, in May 1919, A. P. Davis believed substantial legislation would pass and thanked Mead for his help on the matter with Congress. But even the hallowed cause of rewarding veterans could not prompt Congress to

4.19. Alfred Fincher in the depression years chose to exercise his option to take up a Veteran's Homestead. This is the family's temporary camp nine days after arrival on the Vale Project. September 1936.

underwrite the settlement of veterans on reclamation projects. Free water as well as free land was too much to ask.[60]

In February 1920, Congress belatedly agreed to grant veterans a 60 day (later extended to 90 days) preference right when they applied for homesteads on the public lands, including reclamation projects. Hardly a major giveaway or significant benefit program for veterans, it was an invitation to gamble on establishing a farm on reclaimed lands and tie one's future to the fortunes of the troubled reclamation projects. The response of veterans was not overwhelming. Still, from 1920 to 1922 lands opened to settlement went almost exclusively to veterans. Most of the new farms were opened on the Klamath Project, which straddled California's northern border with Oregon and on the Shoshone and North Platte Projects in Wyoming. Indicative of the slow pace of project development in the interwar years (1920 to 1940), only 1,331 farms were made available for 10,875 applicants who vied for farms largely by lottery. [61]

A 1924 investigation into the plight of veterans revealed that they fared no better than others who came to the projects. This was not a surprise

4.20. World War I veterans at Torrington, Wyoming, wait to see whether their name will be drawn, allowing them to homestead on the North Platte Project, Nebraska and Wyoming. September 9, 1921.

because the law granted veterans no special advantage beyond an opportunity to select a homestead ahead of other applicants. For veterans and non-veterans alike, the report cited the usual reasons for failure: lack of capital, inexperience in farming, the poor quality of the lands settled, distance from markets, and disillusionment with farming. In the case of veterans, poor health related to war injuries often figured in the failures. Many quickly sold their homesteads after proving up. They regarded their presence on the projects as a speculative venture from the outset. Those who took up claims on the Klamath Project, however, proved the most successful and persistent in their landownership. On Klamath, 65 percent of those who gained a patent to their lands were still on them in 1944, while on the North Platte Project, only 19 percent remained. A major determinant was the quality of lands within a project. The Klamath met this standard and became the best example of veterans winning the gamble to make a go of it on a government irrigated homestead.

In 1944, when Congress again considered a soldiers' settlement bill – this time for the post World War II period – a report noted that, of the 1,311 new homesteaders on reclamation projects in the interwar period, nearly 60 percent obtained patents. Of the original 1,311 claimants, 45 percent still possessed those farms in 1944. Reclamation resettlement historian Brian Q. Cannon sees this as "an impressive rate considering the economic volatility of the 1920s and 1930s." Still, not half (46 percent) of these owners worked the land themselves. The percentage was much higher on projects with more fertile land, as was the case with the Klamath Project.[62]

Interwar Realities

The Reclamation Service faced a changing administrative landscape after 1920. In ill health, Secretary Lane retired in February 1920 and died in 1921, as the Democratic Administration of President Wilson faded into a postwar period of Republican Party ascendancy. The fall elections brought the presidency of Warren G. Harding. Harding promised "normalcy" to Americans who welcomed his bland pronouncements after the frenzied months of war and the heated reformist idealism of the Progressive years. These reforms began with Theodore Roosevelt's presidency in 1901. They included the Reclamation Act and other conservation measures as well as greater governmental regulatory powers over the economy, and they culminated with the wartime crusade to "Save the World for Democracy."

Normalcy also meant a return to sound business practices as the nation embraced the conservative values of the Republican Party in the 1920s, including balanced budgets, the retirement of war debt, reduced taxes, and less government regulation of corporations.

After the upswing of agricultural prices in World War I, prices turned down sharply by 1919, and Director A. P Davis declared the Reclamation Service had succeeded admirably in carrying out the purposes of the Reclamation Act by establishing farms and homes in the arid lands. He noted that the government had spent $122 million in the building of reclamation projects that had created over a half a billion dollars of wealth in western states in the form of land and crop values. The growth of accompanying businesses connected to this wealth was incalculable. His conclusion was that Uncle Sam had made a wise investment and that the estimated $49 million to complete work on existing projects was only one-half the annual value of crops raised on the projects. But the postwar years found the Reclamation Service under close scrutiny. [63]

Western reclamation projects that could not pay their way were a thorn in the side of the cost-conscious Congress and Republican Administration. The 1920 Congress provided that 52½ percent of proceeds from the Oil Leasing Act on public lands go to the Reclamation Service. And the Federal Water Power Act of that year earmarked 50 percent of the licensing fees for hydroelectric plants to go to the Reclamation Fund. Still, the Reclamation Service grew poorer each year. By 1921, the postwar agricultural depression erased many of the gains made during the war. To counter the bad news coming in from the projects, the Reclamation Service claimed that it had increased the value of lands on its projects by $350 million and by $100 million on private projects served by government water under the Warren Act. Added together, the $450 million figure compared favorably with the total of $130,742,488 the government had invested into reclamation projects. But the land values did nothing to repay the government. By 1921, the government had received only $10,677,250 in repayment on construction costs, and most of that had been generated by power revenues. As the land historian Paul W. Gates noted: "Reclamation Service officials … refused to face the fact that the Service was performing a task assigned it, perhaps on the false assumption that it would pay off in returned dollars, and when they recognized that it could not they resorted to what amounts to circumlocution to prove good results." [64]

The major problem remained the inability, if not outright refusal, of project water users to retire their debts to the Reclamation Fund. Despite Congress's extended repayment periods and graduated payment schedules, most projects struggled to meet their financial commitments. The postwar agricultural depression (1919-1921) only compounded the problems on many of the faltering projects. Crop values fell by nearly 50 percent from 1919 to 1922. This prompted a comparable fall in land prices, but as both crop and land prices fell, the cost of irrigation rose. Early estimates (1902-07) suggested a figure of $31.00 per acre for construction costs on reclamation projects, but by 1923 the Reclamation Service estimated that the construction costs on projects under construction would average $84.00 an acre. The projects on the northern Great Plains, in Wyoming, Montana, and the Dakotas seemed mired in failure.[65]

Scandal compounded these problems. In what came to be known as the Teapot Dome scandal, Secretary of the Interior Albert Fall faced charges of accepting bribes from oil companies in return for granting leases to the companies for exploration and drilling on public lands. His resignation came in early 1923. Fall paid little attention to the troubles of the Reclamation Service other than to demand that it operate more efficiently. Nothing like the eagerness displayed by Secretary Lane to attack problems occurred in Fall's brief tenure in office. Congress, however, tried to fill the vacuum, as did the leadership of the Reclamation Service. Reclamation leaders continued to assure Congress that its biggest problem was repayment.

Director A. P. Davis remained adamant on this point. Like Newell, he believed that farmers who could not make their repayments were improvident, poor managers, and somehow morally deficient. Those who could not make it on the projects should move on and make way for more competent farmers. After all, the Reclamation experience suggested that it took two or three waves of unsuccessful settlers before a competent farmer could succeed.

By 1922, the Reclamation Service reported a 40 percent delinquency figure on construction payments. Some eastern newspapers told the public that Uncle Sam would have to write off the entire debt of western farmers owed to the Reclamation Fund. For Congressman Mondell of Wyoming, a champion of reclamation from its beginnings, such talk doomed any further appropriations from Congress. He opposed any postponement of payments because further delays risked complete disillusionment in Congress with the entire proposition of reclamation. Still, Mondell realized that many western representatives could not speak against postponement for fear of offending constituents. Their talk about "repudiation" of the debts was completely irresponsible, in his view, and did great injury to the cause of reclamation. Mondell insisted that the Reclamation Service should "lay down the law" on these matters. He believed that the movement for additional concessions was more in the interests of "the banker and the town man" rather than the poor farmer. Bankers would be in a better position to get repayment on their loans to farmers if payments to the government were postponed, and many town men held project lands for speculation and wished to avoid payments. In fact, failure to enforce repayment encouraged speculation. When there was talk of forgiving the payments, responsible farmers feared making payments while their neighbors resisted.[66]

In an insightful letter (insightful into the minds of the Reclamation leadership) to Senator George Norris of Nebraska on January 10, 1922, Director Davis stressed that the "moral obligation" of repayment should be honored in the light of pledges by western congressional delegations that all of the money spent on irrigation projects would be returned to the treasury. Davis, however, admitted that there were "industrious and deserving" farmers on almost all of the projects who would be ruined if the existing repayment laws were literally enforced. He wanted legislation that would protect the deserving, extract payments from those who could pay, and hurry the departure from the projects of the hopeless cases that could never make good.[67] An impatient House Committee on Irrigation of Arid Lands wrote to Secretary Fall at the end of 1922 requesting information on the repayment question. Chairman of the Committee, Addison T. Smith, of Idaho, noted the contrasting experiences on many of the projects: some successfully met repayment obligations while others failed utterly. Requests to defer collection were constant from some projects. Citing the five year special extension for the beginning of payments on the Uncompahgre Project in Colorado, the chairman said that it had left the project in a worse position than before. "The mere extension of time can be of little permanent value in such cases," he said, "it only serves to put off the day of settlement and piles up added deficits." Against those who advocated postponement, Smith argued that repayment was a check on speculation and tenant farming. In the long run, a policy that demanded adherence to repayment schedules would promote a larger number of self-supporting farm homes. The Congressman expressed frustration at the continued demand each year for new legislation to extend payment schedules or to cancel the debts entirely and declared that the repayment dilemma stood in the way of legislation to continue or expand reclamation as a national policy.[68]

Western congressional representatives found themselves caught in a difficult position about what to do with the entire reclamation program. They were under pressure from constituents, especially water users' associations, to extend payments or declare moratoriums. But they also realized that relief measures would bring Reclamation under even greater suspicion in the eyes of their midwestern, eastern, and even southern colleagues, closing off any possibility of obtaining additional money for water development in the West. The engineering cadre that led the Reclamation Service seemed to hew too strictly to the doctrine of repayment on the grounds of "moral obligation." Western congressmen, however, knew forcing repayment might cause the projects to default on a wide scale, and this would surely seal the doom of Reclamation.

It was Congress that gave ground. By March 1922, Congress granted a one-year extension on all payments for construction and operation costs on government projects. This also applied, in the opinion of Director Davis, to Indian projects in which the Reclamation Service had been involved since 1907. But the one-year moratorium could not be extended to lands within those projects that Indians no longer owned. Suspension of payments to those on Indian lands would be, according to him, "in keeping with the spirit of the relief act of March 31, 1922." This was the first relief act Congress passed in the 1920s, and it would not be the last in the opinion of Davis. Still, he was hopeful that project financial conditions would improve to avert future relief laws.[69] In addition, Congress permitted the Reclamation Service, in 1922, to enter into contracts with irrigation districts that made the districts responsible for running the projects, collecting construction and operation costs, and, most importantly, removing the government as the holder of the first mortgage on lands where construction costs had not been repaid. This opened virtually all lands on the projects to the benefits of loans from the Federal Farm Land Banks, created in 1916. Local water districts and associations quickly responded to the provision permitting them to take control of operations. By the end of the 1920s, government reclamation officials administered only the Rio Grande, Carlsbad, Yuma, Orland, and Klamath Projects.[70]

In spite of the 1922 legislation, the leadership of the Reclamation Service, Congress itself, and the office of the Secretary of the Interior were all, in one way or another, at an impasse in early 1923. The Reclamation Service faced a limited future with little revenue coming in from either public land sales or mineral leases. Water users continued to resist the repayment of construction and other costs. While Congress bowed to pressures for debt relief, western representatives realized there would be no infusion of money for new programs under these conditions, and the Department of the Interior was paralyzed by Secretary Fall's rumored departure. Adding to the confusion, President Harding died in San Francisco on August 2, 1923. With scandal in the Department of the Interior, the Veterans Administration, and elsewhere in the government there was little that smacked of normalcy in the everyday affairs of the Administration except the intentions and efforts to achieve balanced budgets and to curtail spending to lower taxes.

Difficult economic years along with a tight-fisted Congress and Administration curbed new initiatives for Reclamation. With its hands tightly around the purse strings, Congress was puzzled over how to handle the Reclamation Service's financial crises. Director Davis's solution was simple: it was time for the water users to pay up and come to the aid of the Reclamation Service instead of the Reclamation Service coming to their aid. Yet, his direct and clear answer was not without the caveat that some industrious, honest farmers needed deferred payments and were in genuine need of relief.[71]

By early spring 1923, the Teapot Dome scandal forced Secretary Fall from office. In October, the Senate opened hearings on charges that he accepted bribes from oil company officials in return for oil leases from naval oil reserves in Elk Hills, California, and near Teapot Dome in Wyoming — after finagling the transfer of the reserves to the Interior Department. The affair, famously known as the Teapot Dome scandal, came to symbolize the corruption of the Harding Administration. Fall was the first cabinet member ever to be convicted of crimes and sent to jail for his misdeeds.[72] President Harding appointed his Postmaster General, Hubert Work, as the new Secretary of the Interior. Work, a practicing physician in Colorado, brought political skills to the

4.21. Secretary of the Interior Hubert Work served 1922-1928.

Department of the Interior, a determination to salvage its reputation, and a charge to address the financial woes of the Reclamation Service. A professional himself, he was not overawed by the professional credentials of the engineers who ran the Reclamation Service, that is to say, the reputation and achievements of Director Arthur Powell Davis.

As a westerner, a physician, and a person prominent in Republican Party politics, Work quickly confronted the leadership of the Reclamation Service. In April, he established the office of "Reclamation Field Commissioner" to represent his office in the various projects with the purpose of improving business and agricultural practices. He and other critics believed that Director Davis looked at federal reclamation solely as an engineering job. Consequently, the engineering-only mentality resulted in an overemphasis upon construction, ditches, drainage, operations, and maintenance questions while neglecting the economic, social, and community aspects of the settlement process.

Although there had been intimations of Davis's dismissal under Secretary Fall, the understandably confused conditions in the Department of the Interior prevented decisive actions. The new Secretary was in a position to put intentions into action. In late June, Secretary Work asked for Davis's resignation, but wanted the resignation letter framed to indicate it was not a "firing." Davis refused. The announcement came from the Secretary's office in what one historian called a "deliciously ambiguous press release," in which the Secretary praised the contributions of engineers to the construction of the great works of reclamation, but declared it was now time for the Reclamation Service to follow sound business principles. On June 20, he abolished the Office of Director of the Reclamation Service. At the same time, Secretary Work ordered that the name be changed to Bureau of Reclamation and appointed David W. Davis, former governor of Idaho and businessman, as "Commissioner" of the Bureau of Reclamation.[73]

4.22. David W. Davis, Commissioner of the Bureau of Reclamation from 1923-1924.

A. P. Davis, chagrined over his removal, refused, like Newell before him, to stay on with the Reclamation Bureau as a consulting engineer. After all, he deserved better. His many accomplishments included Shoshone, Arrowrock, and Elephant Butte Dams as well as the four-mile-long Strawberry Tunnel and the six-mile-long Gunnison Tunnel. He had also pushed for the damming of the Colorado in a large project to harness the power of that river and its water supply. But his progressive credentials made him suspect in the eyes of the business community. That is to say, private power interests suspected that he favored the national development of the Colorado and a national monopoly of the river's power. This alone was a major argument for his replacement by a man with "business acumen," if the Colorado River project went forward in the immediate future. Moreover, those who backed the development of a Columbia River project suspected him of ignoring their interests because, to them, he put the Colorado River before the needs of the Pacific Northwest.[74]

From the projects themselves, Secretary Work received plenty of support for his dismissal of A. P. Davis. Water users' associations constantly discounted Davis's belief that meeting obligations was essential to the future health of the entire reclamation enterprise. A particular example of animosity to Davis came from the Pecos Water Users Association on the troubled Carlsbad Project in New Mexico. Its directors passed a series of resolutions that applauded the "retirement" of Director A. P. Davis. It noted that protests over the resignation came almost entirely from middle and eastern states "not familiar with the merits of the case." Director Davis's departure was a part of the welcomed reorganization of the Reclamation Service by Secretary Work "upon a strictly business basis to which engineering and construction must inevitably be subordinated." Furthermore, the Pecos Water Users Association expressed confidence in the new commissioner, D. W. Davis, and in his effort to move the Reclamation Service "along thoroughly practical lines, which in no way conflict with sound engineering but are absolutely necessary if engineering and construction work are to be properly financed and continued." Finally, it declared that "the prime purpose of reclamation is not engineering design and construction of vast irrigation works, but this is all incidental and subordinate to prosperous, contented and happy homes in desert places, through the complete co-operation of the government 'to enable,' as former President Roosevelt expressed it, 'the people in the local communities to help themselves.'"

For his part, A. P. Davis was not shy about embracing the support and protests of the American Society of Civil Engineers (ASCE) over his ousting. He emphasized that his dismissal occurred primarily because of pressure from water users' associations and other corporate interests that stood to benefit from continued repayment delays. On the other hand, the Pecos Water Users Association denounced these claims as "contemptible propaganda" and denied that private corporate interests exercised any undue influence against any policy or individual in the Interior Department. The Pecos Water Users' Association was a prominent example of what Davis and Reclamation Service officials criticized about such associations. Its members resisted repayment and disputed construction, operation, and maintenance charges while demanding that the Reclamation Service expand water delivery and storage facilities for lands with very little chance that they would soon be occupied.[75]

Secretary Work felt compelled to reply to the protests of the ASCE. He noted that the Reclamation law placed the entire responsibility for the administration, construction, and operation of irrigation projects on the Secretary of the Interior. In recounting the history of Reclamation, he blamed the Reclamation Service for a preoccupation with engineering. The completion of construction projects, however, raised new issues, and these problems were best handled "by a practical business man, familiar with conditions peculiar to irrigation in the West." The emphasis now, Work believed, should be on helping farmers and breaking up large landholdings to get more settlers on the projects. In addition, Reclamation should promote creameries, sugar factories, and other enterprises and encourage the diversification of crops. It should cooperate with the farmers and tenants on project farms to help them process and market their products.

Work sought efficiency in every phase and aspect of reclamation. Change must occur, emphasized Secretary Work, "for unless improvement can be brought about, many projects will be abandoned entirely by settlers, some have already gone, the Government not only will lose millions of dollars invested, but the settlers themselves will lose time, labor, and money already placed by them on their farms." Then he struck his major theme: "Although it is primarily essential to construct dams and ditches, these are not alone enough to secure successful farming to settlers, for whom reclamation was instituted." Since most of the reclamation projects were not prosperous, "business acumen" was more necessary than engineering at this point to facilitate the reimbursement of the government for the millions of dollars advanced for the reclamation of western lands, and this necessitated reorganization of federal reclamation and Reclamation's leadership.[76]

With the reclamation program at a crossroads, Secretary Work appointed a commission to study new directions for Reclamation. Called the Fact Finders' Commission,[77] it was made up of six members: Chairman and ex-Governor of Arizona, Thomas Campbell; ex-Secretary of the Interior under President Roosevelt, James Garfield; Elwood Mead (Secretary Work requested Mead's service while he was on his most recent assignment in Australia); Julius Barnes, President of the National Chamber of Commerce; John Widtsoe, professor of irrigation studies at Utah Agricultural College [now Utah State University in Logan]; and Oscar E. Bradfute, president of the American Farm Bureau. A. P. Davis later commented that Secretary Work had appointed "a number of politicians, upon whom he apparently relied to secure the kind of report he wanted." This was unfair. Only the

4.23. The Committee of Special Advisers on Reclamation, popularly known as the Fact Finders' Committee.

ex-Governor of Arizona appeared to have political ties. Davis did express confidence in Mead, Garfield, and Widtsoe, but was dismissive of the others, noting that Barnes "knew nothing about the irrigation work." [78]

The commission held its first meeting on October 15, 1923. In part, Secretary Work created the commission to fend off criticism in Congress surrounding the dismissal of Director Davis. The Department of the Interior had already suffered embarrassment in the wake of the Teapot Dome affair. For the Bureau of Reclamation and its longtime employees, the work

The Special Advisory Committee at Salt Lake Hearings, Left to right: Hon. James E. Garfield, Hon. Thomas E. Campbell, Dr. Elwood Mead, and Dr. John A. Widtsoe.

4.24. Members of the Committee of Special Advisers on Reclamation at a hearing in Salt Lake City. *New Reclamation Era,* March 1924.

of the commission threatened "a shake up" in the conduct of Bureau affairs. Some within Reclamation called it the "Fault Finders Commission" and saw no good coming of its investigations. When Mead returned from Australia in late December, he actively pursued the investigations, traveling widely to projects in the West. He asked many questions of project managers and settlers alike. The commission took testimony from officials in both the Department of the Interior and the Department of Agriculture, including a long session with A. P. Davis. The recommendations emerged over a seven month period, with the final report issued in April 1924.

The Fact Finders' Report echoed many recommendations of a report Mead had made to Secretary Lane in 1915, according to historian Pisani. The Fact Finders called for less emphasis on hydraulic engineering and more on how to settle the projects. The Commission made sixty-six

68TH CONGRESS
1st Session | SENATE | DOCUMENT
No. 92

FEDERAL RECLAMATION BY IRRIGATION

MESSAGE

FROM

THE PRESIDENT OF THE UNITED STATES

TRANSMITTING

A REPORT SUBMITTED TO THE SECRETARY OF THE INTERIOR BY THE COMMITTEE OF SPECIAL ADVISERS ON RECLAMATION

APRIL 21, 1924.—Read; referred to the Committee on Irrigation and Reclamation and ordered to be printed with illustrations

APRIL 24 (calendar day, MAY 2), 1924.—Action of April 21, 1924, rescinded and document ordered to be printed without illustrations

WASHINGTON
GOVERNMENT PRINTING OFFICE
1924

4.25. The report "Federal Reclamation by Irrigation" is popularly known as the "Fact Finders' Report."

recommendations. It advocated classifying project lands, turning over control of projects to water users as quickly as feasible, restricting future settlement to experienced farmers, and extending repayment periods. The report reflected the thinking of agricultural economists who emphasized farm efficiency, land-use planning, and even social engineering.[79]

Shortly after issuance of the Fact Finders' Report, on April 3, 1924, Elwood Mead succeeded D. W. Davis as Commissioner of Reclamation. The rapid turnover in leadership certainly had something to do with the withering criticism heaped upon Secretary Work over his replacement of an engineer at the head of Reclamation with an Idaho politician. Davis assumed a position as "Director of Finance" overseeing settler repayment. If the Fact Finders' Report can be taken as a guide to Mead's impending administration, it suggested that the Bureau of Reclamation would pay more attention to the practical affairs of making a living on the projects. These involved farming methods, soil studies, land classification, crop selection, access to agricultural loan programs, and a screening of those permitted to take claims on the projects. All of this was a far cry from the original tasks of hydraulic engineering. The new directions accurately mirrored Mead's vision, gleaned from a career in developing planned irrigated agricultural settlements in California, Australia, and the Middle East, where Mead offered advice to Jewish resettlement groups in Palestine. These latter assignments called for the creation of communities as well as irrigation systems.[80]

4.26. Reclamation published this portrait of Elwood Mead in the May 1924 issue of *New Reclamation Era*.

For the government to wash its hands of the economic problems facing settlers on the projects and let the forces of nature and economics take their course might well pose the risk of abandoning the now almost

$200,000,000 invested in irrigation by Congress. The best course appeared to be accommodation. Professor Pisani writes: "Having made an economic commitment to develop the West, the federal government could not back out. Simply to stay afloat, federal reclamation would have to go far beyond the objectives of the Reclamation Act of 1902."[81] Still, a penurious Congress imposed limits and was in no mood to move aggressively in expanding the scope of reclamation. It was willing to grant further repayment delays, and, toward the end of 1924, Congress enacted the "Second Deficiency Appropriation Act" that extended all repayments to twenty years. Moreover, charges were now based upon land production values which permitted inferior lands to pay lower construction charges than productive land. Two years later, in 1926, came the "Omnibus Adjustment Act" which entirely excluded nonproductive agricultural lands from the repayment schedule. It extended the repayment period of twenty years in the previous act to a maximum period of forty years, provided that settlers agreed to sell any lands they owned in excess of the 160 acre limit.

While Congress was generous in the revision of repayment provisions, it took little interest in the other recommendations of the Fact Finders' Report, such as loans, education, and community development. Therefore, the reclamation effort remained rudderless in the stymied agricultural economy of the mid-1920s. Yet, other plans were afoot.

The Reclamation Service had long studied the possibility of developing the lower Colorado River. And from the Pacific Northwest came voices advocating the utilization of the power and waters of the Columbia River. Both rivers presented opportunities for the Bureau of Reclamation. In 1922, a major hurdle had been overcome when the states on the upper Colorado River signed the Colorado River Compact, agreeing to a division of water and power between upper and lower basin states, should a large dam be built. This offered protection to the upper Colorado River Basin states against threats by interests in the Lower Colorado River Basin states (most prominently California interests) to claim the lion's share of the river's water through dam and diversion projects. These water uses could assert irrevocable water claims on the basis of the West's long standing prior appropriation water law legacy.

Endnotes:

[1] Lampen, *Economic and Social Aspects,* 54-5, 61-2.
[2] U.S. Department of the Interior, U.S. Reclamation Service, *Eleventh Annual Report of the Reclamation Service, 1911-1912* (Washington, D.C.: Government Printing Office, 1913), 1-3.
[3] Carpenter, *The Forging of Bureaucratic Autonomy*, 344.
[4] A. P. Davis, "Memoirs of Reclamation Service Abridged, 1913-15," typescript, Davis Papers, Box 13, 4-5.
[5] F. H. Newell to Project Engineers and Managers, November 12, 1912, RG 115, Entry 3, Box 251, NARA, Denver.
[6] F. H. Newell to Project Manager, Sunnyside, Washington, December 10, 1912, RG 115, Entry 3, Box 241, NARA, Denver.
[7] "They Have 'em in Australia, Too," *Reclamation Record*, 6 (January 1915), 20.
[8] George W. Norris to Secretary of the Interior, January 5, 1914, RG 115, Entry 3, Box 88, NARA, Denver; A. P. Davis, "Memoirs," Davis Papers, 5-6.
[9] A. P. Davis, "Memoirs," Davis Papers, 7; Pisani, *Water and American Government,* 117; Franklin K. Lane to A.P. Davis (December 17,1914), RG 115, Entry 3, Box 102, NARA, Denver.
[10] "Abstract of Hearings before Secretary Lane (May 1-17, 1913)," RG 115, Entry 3, Box 193, NARA, Denver; "Activities of Water Users' Associations," *Reclamation Record,* 5 (February, 1914), 44-8; Roy E. Huffman, *Irrigation Development and Public Water Policy* (New York: The Ronald Press Company, 1953), 83.
[11] William S. Cone, Superintendent of Construction, "The Salt River Valley Power Situation," *Reclamation Record*, 5, (May 1914), 176.
[12] "How National Forest Administration Benefits Water Users: The Relation of Grazing to Stream Flow," prepared by the Forest Service, *Reclamation Record*, 7 (May 1916), 223-4.
[13] William D. Rowley, *U.S. Forest Service Grazing and Rangelands* (College Station: Texas A & M University Press, 1985), 64; F. H. Newell to Morris Bien, February 26, 1914, RG 115, Entry 3, Box 86, NARA, Denver; C. J. Blanchard, "Current Comments from Projects: Idaho, Minidoka," *Reclamation Record*, 5 (March 1914), 80, notes that only cattle owned on the Project will be admitted to the Minidoka National Forest by agreement with the Forest Service.
[14] "Comparative Cost of Public and Private Irrigation Projects," *Reclamation Record*, 5 (February 1914), 56.
[15] Pisani, *Water and American Government*, 117.
[16] Pisani, *Water in American Government*, 116.
[17] "Minutes of Conference of Supervising Engineers and Others of the Reclamation Service at Lake Tahoe, California, September 4-8, 1913," RG 115, Entry 3, Box 103, NARA, Denver; F. H. Newell to H. B. Walker, Kansas State Agricultural College, January 9, 1913, RG 115, Entry 3, Box 251, NARA, Denver; Pisani, *Water and American Government*, 116; Kluger, *Turning on Water*, 71.
[18] Newell, "Memoirs."
[19] Pisani, *Water and American Government*, 121.
[20] Franklin K. Lane to A.P. Davis (December 17,1914), RG 115, Entry 3, Box 102, NARA, Denver.
[21] Franklin K. Lane to A. P. Davis (December 17, 1914), RG 115, Entry 3, Box 102, NARA, Denver; Pisani, *Water and American Government*, 118-9; Golzé, *Reclamation in the United States*, 27; Kluger, *Turning on Water*, 75.
[22] "Report of Central Board of Review," *Reclamation Record*, 7 (July 1916), 298; see also Central Review Board to Franklin K. Lane, September 27, 1915, Carton 18, Elwood Mead Papers, Bancroft Library, University of California, Berkeley, hereafter cited as Mead Papers.

[23] Pisani, *Water and American Government*, 274; H. G. Tyson, Jr., "A Reply to Comptroller Ryan's April Article," *Reclamation Record*, 5 (June 1914), 211-3; Abel Ady, "The Extension Bill," *Reclamation Record*, 5 (September, 1914), 306.
[24] C. B. Randolph, Omak, Washington, "Cooperative Farm Organizations," *Reclamation Record*, 5 (December 1914), 470; George W. Lee, "Cooperation on the Okanogan Project," *Reclamation Record*, 7 (December 1916), 557-9; C. J. Blanchard, "Co-operation on Reclamation Projects," *Reclamation Record*, 5 (February 1914), 42.
[25] Wayne D. Rasmussen, *Taking the University to the People: Seventy-Five Years of Cooperative Extension* (Ames: Iowa State University Press, 1989).
[26] Townley, "Soil Saturation Problems," 280; Hurt, *American Agriculture*, 263; *Maxwell's Talisman* 6 (July 1906), 5-6 [a publication of George Maxwells's National Irrigation Association largely supported by railroad money]; R. E. Bassett, Secretary of Immigration, Elephant Butte Water Users' Association, "Advertising and Immigration Work," *Reclamation Record*, 5 (October 1914), 374-5.
[27] P. W. Dent, "Community Ditches: Their Rights and Powers," *Reclamation Record*, 6 (August 1915), 371-4.
[28] U.S. Department of the Interior, U.S. Reclamation Service, *Fifteenth Annual Report of the Reclamation Service, 1915-1916* (Washington, D.C.: Government Printing Office, 1916), 11; Pelz, *Federal Reclamation and Related Laws Annotated*, volume I, 197.
[29] C. P. Wickersham, Tieton Project Farmer, "Advice From an Optimist," *Reclamation Record*, 5 (September 1914), 323-4.
[30] B. E. Stoutmyer, "Irrigation Districts: Their Relation to the Reclamation Service," *Reclamation Record*, 5 (October 1914), 861-3.
[31] *Fifteenth Annual Report*, 11.
[32] John Wesley Powell, "The Irrigable Lands of the Arid Regions" *Century Magazine*, 39 (March 1890) identified "duty of water" as the amount of water needed to serve an acre of land and different crops or uses could have different duties for water to perform; other factors bearing on the "the duty of water" might be soil types, slope, and climate, 777; E. B. House, "The Duty of Water," *Reclamation Record*, 5 (October 1914), 355; "Increase the Duty of Water," *Reclamation Record*, 5 (May 1914), 160.
[33] T. N. Carver, Director of the Rural Organization Service of the Department of Agriculture, "The Probable Effects of the European War on American Agriculture," *Reclamation Record*, 5 (November 1914), 405-6.
[34] S. B. Williamson to Franklin K. Lane, April 7, 1915, Entry 3, RG 115, NARA, Denver.
[35] *Fifteenth Annual Report,* 771-3.
[36] Kluger, *Turning on Water*, 85-101; Elwood Mead, *Helping Men Own Farms: A Practical Discussion of Government Aid in Land Settlement* (New York: The Macmillan Company, 1920); Elwood Mead to A. P. Davis, July 29, 1915, Box 6, Davis Papers.
[37] Elwood Mead to Secretary of the Interior Franklin K. Lane, December 6, 1915; A. P. Davis to Senator Henry F. Ashurst, October 6, 1915; Will R. King statement, December 29, 1915, RG 115, Entry 3, Box 6, NARA, Denver.
[38] Brian Q. Cannon, "Water and Economic Opportunity: Homesteaders, Speculators, and the U.S. Reclamation Service, 1904-1924," *Agricultural History* 76 (Spring 2002), 199; Pelz, *Federal Reclamation and Related Laws Annotated*, volume I, 177, 183.
[39] Pamphlet by Fulton H. Sears, "A System of Rural Credits adapted to Federal Reclamation Project," (Fallon, Nevada: 1916), RG 115, Entry 3, Box 6, NARA, Denver; U.S. Department of the Interior, U.S. Reclamation Service, *Sixteenth Annual Report of the Reclamation Service, 1916-1917* (Washington, D.C.: Government Printing Office, 1917), 13-4.
[40] Golzé, *Reclamation in the United States*, 27; Lampen, *Economic and Social Aspects*, 69.
[41] Kluger, *Turning on Water*, 74-7; Elwood Mead to Franklin K. Lane, December 6, 1915, RG 115, Entry 3, Box 6, NARA, Denver; Brookings Institution, *U.S. Reclamation Service*, 68.

[42] "OFFICES," circular, November 22, 1915; Franklin K. Lane to A. P. Davis, November 22, 1915, RG 115, Entry 3, Box 102, NARA, Denver; F. H. Newell to A. P. Davis, February 7, 1916, Box 6, File - "Frederick Newell," Davis Papers.
[43] A. P. Davis to Chief of Construction in Denver, December 16, 1916, RG 115, Entry 3, Box 102, NARA, Denver; Rocca, *America's Master Dam Builder*, 55.
[44] J. B. Lippincott to Francis G. Newlands, November 2, 1915, Davis Papers, Box 6.
[45] "Description of Reclamation Exhibit, Panama-Pacific International Exposition, San Francisco, Cal.," *Reclamation Record*, 6 (June 1915), 257-8.
[46] A. P. Davis to Secretary Franklin K. Lane, July 27, 1916, RG 115, Entry 3, Box 102, NARA, Denver; Blanchard's work for agencies other than the Reclamation Service began as early as 1903. Pinchot made arrangements with Newell to do an article on the forest reserves for the Idaho newspapers, Harold K. Steen, ed., *The Conservation Diaries of Gifford Pinchot* (Durham, North Carolina: Forest History Society, Pinchot Institute for Conservation, 2001), 108.
[47] Pisani, *Water and American Government*, 121-2, 124; F. G. Hough to Mr. Cole, December 13, 1915, Box 13, file, "Memoirs of Reclamation Service, 1913-1916," Davis Papers; *Sixteenth Annual Report*, 16; U.S. Department of the Interior, U.S. Reclamation Service, *Twentieth Annual Report of the Reclamation Service, 1920-1921* (Washington, D.C.: Government Printing Office, 1921), 8.
[48] Elwood Mead to A. P. Davis, August 20, 1917; A. P. Davis to Elwood Mead, August 31, 1917, Davis Papers, Box 6, folder, "Elwood Mead"; A. P. Davis to Secretary of the Interior, July 20, 1917, RG 115, Entry 3, Box 102, NARA, Denver.
[49] James, *Reclaiming the Arid West*, xx, 12.
[50] "The Hospital at Neuilly, France," *Reclamation Record*, 8 (September 1917), 406; "American Ambulance Hospital in Paris (Neuilly)," *Reclamation Record*, 8 (November 1917), 503-4; "Increased Food Production on Reclamation Projects," *Reclamation Record*, 8 (December 1917), 547; Kluger, *Turning on Water*, "Agriculture Department officials blocked the plan. They not only objected to another department going into farming, but they questioned whether such an activity was a 'desirable one for any branch of the Government,'" 80-1.
[51] Frank Weymouth to A. P. Davis, November 17, 1917, RG 115, Entry 3, Box 102, NARA, Denver; U.S. Department of the Interior, U.S. Reclamation Service, *Seventeenth Annual Report of the Reclamation Service, 1917-1918* (Washington, D.C.: Government Printing Office, 1918), 441; *Sixteenth Annual Report*, 7, 17; U.S. Department of the Interior, U.S. Reclamation Service, *Eighteenth Annual Report of the Reclamation Service, 1918-1919* (Washington, D.C.: Government Printing Office, 1919), 10, 25.
[52] Kluger, *Turning on Water*, 79-80.
[53] Lampen, *Economic and Social Aspects*, 55.
[54] A. P. Davis to Gifford Pinchot, May 5, 1918, Davis Papers, Box 6, Folder - "Gifford Pinchot Corres."
[55] D. W. Cole, "Extension of Time for Payment of Water Charges," *Reclamation Record*, 9 (July 1918), 329-30.
[56] *Fifteenth Annual Report*, 46; *Eighteenth Annual Report*, 41-2; Brookings Institution, *U.S. Reclamation Service*, 56-7.
[57] Smythe's memorandum cited in Pisani, *Water and American Government*, 124-25.
[58] C. J. Blanchard, "Give Our Fighting Men Their Opportunity on the Land," *Reclamation Record*, 10 (February 1919), 50; Senate, *Federal Reclamation by Irrigation*, 97 as cited by Brian Q. Cannon in unpublished manuscript, "Creating a 'New Frontier Opportunity: World War II Veterans and the Campaign for Western Homesteads," read at the Pacific Coast Branch of the American Historical Association, Honolulu, Hawaii, (August 2004).
[59] Memorandum for Director A. P. Davis from W. I. Swanton reporting on Secretary Franklin K. Lane's testimony to Congress, January 9, 1919, R.G. 115, Entry 3, Box 93, NARA, Denver; *Fifteenth Annual Report*, 19-21.

[60] Davis to Mead, May 31, 1919, Box 6, File "Elwood Mead," Davis Papers.
[61] U.S. Congress, House, Committee on Irrigation and Reclamation, *Settlement of Returning Veterans on Farms in Reclamation Projects*, 79th Congress, 1st session, April 12, 1945, 97.
[62] Andrew Weiss, "Special Report on 80 Soldier Entries of March 5, 1920, Fort Laramie Division, North Platte Project," unpublished typescript, 1924, 151-52, RG 115, Box 846, General Administrative and Project Records, 1919-1945, NARA, Denver, cited in Cannon, "Creating a New Frontier of Opportunity," 3.
[63] Arthur P. Davis, "Results of National Irrigation," *Reclamation Record*, 10 (December 1919), 546-7; as noted by Pisani, *Water and American Government*, 124; see also Davis's statement in U.S. Congress, House, Committee on Appropriations, Hearings on the Reclamation Service before the Committee on Appropriations, December 23, 1921, 67th Congress, 2nd session (Washington, D.C.: Government Printing Office, 1922), 608-23.
[64] Pisani, *Water and American Government*, 132; *Twentieth Annual Report of the Reclamation Service*, 8; Gates, *History of Public Land Law Development*, 673.
[65] Pisani, *Water and American Government*, 129-30.
[66] J. B. Beadle, Assistant to the Director, to A. P. Davis, December 7, 1921, RG 115, Entry 7, Box 177, NARA, Denver.
[67] A. P. Davis to George Norris, February 1, 1922, RG 115, Entry 7, Box 177, NARA, Denver.
[68] Addison T. Smith to Albert B. Fall, December 5, 1922, RG 115, Entry 7, Box 177, NARA, Denver.
[69] A. P. Davis to Francis M. Goodwin, Assistant Secretary of the Interior, April 12, 1922, RG 115, Entry 7, Box 177, NARA, Denver.
[70] Pisani, *Water and American Government*, 134, 150.
[71] "Will Not Deprive Striving Settler of Land, Fall Says," *Yakima Daily Republic*, February 6, 1922.
[72] General information on the Teapot Dome affair may be found in Burl Noggle, *Teapot Dome: Oil and Politics in the 1920s* (New York: Norton, 1965).
[73] "Department of the Interior Memorandum for the Press," announced Davis's departure as Director, August 15, 1923, Box 4, Davis Papers; "Davis Memoir," describes his refusal to resign, Box 4, Davis Papers; Davis, "Development of Reclamation," 75-77, Box 9, Davis Papers; Pisani, *Water and American Government*, 138.
[74] Department of the Interior Memorandum for the Press, August 15, 1923, Davis Papers, Box 4.
[75] "Pecos Water Users Association Endorses Reorganization by Secretary of the Interior Work," Davis Papers, Box 4; Stephen Bogener, *Ditches Across the Desert: Irrigation in the Lower Pecos Valley* (Lubbock: Texas Tech University Press, 2003), 214.
[76] Department of the Interior Memorandum for the Press, August 15, 1923, Box 4, Davis Papers.
[77] The "Fact Finders' Commission" became the popular designation for the "Committee of Special Advisers on Reclamation." The committee eventually submitted its report which was titled "Federal Reclamation by Irrigation," but it quickly became known as the "Fact Finders' Report." Find the report at U.S. Congress, Senate, *Federal Reclamation by Irrigation*, Senate Document No. 92, 68th Congress, 1st Session, Serial Number 8238, 1924.
[78] "Development of Reclamation," Davis Papers, Box 9; Hubert Work to Elwood Mead, September 8, and December 14, 1923, Mead Papers, Box 3.
[79] Pisani, *Water and American Government*, 140; Kluger, *Turning on Water*, 11-2.
[80] Kluger, *Turning on Water*, 110-1; Elwood Mead, *Agricultural Development in Palestine: Report to the Zionist Executive* (London, U.K.: Zionist Executive, 1924).
[81] Pisani, *Water and American Government*, 142.

4.27. Using a steam tractor to break up sage lands on the Yakima Project, 1929.

4.28. A government construction camp on the Blackfeet Project in Montana in 1911. Indian construction workers are camped in the background.

4.29. Reclamation built Lahontan Dam on the Newlands Project between 1911 and 1915. This is the left spillway overflowing.

4.30. The Owl Creek Trestle before closing the embankment on Belle Fourche Dam. Belle Fourche Project, South Dakota. May 1909.

CHAPTER 5

CHARTING A NEW FUTURE

Introduction

Elwood Mead presided over a remarkable transition in his years as Commissioner from 1924 until his death in 1936. What occurred was much more than an effort by Mead to tweak the organizational structure to achieve the goals of family farms. With the Boulder Canyon Project Act of 1928 and the subsequent construction of Hoover Dam, the Bureau of Reclamation emerged as a supplier of hydroelectricity and municipal water to the urban West. Reclamation joined American population trends of the twentieth century by following those who fled rural America to the city. A movement originally conceived to revitalize and make possible rural agricultural life in the arid lands of the West now built the cities that depopulated the countryside. The Bureau of Reclamation charted a new course when it lent itself to the expansive urbanization of southern California and to the growth of large agricultural enterprises in that state.

Long a formidable critic of the Reclamation Service, Mead now assumed its leadership. He brought to Reclamation a lifetime of work in arid land irrigation, both in the United States and Australia. From 1899 to 1907, when he headed the Office of Irrigation Investigation in the Department of Agriculture, he had watched with alarm, and probably some envy, as hydrological engineers in the U.S. Geological Survey joined with western irrigation enthusiasts to launch a federal irrigation program in the Department of the Interior. While water from engineering works (dams, canals, and ditches) became the fundamental first step in the irrigation of arid lands, it also followed that the success of the projects depended on agricultural planning and community building. And no one knew more about those subjects than Mead.

From the early days of federal reclamation, Mead expressed misgivings about Reclamation's focus on engineering rather than soils, crops, credit, and markets, all of which successful agricultural communities required. His criticisms opened a wide breach between him and the early leadership of the Reclamation Service. Now, after almost a quarter of a century during which Mead gained a worldwide reputation in irrigation matters, he returned to

Washington to save Reclamation from its critics and from itself. His devotion to *Helping Men Own Farms*, the title of his 1920 book, was now to take him into the realm of building infrastructure for power and water supplies that, in large part, supported the urbanization of the West.[1]

Mead Takes the Helm

Mead was an engineer by training and background, and his outlook was framed by rural midwestern farm life after the Civil War. Born in 1858, he grew up in southern Indiana where he saw freeholds replaced by farm tenancy, with a resulting decline in community and civic spirit. His higher education occurred in the struggling new colleges of the Midwest. From Purdue, he earned an undergraduate degree in 1882, and, after a brief period teaching at Colorado State College, he took a degree in civil engineering from Iowa Agricultural College and returned to Purdue for a Master of Science degree in 1884. Mead believed in the ability of science and engineering to shape the natural environment for the benefit of human society and the "wise use" of commodities and the land. Farmers who owned their own land, he believed, were better farmers and citizens than renters. Mead believed it was incumbent upon the American government, as a matter of policy toward its public domain, to assist in the creation of small farms. This meant, if necessary, "a beneficial paternalism" and the subsidization of irrigation technology and engineering to serve the cause of the common man.[2]

On the other hand, Mead showed remarkable adaptability in his career. According to his biographers, he combined wide experience with a breadth of reading and education that kept him current with developments far beyond the confines of hydrological engineering. In his early work in Wyoming and Colorado, he presided over the development of administrative water law in the West. He quickly grasped the problems that agriculture faced under conditions of irrigation during his work with the USDA and his years working on irrigation projects in Australia and California. His involvement in promoting irrigation colonies drew him to rural sociology and economics. To him, farmers on irrigation projects were "little men" who needed governmental guidance, support, and protection from the big businesses that controlled transportation, credit sources, and marketing. This paternalistic view of rural sociology and economics displayed itself throughout his career.[3] Mead's adaptability moved him beyond the soft idealism of building family farms to the challenges of supplying water and power to factories and the large farms of California's Central Valley. Water for power, water for sprawling municipalities and agribusiness corporations, and water for easier navigation of waterways translated into the hard-headed questions pertaining to the industrialization of the West.

Mead led Reclamation, or should it be said that he followed the lead of the former head of the Reclamation Service, A. P. Davis, into the development of the lower Colorado River with a high dam. From its nadir in the mid-1920s, the Bureau of Reclamation and Commissioner Mead (whether they knew it or not) stood poised to open new vistas in a vast industrialization of western water and rivers in lock step with the expansion of the role of the federal government in American life. Under Mead's leadership, Reclamation (1924 -1936) became, in the words of historian Norris Hundley, "the world's preeminent builder of massive water projects." In the process, despite the persistent rhetoric of Reclamation, the cause of the family farm failed to keep pace with the exciting enterprises of managing great rivers and delivering water and hydroelectricity to a new urban West.[4]

The Fall-Davis Report of 1922, named after the discredited Secretary of the Interior Albert Fall and the Director of the Reclamation Service, laid out comprehensive plans to develop the lower Colorado River. In 1920, Congress had authorized the study to determine the feasibility of a large dam on the Colorado and an All-American Canal to serve the Imperial Valley with the waters of the Colorado.[5] The Fall-Davis Report declared in favor of the All-American Canal. A successful canal, however, would require construction of the world's most ambitious dam in order to harness the rushing waters of the Colorado and create an immense reservoir. (The reservoir eventually assumed the name, Lake Mead). The water that Hoover Dam stored went largely to already existing farms, with a significant portion going to urban development in coastal southern California. The hydroelectric power that Hoover produced, as well as the recreation provided by the lake, served mostly an urban and commercial West. Electrical power sustained the growth of southern California and, eventually, nearby Las Vegas with all of its novelty and detachment from the land. Most of Lake Mead's water did reach the land, but this occurred primarily in the Imperial Valley, where agribusiness eventually predominated. [6]

But this is getting ahead of the story. While the Boulder Canyon dam project proposal marked time in Congress and elsewhere, the unsettled affairs within Reclamation as well as the Department of the Interior prompted speculations in the press. In January 1923, *The New Republic* suggested that Mead would be an excellent choice for Secretary of the Interior, given the pressing need to solve the problems of the Reclamation Service. After President Harding's death in August 1923, President Coolidge chose Work as Secretary of the Interior to replace Fall. Work quickly appointed D. W. Davis as "Commissioner," but his rapid replacement with Mead had an air of desperation about it. Secretary of the Interior Work presided over a Department of the Interior that possessed little public confidence after the resignation of Secretary Fall, and, certainly, Reclamation was adrift after A. P. Davis's departure. In light of the Fact Finders' Report, Secretary Work's appointment of one of the keenest critics of government reclamation and the chair of the Fact Finders' Commission appeared shrewd. The Secretary saw Mead as the best hope to defuse criticism of a bureau that might face bankruptcy if Congress heeded the advice and criticisms of the Department of Agriculture and the eastern press.[7]

Not everyone agreed. Among the engineers who had run the Reclamation Service from its beginning, 1924 was a low point in the internal morale of Reclamation. Engineers and their contributions appeared to be under attack. As one critic put it, the engineers of the Reclamation Service had built admirable structures, but their accomplishments were tantamount to building factories equipped with machinery but with no workers to work in them and no products of value to produce. This view struck close to Mead's own assumption that simply building irrigation works had not and would not attract settlers. More than engineering was required if government reclamation was to be salvaged. "If we are to think the problem of reclamation through and do all that success requires," he said, "we must give as much attention to the selection of settlers, to providing credit for the improvement of farms, and aid and advice to farmers in development as we have been giving in the past to planning and financing of irrigation works."[8]

The disassociation of Arthur Powell Davis from the affairs of Reclamation was another indication of the retreat of the engineers. In late 1924, Davis, in a letter to Mead, remarked that he seldom heard anything about the operations of Reclamation. He related that his life was kept extremely busy with his work as Chief Engineer and General Manager of East Bay Municipal Utility District in Oakland, California, and a recent trip to Turkistan, in the Soviet Union, to advise on irrigation issues.[9] Davis's detachment from a government service to which he had devoted most of his life added to the low morale amongst engineers within Reclamation in 1924.

Major engineering journals also voiced discouragement, particularly during the months when D. W. Davis held the top post from June 1923 to April 1924.

Still, A. P. Davis was not totally detached. In a personal letter to the editor of *Engineering News-Record*, he noted that Secretary Work's sole purpose in appointing the Fact Finders' Commission was to find incompetence in the Reclamation Service, but it failed to reveal the kind of inadequacies the Secretary wanted to explore. In the end, the report of the commission recommended that an engineer head the bureau as had been the practice in the past. The Secretary complied, in the opinion of Davis, and removed the politician D. W. Davis when Mead agreed to serve. Arthur Powell Davis, however, continued to be disappointed. He wrote: "I knew Dr. Mead's qualifications and believed he would have the strength to prevent further malicious action, but it seems I was mistaken." He called Mead's appointment by Secretary Work "a smoke screen" to cover up the Secretary's continuing mischief "in disorganizing the service." He noted that Work's first act was one "of revenge," when he directed that all the engineers in charge of projects have their titles changed from "Project Manager" to "Superintendent of Ditches."

The removal of some of the long-time employees in the top echelons of Reclamation, including Chief Counsel Morris Bien, according to Davis, only made way for the appointment of "incompetents" loyal to Secretary Work. Davis predicted that the next man to leave would be Chief Engineer Frank E. Weymouth. As part of his assault on the engineers, Davis believed, Secretary Work was conducting a systematic campaign to undermine the chief engineer. Should Weymouth be discharged, Davis said there would be "utter collapse of the enthusiasm and efficiency that has heretofore characterized the service." He believed that the Secretary saw the engineers within Reclamation as having greater loyalty to the older and now departed leadership than to the Secretary of the Interior. He was particularly suspicious of what he believed were Secretary Work's attempts to squelch his longstanding recommendations on what to do with the Boulder Canyon Project, namely, build the big dam. The engineers continued to support his earlier studies and reports in spite of Secretary Work's threats to dismiss them if they continued to back his position.

These opinions, of course, came from a former director who resented his forced departure from the Reclamation Service. Admittedly, Secretary

Work was no friend of public power and Davis's recommendation for one big dam appeared to take a major step in the direction of providing public power to the entire Southwest. The fear was that a big dam built with public funds inevitably meant cheap public power and the establishment of a cost of production yardstick to measure profits and the fairness of prices charged consumers by private companies. Secretary Work was not necessarily averse to building power facilities with government funds, but his words and actions opposed the distribution and sale of the power by public authorities. It is safe to say that in the bitter fight that developed in the 1920s between public and private power, Secretary Work generally fell into the camp of private power and suspected grand schemes of river development that included major public power generating plants. [10]

As Davis predicted, Chief Engineer Weymouth resigned in September 1924. The resignation confirmed a growing rift between the engineering profession and Reclamation that developed after Davis's firing. Davis perceived that Secretary Work sought other dismissals among engineers who harbored a continuing loyalty to him. Hurried along by Secretary Work, the departure of the old guard within Reclamation was at hand. The resignations provoked a flurry of comments from engineering publications, partly encouraged by Davis, which attacked Work's policies that even Commissioner Mead seemed to endorse.[11]

When Mead became Commissioner of Reclamation in 1924, he turned his attention to the more pressing problems of the projects rather than to Boulder Canyon. His familiarity with the ills of the Reclamation Service appeared in his report on the Reclamation Service in 1915. It concurred with other commentaries of the time that the Reclamation Service suffered from overbuilding, land speculation, and despairing, resentful settlers. Amidst all these problems, only 7 percent of the land under irrigation in the West could be credited to government reclamation efforts. As he undertook to defend Reclamation, Mead faced a difficult task. His first objective was to temper many of the criticisms he himself had lodged against the Reclamation Service in the past. He now praised the Bureau of Reclamation program for raising land values, creating new business opportunities, and increasing the population of western states and communities. He never doubted the legitimacy of federal reclamation, only the way it had been administered.

Now, the new Commissioner stood ready at last to put his ideas into play. Many believed he would spend his energies relieving settlers of unwarranted debt, scaling back costs, fostering cooperatives, making credit available, and generally offering scientific agricultural advice for the success of the projects. While he successfully pursued many of these goals, larger opportunities presented themselves for the development of water resources beyond irrigated agricultural projects.

Beyond the Fact Finders' Report: The Colorado Question

While a larger future for Reclamation started to take shape on the lower Colorado and even in the Columbia River Basin, the Fact Finders' Report continued to direct Reclamation's attention to the irrigation projects. It pointed to past failures and present problems while recommending more efficient administration, improving life on the projects, and repaying long term loans. All in all, the new/old directions outlined in the Fact Finders' Report were not very exciting when contrasted to growing opportunities about to be offered to Reclamation in the development of river basins in grand, new projects. In 1922, California Congressman Phil Swing and California's U.S. Senator Hiram Johnson began introducing various versions of the Boulder Canyon Project bill in Congress. By 1928, a combination of events resulted in Congress finally approving the project that featured a large dam on the Colorado River. The Boulder Canyon Project was not a complete break with the past. In most respects, it rested upon past accomplishments of the Reclamation Service – dam building, the production of hydroelectricity, and the delivery of water. The difference was in the size of the new structures proposed; the integration of their functions into irrigation, flood control, and urban water supplies; and, most importantly, the focus on the production and sale of hydroelectricity to pay for the project.

5.1. Senator Hiram Johnson of California. Courtesy of U.S. Senate Historical Office.

From a broader perspective, the Bureau of Reclamation assumed the role of a multiple-use, river basin planner and developer. As early as 1907,

President Roosevelt brought together advocates of planned river developments with the appointment of the Inland Waterways Commission. It was another Progressive Era commission that promised much but expired in 1909. The construction of Hoover Dam introduced an era of enormous dams, gargantuan hydroelectricity production, and water deliveries via aqueducts from reservoirs over great distances — to cities as well as farms. Water from Lake Mead reached Los Angeles in 1941 via Parker Dam and the Colorado River Aqueduct. Not only did water delivery occur over greater distances than previously, the transmission of electricity via high-tension power lines became practicable, enabling dams in remote locations to deliver power to distant urban and manufacturing centers. This larger playing field for Reclamation, growing out of its capacity to build huge dams, extended to flood control and navigation of inland waters plus the added bonus of recreation on the waters of huge reservoirs enjoyed chiefly by urban visitors. [12]

Arthur Powell Davis could only regret his absence from this larger playing field starting with the successful development of the lower Colorado River Basin. From the beginning of his career in the late nineteenth century, Davis had studied the Colorado River. As early as 1902, he headed a team in the new Reclamation Service to study the possibilities of the lower Colorado River for development, primarily in terms of irrigation. Davis's departure from Reclamation in 1923 meant he would be denied the opportunity to carry out his expanded vision for development of the Colorado River centered on the construction of one monumental dam that could serve multiple purposes. He had long planned for this event and assumed he was the natural heir to any project that tapped the energy of the Colorado River. His uncle, John Wesley Powell, had initiated the first steps toward conquering the river in the exploration of its canyons in 1869 and 1871-72. Powell's account of the first trip was published in 1875 as *Exploration of the Colorado River of the West*, and it was widely read. By the twentieth century, other developments brought the Colorado to the public's attention, particularly in California. The bed of an ancient inland sea, the Salton Sea, in southern California, into which the Colorado once flowed, became a major issue in the control of the Colorado. Extensive flooding at the turn of the century raised the need for flood control if the Imperial Valley was to prosper.

The Colorado River and Southern California

Opening the resources of the Colorado River to southern California proved to be a complicated matter. In 1913, the Reclamation Service had abandoned a planned reclamation project in the Owens Valley on the east side of California's southern Sierra Nevada in favor of Los Angeles's aqueduct to the city. Now, for southern California to go farther afield in its search for water, to the Colorado River, was more complicated but not impossible.

5.2. The Colorado River Basin showing the upper and lower basins.

The Colorado was the great river of the Southwest. In volume, it did not compare with the Columbia, the Mississippi, or many lesser rivers in the eastern United States. But its limited waters are practically the only available supply in the arid Southwest. Its watershed reaches into seven states: Colorado, Wyoming, Utah, Arizona, Nevada, New Mexico, and a small portion of California. California contributes very little water to the river — the greatest accumulations come from snowpack on the western slope of the central Rocky Mountains in Colorado.

The seven states of the Colorado River Basin all had claims to the great river of the West. It would be a formidable accomplishment to bring all of them to the table, let alone agree on how to share the river. Despite long odds against it, the Colorado River Compact was signed in 1922 through the efforts of Secretary of Commerce Herbert Hoover and many state officials,

5.3. Federal and state representatives at a meeting of the Colorado River Compact Commission, north of Santa Fe, New Mexico, at Bishop's Lodge. Left to right: W. S. Norviel, Commissioner for Arizona; Arthur P. Davis, Director, Reclamation Service; Ottamar Hamele, Chief Counsel, Reclamation Service; Herbert Hoover, Secretary of Commerce and Chairman of Commission; Clarence C. Stetson, Executive Secretary of Commission; L. Ward Bannister, Attorney, of Colorado; Richard E. Sloan, Attorney, of Arizona; Edward Clarke, Commissioner for Nevada; C. P. Squires, Commissioner for Nevada; James R. Scrugham, Commissioner for Nevada; William F. Mills, former Mayor of Denver; R. E. Caldwell, Commissioner for Utah; W. F. McClure, Commissioner for California; R. F. McKisick, Deputy Attorney General of California; Delph E. Carpenter, Commissioner for Colorado; R. J. Meeker, Assistant State Engineer of Colorado; Stephen B. Davis, Jr., Commissioner for New Mexico; J. S. Nickerson, President, Imperial Irrigation District of California; Frank C. Emerson, Commissioner for Wyoming; Charles May, State Engineer of New Mexico; Merritt C. Mechem, Governor of New Mexico; T. C. Yeager, Attorney for Coachella Valley Irrigation District of California. November 24, 1922.

notably Delph Carpenter of Colorado. Arizona, however, refused to sign. Water lawyer Carpenter, from his perspective far up on the headwaters of the river, recognized that an early agreement was necessary to protect water rights of states in the upper Colorado Basin. A compact agreement among states would also avoid endless litigation and prevent California from asserting the widely-recognized western water law of prior appropriation over the waters of the Colorado should California interests (either private or public) move rapidly to build dams and divert water on the lower Colorado River. The *Wyoming v. Colorado* decision by the U.S. Supreme Court in 1922 sustained the argument that states could assert prior appropriation water claims should they or their citizens be first to utilize water. This further confirmed the wisdom of the upper Colorado Basin states in joining the Colorado River Compact.[13]

Southern California interests early saw the possibilities of water from the Colorado. Under the rules of prior appropriation water law, tapping the river's water for agricultural and urban centers in the southern part of the state at an early date could claim much of the river, denying states upstream water that originated within their borders. Southern California land barons, among them George Chaffey, in the 1890s, contemplated the growth of an agricultural empire in the lands of the ancient lake bed called the Salton Sink or the Colorado Desert. At one time the Colorado River had flowed into the sink, but it turned eventually toward the sea, finding a way to the Gulf of California and leaving behind a broad fertile plain that developers named the Imperial Valley of California to avoid such unflattering terms as "sink" or "desert."

To convert the fertile lacustrine soils deposited there in eons past from desert into a bucolic valley, water from the Colorado was necessary. In 1892, Charles Robinson Rockwood saw what others had seen before him. All that was needed for the valley to flourish was a canal through the natural levee the Colorado had built up when it bypassed the Salton Sink and turned on a new course toward the Gulf of California. Headworks breaching the natural levee could divert water down a natural channel, the New River, into the awaiting soils. By 1901, Chaffey's California Development Company (CDC) cut through the natural levee and diverted river water through a canal and the ancient river bed into the Imperial Valley. Settlers immediately appeared in response to a grand publicity drive that advertised the advantages of the valley: fertile land, connections to rail transportation, and water from the Colorado. In 1900, an article by William E. Smythe, in *Sunset Magazine*, sketched a glorious future for the Colorado Delta and its adjacent valleys which both the United States and Mexico would enjoy.[14]

In the background, however, major threats to the valley's water supply portended disaster. The first was legal. The newly established Reclamation Service questioned the right to divert a river for which it already had great plans and over which it claimed a control that rested, in theory at least, on the navigable character and history of the lower Colorado. Another threat was from nature. By 1904, the first four miles of the sixty-mile-long canal that followed a once living channel to the valley became clogged with silt. (Not unwittingly had the Colorado been given its name — red- or reddish-river). This was a river of mud, and the silt contributed to a rich and diversified ecosystem in the sprawling delta streamlets before they reached the Sea of Cortez (Gulf of California).[15] Yet the volume of silt proved the undoing of the first diversion canal from the river to the Imperial Valley. By 1904, the canal had lured 7,000 settlers to the valley. To foil both the government's challenge to its use of Colorado water, and to solve the problem of its clogged canal, the CDC decided to divert water from a point along the river's banks in Mexico. In its rush to make the diversion, the company failed to build an adequate headgate to close off the flow of the river when it flooded.

Diverting water at a point on the river in Mexico presented hazards. The canal from Mexico in its route north to the Imperial Valley flowed at a steeper incline than the grade of the river channel toward its outlet in the Gulf of California. Moreover, the Colorado River carried very heavy flows at the very time the CDC chose to cut the new diversion. The months leading up to the 1905 flood were "a particularly dangerous period for tinkering with the river ... the CDC could hardly have selected a worse time to demonstrate its ineptitude," writes William deBuys in *Salt Dreams: Land & Water in Low-Down California* (1999). The river hit flood-stage during the winter of

5.4. The New River cutting its bed back toward the Colorado River from the Salton Sea during the break of 1905-1906 (see map on page 267).

1904-05 and "monumental volumes of snow and rain far upstream" brought flooding to the lower Colorado in early 1905 and again in November of 1905. The river eroded out the headgates, and all its water drained into the valley forming, again, the large Salton Sea and threatened the livelihood of the entire valley. The Southern Pacific Railroad mounted a gigantic effort to plug the inflow from the river and return it to its normal channel. The Southern Pacific stood to reap revenues from carrying the products of the Imperial Valley and from selling the land it owned there.[16]

To rescue the valley, containment of the river was essential. The railroad employed every available freight train in the Southwest to carry rocks and soil to plug the gaping breech in the banks of the Colorado. It took two years and millions of dollars to push the Colorado back into its channel so it could flow south to its delta on the Gulf of California. Over the next couple of decades, problems with the Mexican canal persisted, including mounting demands from Mexico to share in the canal's water. Soon the cry for an all-American canal became a prominent slogan in southern California. But

5.5. In August of 1906 the Southern Pacific Railroad's Salton Station had already been moved three times because of the Colorado River break sending water to the Salton Sink.

5.6. The Southern Pacific Railroad prepared to close the Colorado River break by building a trestle across the break in the levee. August 26, 1906.

5.7. October 20, 1906, the Southern Pacific Railroad was dumping gravel and rock into the Colorado River break.

Reclamation officials, including A. P. Davis, opposed building an expensive all-American canal within the United States unless it was accompanied by the construction of a major flood control dam on the river. Without a major upstream dam the security of all of the irrigation works, including an all-American canal, could not be guaranteed.

5.8. Still dumping rock and gravel on November 1, 1906, at the Colorado River break.

It was no secret that, under Davis's leadership, Reclamation had long planned for this development. Now, the Imperial Valley and Los Angeles saw common interest in harnessing the Colorado. Reclamation also saw opportunity in building what would become, for a time, the highest dam in the world. The Imperial Valley wanted irrigation water and it could only have it on a secure basis if a large dam guaranteed flood control. While Los Angeles and greater southern California stood outside the flood threat from the Colorado, a big dam's reservoir offered the potential for additional municipal water to augment the supply from the Owens Valley and the hydroelectric power necessary to pump Colorado River water over great distances to Los Angeles. In addition, it would provide power for the city's population and industrial growth that seemed to know no

5.9. Large rock in the Colorado River break as seen from downstream on November 1, 1906.

bounds. The decennial census records indicate the growth: 1900 - 102,479; 1910 - 319,198; 1920 - 576,673; 1930 - 1,238,048. While the figures are not spectacular by twenty-first-century standards, the percentage of growth is.[17]

In response to the immediate problems posed by the Mexican Canal, the Imperial Irrigation District, the successor to the CDC, sent lawyer Phil Swing to Washington in 1917 to seek aid from Congress. In that same year, the Reclamation Service and the Imperial Valley water users joined forces to produce a report demonstrating the feasibility of an all-American canal, but its construction required a water storage facility and dam on the Colorado to regulate the flow of the river. By 1919, Swing represented the Imperial Valley as a congressman from southern California. Even as a congressman, he could not move the Reclamation Service to back an all-American canal without an upstream dam. Swing ran into the determined opposition of the Director of the Reclamation Service, A. P. Davis, when he promoted legislation for the construction of a new canal. Davis wanted no piecemeal effort to address the problems of flooding and a stable water supply to the communities of the lower Colorado. Davis stood for the comprehensive solution — the big dam on the Colorado to meet the multiple problems of flood control, irrigation water, municipal water, and hydroelectric power.

Hurdles for the Boulder Canyon Project Legislation

The Colorado River Compact (1922) removed a major legal obstacle to the construction of a big dam on the Colorado. The U.S. Constitution also required Congress's approval of the Compact, but that would not come until Congress passed legislation to build the dam. If California congressmen had anything to say about it, that day would occur very soon. In 1922, Senator Hiram Johnson of California quickly joined forces with Congressman Swing to introduce the first of a series of Swing-Johnson bills authorizing a large dam on the Colorado. The bill generally followed the recommendations of Reclamation's Fall-Davis Report of 1922. To guarantee multiple benefits to the people of the southwestern states, this report, prepared under Davis's direction, recommended building the highest dam in the world in or near Boulder Canyon on the Colorado River.

Arizona was not impressed. It feared the excessive demands of California upon the power and water of the river. These objections, and

Davis's dismissal in 1923, blunted the efforts of the Reclamation Service to lend its strong support in Congress to the Swing-Johnson bills. Nevertheless, by 1924 Secretary Work called for a general policy to deal with the Colorado River. He wanted no "temporary expedients" and recommended a comprehensive plan of development to meet "immediate necessities and those of succeeding generations." The Secretary's views, as enthusiastically reported in the *New Reclamation Era*, reflected the long-standing position of the Bureau of Reclamation that the development of the Colorado should be comprehensive and center around one enormous dam. This clashed with the vision of a series of smaller dams recommended by private power companies led by Southern California Edison Company. If the enmity of private power companies was the price of obtaining the large, comprehensive project, both the Department of the Interior and the Bureau of Reclamation were prepared to pay it.[18] Despite promises that the sale of hydroelectric revenues would pay off the cost of the dam, Congress was wary of spending money on a far western water project whose major benefits seemed to accrue to only one state. It was also reluctant to involve the federal government in the business of producing electrical power. In the view of President Calvin Coolidge, "After all, the chief business of the American people is business," and government should stay out of the affairs of the economy and business as much as possible. Endorsing this outlook, powerful interest groups formed around private power and mounted vigorous campaigns in this decade against any federal water projects that might lead to the production and sale of power by the federal government. The "power lobby" became one of the major forces that blocked the Swing-Johnson bills. Nor did southern California speak with one positive voice in favor of the Boulder Canyon Project. The influential publisher of the *Los Angeles Times*, Harry Chandler, whose land syndicate owned 830,000 acres of irrigated land in Mexico, feared that the proposed project would curtail water to Mexican lands.

By and large, however, the bitterest rivalry occurred in Congress. The private power camp saw the big dam as a precedent that would depress power prices and result in unfair competition from government. Reclamation, on the other hand, shared the fear of many in the states of the upper Colorado River Basin (who were not in the pocket of private power, i.e., Delph Carpenter) that, if Congress did not act, private power companies and even Los Angeles's Municipal Department of Water and Power would build a series of small dams on the lower Colorado, monopolizing the river.

Mead's Continuing Challenges

In the midst of the struggles over the future development of the lower Colorado, Mead faced the daunting task of shoring up Reclamation finances, conforming to the general policy goals of his boss, the Secretary of the Interior, and implementing recommendations of the Fact Finders' Commission. Mead desperately wanted to provide loans and other forms of agricultural assistance to the farmers on the projects, but delays in repayment of construction costs left Reclamation with little income. The only remedy, he believed, was to insist upon repayment with no further postponements. Of course, special arrangements must be made for the poorer projects, some of which could even be closed and sold off.[19]

Regardless of Mead's views about enforcing repayment discipline upon the projects, Congress did what it wanted. In 1926, it cancelled debt for North Dakota's Williston Project and Buford-Trenton Project. Indeed, the Williston Project was abandoned. It was on this project that the Bureau of Reclamation had owned and operated a coal mine. The coal supplied a powerplant that pumped water from the Missouri River because the fall from the river was insufficient to permit a diversion dam to work properly.

5.10. A Reclamation photographer visited the Williston Project in North Dakota, on July 4, 1910, and took several pictures of the powerplant and coal mine. Reclamation later abandoned the project.

Although Reclamation suffered withering criticism in 1924, Congress voted over seven million dollars for improvements in reclamation facilities and between 1924 and 1926 authorized five new projects and the expansion of three old ones at a cost of $60 million. Even the Fact Finders' Commission had proposed six new projects.[20] A decision by the U.S. Attorney General, however, ruled that the Reclamation Act of 1924 had allowed the Secretary of the Interior to reject recommended projects approved by Congress. Repayments to the Reclamation Fund continued to be unimpressive, and new sources of revenue from oil and gas leases on public lands failed to generate much income for the fund. When Congress voted for projects, it knew that appropriations were to come from the Reclamation Fund. This financial reality placed at least some limits upon capricious congressional actions when it came to approving new projects.[21]

The Fact Finders' Commission recognized that the overly ambitious programs of the early years spelled bankruptcy for the Reclamation Fund. The Reclamation Act directed that revenues from public land sales in the western states be used for construction of projects within the states in which the sales occurred. The result was the start up of numerous projects throughout the West in a short period of time. Projects for every western public land state also met with approval from the political forces on the local and national levels. Hindsight suggests that launching a few model projects would have been better than undertaking a multitude of projects throughout the West. According to the early critics of the Reclamation Service, a few projects of a largely experimental character would have provided lessons and guidance for future projects. But the past could not be remade. The Commission now emphasized that new projects should not be authorized until full information was secured concerning water supply, engineering features, soil, climate, transportation, markets, land prices, the per acre cost of development, and other factors upon which the success of a project must depend. The report's recommendations sounded much like closing the barn door after the horse ran off.[22]

In the first years of his tenure, Mead performed a difficult balancing act between the demand for additional projects and restrictions of Reclamation's limited budget. As the author of much of the Fact Finders' Report, Mead wrote many of his own long-held convictions about reclamation into it. He saw the importance of directing efforts toward improving life on the present projects rather than bringing new projects on line. People already on the projects needed help in terms of farm credit, scientific advice, and community improvements. The land grant colleges and the extension services, as well as the Bureau of Reclamation, could make common cause during the extended farm depression of the 1920s. If better ways were found to forecast what crops to plant and how to get them to market, farmers would receive a better return. This was true for all farmers, not just farmers on reclamation projects. It became increasingly clear that the reclamation program had drawn the federal government deeper into the farm crisis of the 1920s. In the end, even efficiency was a false promise. More commodities for already saturated markets spelled only further trouble. Agricultural economist Rexford G. Tugwell remarked in 1928 that, "Efficiency digs its own grave."[23]

Still, many looked to economics for clues to solving the farm depression. When it came to applying economic analyses to Reclamation's experience, its shortcomings in planning presented a painful picture. Mead had criticized the Reclamation Service in the past, but he did not appreciate the economic critiques of the organization he now headed. While Secretary Work's dismissal of A. P. Davis and his replacement with a non-engineer, D. W. Davis, brought criticism from the engineering community, the economists now took their turn at finding fault. Dispirited, Mead observed that a group of professionals, i.e., economists, whom he hoped would help him reform Reclamation seemed to be turning into destructive critics.

Some were associated with the Department of Agriculture. In the 1920s, the Secretaries of Agriculture criticized western reclamation and the cultivation of new lands because the program seemingly contributed to crop surpluses. In a study for the USDA that appeared in late summer 1924, R. P. Teele, the associate Agricultural Economist for the Bureau of Agricultural Economics declared, "From the standpoint of the investor in reclamation enterprises, reclamation by irrigation in the United States had always been a failure." By enterprises, Teele meant private investment, and he concluded that the inability of private capital to profit from irrigation enterprises was the principal reason government embarked on a program of arid land irrigation. Even Mead admitted that, as the history of the national Reclamation Act unfolded, "the act has been largely used as a lifesaver for bankrupt private projects."[24]

From Teele's point of view, the act seemed to be doomed as a losing proposition from the outset. As an economist, he objected to the subsidies to public land irrigation projects in the form of free land and no interest charges on the cost of irrigation works. The land subsidy was immaterial because, without the irrigation works, the land was of little value. The no-interest provision on the repayment of the irrigation works and the lengthening periods of repayment offered by Congress made the subsidies even

5.11. Looking north from the outlet tower of Hondo Reservoir in January 1907. Reclamation later abandoned the project.

larger. All of this represented a serious loss to the public purse because the Treasury must pay interest on the money it extended to Reclamation. Teele gave full attention to total cost and said that the cost of these works to the government "included interest as well as the money actually spent on the work." For each project, he provided figures of reported investment to build the project (which were to be paid from sales of public lands) and then what he considered a "corrected investment" to reflect the interest at 4 percent. In many cases, the cost was half again as much and close to double the reported investment. He also noted the abandonment of the

5.12. The interior of the Garden City Project pumping plant, subsequently abandoned.

Garden City Project, in Kansas, and the Hondo Project, in New Mexico, with the imminent closure of the Williston pumping project in North Dakota. Expenses on these projects must be written off by Reclamation because there was no possibility of recovering the investment except for the salvage price when the government abandoned the projects.[25]

The Reclamation Extension Act of 1914 closed off all hope of financial viability for the reclamation program, according to Teele. Extension of the repayment period from ten to twenty years, and a provision that delayed the time when repayment should begin, defied sound economic policy. Teele took a dismal view of agriculture on reclaimed lands because of the cost: "The cost of agricultural production on reclaimed land is not limited to the cost of reclamation works, but includes the cost of establishing new farms, as well as the cost of all community improvements and institutions, such as railways, highways, schools, churches, etc." He charged that Reclamation publications did not present valid statistics on expenditures and benefits: "It is not uncommon to see comparisons between [the] cost of reclamation works and land values or gross crop values, with the implication, if not the statement, that these values can be credited to reclamation alone, and that the difference is a measure of the profit of reclamation." In reference to the state projects in California with which Dr. Mead had been associated, Teele noted that providing loans for farm development aided in bringing land into production, but he cautioned, "This will not be effective if reclamation works are over-built to any considerable extent." The term "over-built" referred to bringing lands under ditch for which markets could not provide adequate revenue to justify the crops produced, especially in the depressed markets of the 1920s. World War I caused the overdevelopment of American agriculture and the problem now was to decrease production. Under these conditions, the goal should be shutting down projects, "unless some strong reasons for departure from this general policy are shown." Teele observed a disturbing difference between private and public investment in reclamation. Under private initiative, funds usually became available when the need for work increased. "On the other hand, the demand for Government construction comes when it appears that it will not pay anyone else to do it," wrote Teele. Under these circumstances, reclamation occurred because government subsidized it. Economists of Teele's stripe opposed subsidies because they encouraged the inefficient use of resources and diverted resources from more productive enterprises.[26]

In an attempt to present a balanced picture, Teele listed the advantages and disadvantages of reclamation under subsidized conditions:

- It makes possible the selection of arenas to be reclaimed, in accordance with an established policy of expansion of the agricultural areas.
- It makes possible the obtaining of funds at low rates of interest, and thus decreases the cost of reclamation.
- It gives assurance as to the sufficiency of water supply, stability of reclamation works, and ability to carry out any contracts made.
- It assures leniency in collections in case of adversity, an advantage not unmixed with disadvantages.

The disadvantages, however, seemed to outweigh the advantages:

- The selection of areas to be reclaimed is likely to be governed by political considerations resulting from pressure by the local land-owning and business interests directly benefited by expenditure of the funds.
- The demand for Government expenditures for reclamation has little relation to the real need for the land to be reclaimed.
- The ease of obtaining funds is likely to lead to reclamation work when and where it is not needed.
- There are likely to be demands for leniency in collecting, when there is not valid reason for it.
- There is a tendency toward extravagant or unnecessary expenditure, because of the lack of incentive for economy in construction as a means of obtaining profits.

Looking at the entire reclamation program, Teele concluded that "Up to the present time the disadvantages have been much more in evidence than the advantages." And the principal reason for this failure was "the subsidy feature of the present system." Public financing often opened the way for basic mistakes. With publicly financed reclamation went the very real possibility that political rather than economic considerations would determine project conditions. For the economist, these points or issues amounted to a formula for failure.[27]

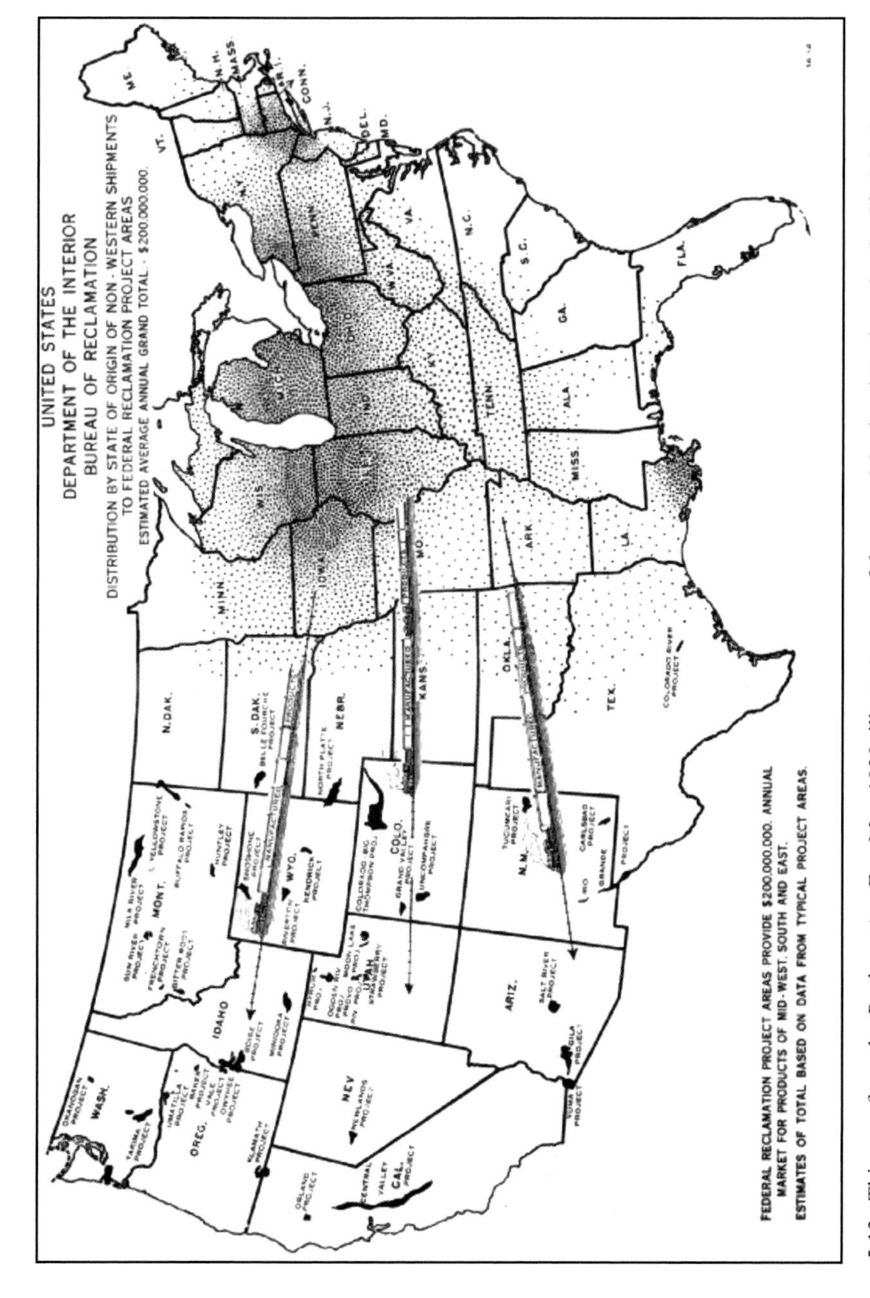

5.13. This map from the *Reclamation Era*, May 1939, illustrates one of the ways Mead and Reclamation justified the importance of Reclamation's programs to the national economy by emphasizing: "Purchasing Power of Federal Reclamation Projects."

By the mid-1920s, critiques of reclamation and its accomplishments were common with the economists leading the charge. The conviction that reclamation could reclaim lost lives, or rejuvenate rural society, or benefit urban workers seemed anachronistic. Nevertheless, as Mead wrote to the Commissioner of Water and Irrigation in Sydney, Australia, in June 1922, "I do not believe in an undertaking by the government if the indirect benefits are not taken into account." The indirect benefits must be factored into any profit and loss sheet, but Mead knew that indirect benefits by themselves could not justify projects. He, too, was upset at the refusal to honor repayment contracts on many of the projects and expressed dismay at the, "Bolshevist attitude that exists on so many of the projects."[28]

In defense of Reclamation, Secretary of the Interior Work dismissed Teele's calculations as, "a waste of effort." In Work's opinion, Teele's paper was an exercise in what might-have-been. Teele's economic analysis offended Mead's missionary zeal and his social goals for reclamation to make homes for small farmers possible . He asserted that Teele refused to consider the expansion of businesses and property values in the western states that, alone, justified the millions of dollars spent by the federal government on the reclamation projects. Teele's study provided the Department of Agriculture with its fundamental critique of reclamation in the 1920s and marked the battleground between Interior and Agriculture until the crisis of the Great Depression.[29]

Mead's Mission to Revive Reclamation

Mead often fell back on the plea that, from the beginning, the national Reclamation Act was an experiment: "No one was to blame. The law was an experiment." He believed that unforeseen conditions prevented settlers from realizing their "agricultural and economic needs" under reclamation. Mead had remodeled Bureau policies. Even the official publication of reclamation, *Reclamation Record*, underwent a transformation with a name change to *New Reclamation Era*. Its coverage shifted to social, economic, and farm improvement topics, and it gave more attention to women on the reclamation projects and to their efforts at social and community improvements. Mead, himself, had always believed that the lack of economic and social assistance "have wrecked some projects," as he put it, "and have created the problems that the Interior Department and Congress are now seeking to solve." He welcomed new legislation that required a thorough soil survey of the lands for each proposed new project, that land should be classified according to earning power, and that annual payments for water rights should be in an amount equal to 5 percent of the average gross crop return. He applauded recent regulations from the Secretary of the Interior that settlers on public land within the projects must have capital of $2,000 and have had at least two years of agricultural or farm experience. He noted that this was the first time that such extensive requirements for settlement on the public lands occurred. The qualification requirements ensured that "whoever undertakes to subdue the desert and bear the burden of its costs should be fitted for the task." Only those who brought resources to the land could enjoy its opportunities for home-makers, and they should have "a lifetime in which to develop and pay for farms and have efficiency and integrity as the watchword in all relations between the water user and the Government." Finally, Mead described reclamation, "as the chief instrument in building up rural civilization in the western third of our country."[30]

The Fact Finders' Report recognized the importance of local control over irrigation projects when it recommended that Reclamation transfer management of the irrigation works to local water users' associations or water districts. By the summer of 1927, with Mead's endorsement, water users' associations operated all or part of sixteen projects. This meant that local people took responsibility for operating the irrigation works and assumed the maintenance costs. The arrangement relieved Reclamation of collecting these operation and maintenance fees. Setting the fee rates and providing for their collection had been a major source of contention. But the government still owned the facilities, and the dams, the gates, and the powerplants — all subject to Reclamation's safety standards via scheduled inspections. Similarly, Reclamation hastened to turn over operation and maintenance of irrigation projects built by it for the Bureau of Indian Affairs. In Montana, this occurred on the Blackfeet, Flathead, and Fort Peck Indian reservations.[31]

As it turned projects over to local associations and limited the number of new projects to be built, the Bureau of Reclamation appeared to embrace the wisdom of downsizing the organization. In response to Congress's approval of six new projects in the spring of 1925, Mead quickly pointed out that the cost of bringing water to them would be higher than on any previous projects. In opposing the projects, Mead fell back on long-standing arguments against irrigation in the arid lands, when he said new investment did not make sense when farms could be developed for far less in the East and under more favorable circumstances for new farmers. He attributed the legislation to pressure from local groups in western states, "biased by their passionate desire for development." To counter the ambitious plans of Congress, the Bureau of Reclamation launched a scaled back ten-year plan in 1926 that projected reclamation income at $97,000,000 annually. As it turned out, even this figure was an optimistic projection. Income from public land sales and mineral leases did not match the spending plans, and repayments fell off with the onset of the Great Depression. In response, Congress granted repayment moratoriums from 1931 through 1933.

Yet the success of Mead's policies cannot be discounted. Project settlers increased their repayments to the Reclamation Fund by over a million dollars from 1928 to 1929. In part, this increase resulted from passage of the Omnibus Adjustment Act of 1926, in which Mead got many of the cost readjustments he had requested from Congress. The Act authorized an economic survey of the operating projects to classify land according to their potential productivity and eliminate unsuitable land from the projects. New contracts enabled repayment over a period as long as forty years, without interest on the investment. In some cases, repayment could be based on a percentage of the average annual crop value, and this extended the payments over an even longer period [ability to pay]. Most important from Mead's point of view, settlers on new land could now be selected on the basis of qualifications of industry, experience, character, and capital. As he put it, "No longer is it possible for a man with a shoestring and a misguided belief that he is a farmer to take up public land on a federal irrigation project. These misfits are weeded out by the project examining boards." Applicants had to possess at least $2,000 in cash or its equivalent in livestock, farming implements, or other assets recognized by the examining board. Mead actually suggested $5,000 to $7,000 as a more realistic figure. On each project there was now an economic adviser to assist in organizing cooperatives for marketing, purchase of farm supplies, and advice on crops to plant. According to Mead, this effort was part of, "placing the economic and social conditions of the projects on as high a plane as the engineering works."

Mead also noted that the 1926 legislation tried to address the problem of speculation on the projects and what he called, "the pyramiding of land prices which have ruined many an otherwise excellent prospect." On the new projects, large areas of land existed in private ownership. This land would now be appraised for its value as undeveloped land without reference to the coming irrigation development. Those values generally ranged from $1 to $15 per acre. Private lands of more than 160 acres could not be sold to settlers at more than the appraised value. Those lands held in tracts of less than 160 acres could be sold for more than appraised value on condition that 50 percent of the selling price in excess of the appraised value be turned over in cash to the irrigation district as a credit to the water right of that particular piece of land. These new rules produced a salutary effect for the "excess" and "incremented value" contracts in holding down the price of land to a figure that a prospective settler could afford. Mead summarized the new measures: (1) thorough classification of the land to be reclaimed under a rating system that classified the land on a 1 to 6 basis with 1 standing for the highest quality, (2) careful selection of settlers, (3) reasonable prices for the land, (4) further extension of the repayment period in some cases, (5) available low interest loans, (6) expert advice on farming operations, and (7) development of cooperative buying and selling.

Because a farmer had 160 acres of land in a project did not mean that the entire acreage could be cultivated and placed under irrigation. The act eliminated construction payments for project land unfit for irrigation agriculture. Overall, construction costs were reduced by almost $14 million, buoying up settler confidence and quieting discontent. The Act also addressed the perennial question of excess lands — land in excess of 160 acres — which already received or had applied for government water. The Act provided for methods of appraisal and sale to avoid destructive land speculation, responding to speculative land prices that had reached levels so high that they discouraged additional settlement and stacked the cards against the success of farmers who had to incur large debt to become established. In this same year (1926), Secretary Work, with the approval of the president, began transferring management (i.e., operations, but not ownership) of the reclamation projects to the water users as soon as "satisfactory purchase contracts were signed." All in all, the 1926 legislation gave Mead what he thought was needed to restore confidence on the projects and gradually build healthy agricultural communities by slowing the pace of any planned expansions in favor of concentrating on the social and economic welfare of the existing projects. Try as he might, Mead never brought under control the practice (denounced by him as an evil) of speculation, which drove land prices ever higher and contributed to the growth of tenancy and farm failures. In 1927, newspaper sources reported tenants farmed one third of the farm units on the projects and many others had been abandoned or foreclosed [See illustration 3.21]. [32]

Yet, in these years prior to the Great Depression, Mead drew great satisfaction from his job as Commissioner. He had intended to stay in Washington only a few years. He was seventy years old in 1928 and required special permission from the Civil Service Commission to work beyond the mandatory retirement age of seventy. His $10,000 a year salary enabled him to live comfortably in Washington, and he and his wife enjoyed the social scene that included frequent invitations from the White House. Mead spent many evenings at the Cosmos Club among the power elite of Washington. Each year he took long tours of western Reclamation projects and traveled abroad for consultations and conferences. He was in Cuba and Haiti in 1926 and, in 1927, he attended a Pan-Pacific conference in Honolulu to explore peaceful relations among nations of the Pacific and economic development. An excerpt from Mead's memo announcing the conference stated:

> It will be an unusual opportunity … to exchange experiences with the other countries around the Pacific in regard to the methods by which unpeopled land may be reclaimed and settled and the prosperity of the people on these reclaimed lands promoted….What the United States Government is doing in reclamation by irrigation will be shown by exhibits and explained by its representatives.
> THE ACCOMPLISHMENTS OF PRIVATE IRRIGATION SHOULD BE AS FULLY PORTRAYED

Later that year, he was in Palestine to study irrigation problems for the Zionist Organization of America. By this time, Mae Schnurr worked in the Commissioner's office taking care of details during his long absences. She served as Acting Commissioner in 1930 and, in subsequent years, when Mead or Assistant Commissioners were away from Washington. By 1929, however,

5.14. Mae Schnurr at her desk at the Bureau of Reclamation.

except for a visit to Mexico City to negotiate an agreement over the water of the Colorado and Rio Grande Rivers, Mead curtailed his travels as he turned his attention to plans for the big dam on the Colorado and problems arising from the Great Depression.[33]

Pondering Dam Design and New Projects in an Era of Constraints

Mead and the Secretary of the Interior managed to pare down the number of new projects proposed by Congress, but authorizations for new dam construction occurred. Mead's priority was to build out and finish existing projects while improving the quality of life of farmers and the profitability of agriculture. The only new projects approved by the Secretary of the Interior were the Owyhee Project in Oregon and Idaho, and the Vale Project, in Oregon. The Owyhee Project included the building of a major dam in the Owyhee River Canyon in southwest Oregon near the Idaho state line. The designs, construction methods, and organization foreshadowed those employed for the larger Hoover Dam. "Owyhee Dam became a proving ground for theories being developed to assist with the design and construction of Hoover Dam," according to the *Project Data* book published by the Bureau of Reclamation. The Trial-Load Method of design used by Reclamation dated back to the late nineteenth century. As early as 1905, an article in *Engineering News* by Reclamation Service consultants George T. Wisner and Edgar T. Wheeler outlined the Trial-Load Method. By applying complicated mathematical calculations in analytical models, they calculated the load factors against a structure's resistance. These calculations allowed the engineers to produce thinner arch designs that de-emphasized the use of massive amounts of material, achieving strength in structural design rather than in mass. The Trial-Load Method, applied to the thin high arch Pathfinder, Buffalo Bill,

5.15. Reclamation used the trial load method to design Gibson Dam on the Sun River Project in Montana, which was completed in 1929. Previously the method was used to ensure the safety of already designed structures like Pathfinder and Buffalo Bill Dams. August 14, 1929.

and Gibson Dams, was refined in Owyhee Dam to combine the structural sophistication of an arch design with the strength of a massive gravity dam. The outcome at Hoover Dam, however, was nothing more than a curved gravity dam in the massive tradition with the Trial-Load Method used to ensure the safety of the dam. Pathfinder and Buffalo Bill dams fall within the structural tradition that emphasized design over mass. They employed the arch to deflect water pressure to the sides of the canyon. Early Reclamation Service/Bureau of Reclamation dams, as well as the later Owyhee and Hoover Dams, were designed to project images of strength and permanency, according to dam historian D.C. Jackson.[34]

The irrigation districts in the Owyhee River region in southeastern Oregon and western Idaho employed a pumping operation from the Owyhee River to irrigate fields, but also saw great possibilities for a dam and reservoir in the deep Owyhee River canyon. For many years, irrigators had petitioned the Bureau of Reclamation for a dam and storage facility. Studies of possible dam sites and dam designs recommended an arch-gravity dam suitable for the narrow canyon. The arch structure was to bear three-fourths of the water load and mass and gravity would handle the rest. Lessons learned from Owyhee Dam proved critical for the future design and construction of Hoover Dam. The thin arch design theory had already been applied in building Pathfinder and Buffalo Bill Dams, and massive concrete/masonry was the rule for early Reclamation Service dam building. For the Boulder Canyon Project, one source notes that "Based upon the Service's experiences with the Roosevelt, Elephant Butte, and Arrowrock dams, it is not surprising that a massive concrete/masonry gravity design attracted the interest of Davis, his Chief Engineer Frank Weymouth, design Engineer John L. Savage, and project engineer Walker Young." Still, as early as 1920, Director Davis showed an interest in a thin arch dam (possibly multiple-arch) across the Colorado in correspondence with European dam engineer and designer Lars Jorgensen, who advocated this design because it combined economic or efficient use of materials (i.e., less) with all the security factors of the massive tradition in dam building design. Most appealing to money-strapped Reclamation, such dams were cheaper to build. [35]

With these two lines of dam design converging in Reclamation in the 1920s, it should come as little surprise that Reclamation engineers attempted to combine thin arch with the massive gravity design for Owyhee Dam. But, ultimately, they chose tradition — the massive gravity design structure with an arch configuration. The arch effect was of little consequence in this design because both Owyhee and Hoover Dams conformed to the basic principles of a massive gravity design. After the failure of St. Francis Dam in March of 1928, Reclamation decided to err on the side of public safety and inspire confidence in the design. Innovation in the construction process, not the design process, marked Owyhee Dam and would be incorporated into the construction of Hoover Dam. Although President Calvin Coolidge approved construction of Owyhee Dam in October 1926, contracts were not awarded until 1928. The system of canals on the project did not reach the remotest irrigation areas until 1939.

5.16. The *Reclamation Era* published a montage of Owyhee Dam images in June 1936. Owyhee Dam was a proving ground for construction techniques used later at Hoover Dam. Note the morning glory spillway in image 4

Projected at 417 feet high, the Owyhee Dam would be the highest concrete dam structure in the world and would employ in its construction a system for cooling concrete that would later be successfully utilized in the building of Hoover Dam. In 28-foot-square sections, 1-inch pipes spaced at 4-foot intervals, with cold river water pumped through them, served to cool the curing concrete, thus preventing heat-induced expansion followed by shrinkage and cracking as it dried and set. This cooling system also facilitated grouting of the sections by providing for fairly uniform, manageable spaces between sections.

However, in contrast to Hoover Dam, hydroelectricity was not a feature of Owyhee Dam. There was no market for the power in southeastern Oregon. Power to build the dam was available from the nearby Idaho Power Company on the Snake River, and, most critically, the Reclamation Fund did not contain the money to build a powerplant that could not show a quick return on the investment. Construction, however, included provisions for power outlet facilities. Generating facilities were not installed until 1985-1993, and then they were installed by water users who sought licenses from the Federal Energy Regulatory Commission. The three powerplants produced only 15,000 kilowatts for local users.

Rivalries Within Government

Both the Department of Agriculture and the Corps of Engineers saw opportunities in the difficulties facing Reclamation. In 1926 the Corps of Engineers identified 180 U.S. rivers in House Document #308 that had navigation and power possibilities. The 308 Document responded to earlier legislation that directed the Corps to identify rivers for intensive study that possessed multipurpose river basin development potential. The Corps's "308 Studies" broadened its scope of activity as it investigated navigation, flood control, and, now, hydroelectric power and even irrigation. The 308 studies increased the possibilities of rivalry between the Bureau of Reclamation and the Corps of Engineers in the management and manipulation of waterways. One of the in-depth studies the Corps chose to undertake was on the Columbia River Basin. The study reached Congress in 1932, and placed the Corps in a strong position to seize the initiative for dam development on that river. Up to this time, the Bureau of Reclamation had shown only passing interest, if not indifference, to the development of the Columbia River despite frequent pleas from local interests in central Washington.[36]

The agricultural depression of the 1920s continued to fuel the Department of Agriculture's smoldering opposition to the Bureau of Reclamation. In 1927, the Secretary of Agriculture, W. M. Jardine, complained that Reclamation had no business bringing new land under cultivation at government expense and with what amounted to "subsidies." To him, a "direct subsidy" for the expansion of agricultural lands was unthinkable at a time when agricultural surpluses had piled up, driving down prices and destroying farm incomes. The eastern press joined the chorus, noting both the failure of projects to pay back loans and the extraordinary gift to western agriculture in terms of interest free loans. Reclamation and Commissioner Mead believed that the Agriculture Department wanted to absorb Reclamation, and in response Reclamation hoisted its usual defenses. It argued that reclamation projects constituted only .5 percent of all the land under cultivation in the United States. Most importantly, many of the crops on the projects were specialty crops that reached markets at times when they did not compete with eastern and midwestern farmers. Also, large government projects were necessary to provide an orderly development to the prime reservoir and power sites in the West. Mead claimed that an estimated 143,000 people living on twenty-four government irrigation projects produced nearly seventy million dollars in agricultural wealth in 1924. There were also over 300,000 people living in towns associated with the projects. He believed that reclamation had saved some western states from total economic collapse. He cited Nevada, with a population of only 42,000 in 1900, as a state rescued by the Truckee-Carson (Newlands) Project. Building the project had stimulated business, created construction jobs, attracted farmers, and established the new town of Fallon. Of course, all these figures were part of Mead's vigorous defense of reclamation against its many detractors who inevitably charged that the government had spent at least $135,000,000 on reclamation with a payback of only $10,000,000 from project farmers.[37]

When Mead defended reclamation as a boon to western economies and small farming enterprises, he appeared to be consigning the Bureau of Reclamation to a narrow future in the service of family farms and irrigated community farm life. This was the traditional rhetoric of federal reclamation, to which he adhered even in the face of impending change. Since the launching of the Reclamation Service, the organization had devoted enormous energy and resources to making its cause known to the public. The *Reclamation Record* became its first publication, to be followed by Mead's *New Reclamation Era* in 1924. Reclamation also had displays at world fairs such as the Panama-Pacific Exposition in 1915. All sought to explain the purposes of federal reclamation and the centrality of the family farm in the mission of federal reclamation. Mead's idealism was no doubt sincere, but events, rather than words, began to shape Reclamation under Mead's leadership. And events carried Reclamation beyond visions of the family farm into a realm of multipurpose large dam building for water and power development.

Reclamation Seeks a Western and Southern Strategy

By the late nineteenth century, Henry Adams saw the newly-invented electric dynamo as the key to modern industrial civilization. From the beginning of the twentieth century, hydropower developed in tandem with the Reclamation Service/Bureau of Reclamation's efforts to develop water resources in the West. By the third decade of the new century, hydroelectricity became the salvation of the Bureau of Reclamation and the vehicle of its expansion. In its earliest construction projects, the Reclamation Service recognized the potential of electrical power from its water storage facilities, but water for irrigation was its highest priority. The priority for hydroelectricity was to generate power for construction: to provide incandescent light, to run motors for concrete mixing, and to move earth. Beyond the construction phase, it saw the possibility of augmenting gravity water delivery systems with electrically-driven pumping stations to lift water into canals and ditches. An early example occurred when the Reclamation Service employed electrical power to supply the "pumping side" of the Minidoka Project in Idaho. Later, before World War II, Reclamation chose a site on the Columbia River for Grand Coulee Dam that required pumping water onto the broad Columbian Plateau as part of the Columbia Basin Project. Farmers on the Salt River Project enjoyed cheap electricity from Roosevelt Dam to pump ground water to their fields, houses, and diversion ditches. This project, alone, told Congress that electricity was a great source of revenue. In Section 5 of the 1906 Town Sites and Power Development Act, Congress authorized the sale of surplus power and directed that resulting revenues pay off the project. When, in 1911, Congress extended project water to private water companies and lands in the Warren Act, the Reclamation Service embraced private irrigation (hitherto shunned) and reaffirmed the practice of hydroelectric power sales. The Salt River Project led the way in taking advantage of this provision in the law.[38]

The pending Swing-Johnson bill, for the big dam on the Colorado, and emerging plans for a Columbia River project in Congress drew a new road map for the future of the Bureau of Reclamation. No one was more eager to follow it than Mead, despite his idealization of the small family farm. With grandiose new projects in the offing, he could move beyond internecine battling with the Department of Agriculture and concentrate on ambitious projects. The road to passage of the Swing-Johnson Bill in 1928 that launched the Boulder Canyon Project, however, had not been easy.

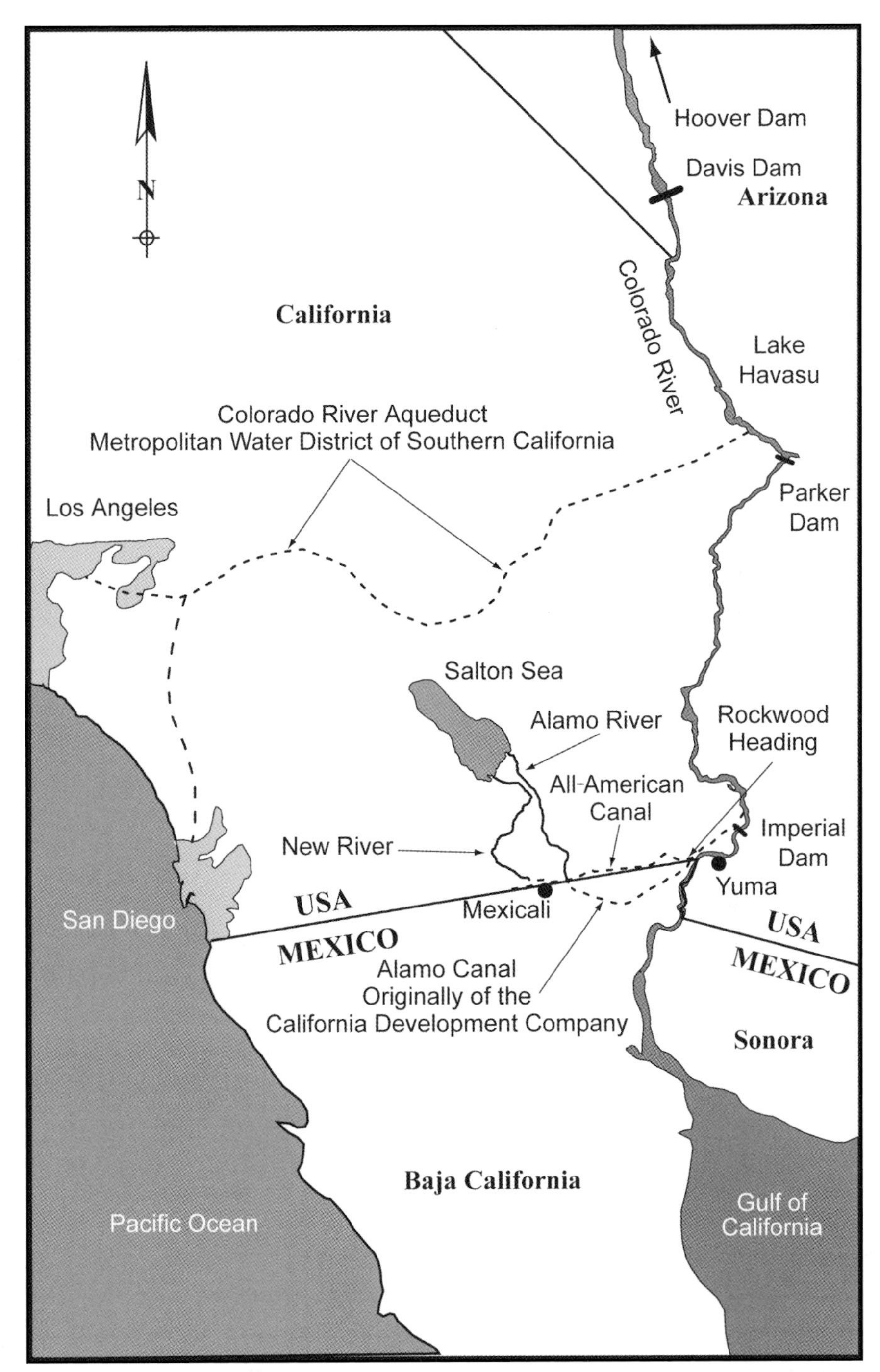

5.17. California stood to gain a great deal when Hoover Dam was in place. With the dam providing flood protection, the California Development Company's Alamo Canal route could be replaced by the All-American Canal, and the Metropolitan Water District of Southern California could plan the Colorado River Aqueduct.

After 1922, Congress considered the project three times before it finally passed. And then Congress granted its approval only after assurances that the stupendous costs would ultimately be reimbursed by the sale of hydroelectricity to the growing southwestern population. Wrangling in Congress, of course, had been preceded by the lengthy negotiations for the achievement of the Colorado River Compact in 1922. And while the compact left several disputes unsettled, namely Arizona's dissent over the amount of water allowed to California, the agreement achieved through the efforts of Herbert Hoover and others appeared firm.

Yet, more immediate events served to push Congress, finally, to approve the Hoover Dam project in December 1928. In the spring of 1927, the Mississippi River wreaked havoc from Cairo, Illinois, to New Orleans in the greatest flood in that river's recorded history. The much-touted levee systems could not withstand the onslaught of the river's high waters. The flood of 1927 killed as many as 1,000 people, drove 700,000 from their homes, and produced property losses of $350,000,000 in 1924 dollars or $4,500,000,000 in 1990 dollars. The waters flooded 27,000 square miles outside the river's channels and touched off a call for aid throughout the country, to which the Red Cross was the principal respondent. President Coolidge took a narrow view of the responsibilities of government, which did not include responses to "acts of God." He resisted the appeals of six governors for federal aid and refused to visit the stricken states. Finally, with New Orleans a "ghost city," the president acted on April 22, 1927. He appointed Commerce Secretary Herbert Hoover to chair a committee of cabinet secretaries to coordinate rescue and relief efforts. Hoover, already known as "the Great Humanitarian" for his work in food relief for Europe in World War I, commanded wide respect. The President also directed Hoover to employ the resources of the Army and Navy to meet the emergency.

Hoover's leadership in the flood relief effort embellished his already shining reputation as a man whose organizational skills saved and uplifted lives. Under Hoover, federal agencies demonstrated a capacity for effective mobilization to meet the catastrophe; the extensive interstate scope of the disaster demonstrated a need for national rather than local and state action. And Congress concluded that it must devise plans to prevent future disasters, including the control of interstate river systems. The African-American population of the Mississippi River Delta suffered the most from the flood. Much of the rescue and relief effort ignored the needs of black people in distress, bypassing them to come to the aid of whites. Hoover and the Relief

Commission rightly received criticism on the issue. More to the point, the Mississippi River Flood of 1927 accelerated change in American society along a broad spectrum from the expansion of the powers of the federal government to speeding up black labor's migration away from the rich cotton producing Delta region of Mississippi to northern cities.[39]

As the effects of the 1927 flood reverberated beyond the Mississippi Valley, backers of the Hoover Dam project took notice. The crisis in the South prompted local support in the South and Midwest for more funds for federal flood control and, more important, broke down the old states-rights arguments against federal power. New congressional appropriations for southern projects increased the chances that support for western water projects might be forthcoming from southern congressmen hitherto indifferent, if not hostile, to appropriations for the further development of western projects.

Although southerners had expressed reservations and even hostility to western reclamation in the early 1920s, Reclamation tried to cultivate southern support by suggesting that it could play a role in revitalizing southern agriculture. Reclamation hoped to resettle farm communities under a program of reclaiming exhausted and abandoned lands and even swamplands. The Department of Agriculture, protective of its southern constituency, usually countered that Reclamation had enough trouble with its western projects and was in no position to undertake work in the South. In fact, at one point the Secretary of Agriculture asked for the closure of the reclamation program because of the failure of many of the projects, the inability of other projects to pay back construction costs, and the persistent charge that the government had no business promoting irrigation in the West when the crops produced could be raised more efficiently in the Midwest or South. The overproduction of staple commodities in those regions and the consequent depressed crop prices compounded the problem. Given the dismal agricultural economy, opponents argued that no new irrigated lands should be opened in any part of the country. Southern politicians opposed the expansion of western projects.

The 1927 Mississippi River flood, however, brought a new consciousness on the part of river communities and people along the Mississippi River of the vital role federal funds played in flood control. Perhaps because its western reclamation projects were under such severe attack, the Bureau of Reclamation continued to look to the South in the late 1920s. After all, reclaiming swampland in the South was cheaper than reclaiming deserts in the West. Mead's experiences in Australia and travels throughout the world left him with an interest in reclamation outside the arid lands. Swamps and cutover lands drew his attention. In 1924, Congress earmarked $15,000 in Reclamation's appropriations to study the development of farm communities in cutover and swamplands. The study committee, headed by Mead, recommended the creation of model farm communities in the South similar to Mead's Durham and Delhi settlements in California. To promote "better rural conditions and a more advanced type of agriculture in the South," Mead called a conference in December of 1927 in Washington, D.C. Representatives from seven southern states met with officials from the Bureau of Reclamation, the Department of the Interior, and agricultural advisors from the USDA. The conference asked for Congress to fund a ten million dollar program to establish demonstration colonies in the South. Congress expressed no interest in the recommendations, particularly with the continued difficulties of the Bureau of Reclamation in the West and, of course, opposition from the Department of Agriculture.[40]

Mead continued to harbor ambitions for the South into the early 1930s. With additional money to study the question, in 1929 he commissioned a report by Dr. E. C. Branson of the University of North Carolina on how to revive the dying rural life of the South. In a work entitled "The Way Out," Branson contended that the southern agricultural crisis was due mainly to tenancy and illiteracy. He concluded that the solution lay in establishing model agricultural colonies that would make farming a profitable business. Such studies offered Reclamation an opportunity to promote community agricultural projects in the depressed regions of the South. Prime farmland of 8,000 to 15,000 acres was easily available for purchase from single owners in the South, noted the Branson article. Some land might have to be drained, and Reclamation should prepare the land completely before establishing group settlements — to avoid the earlier mistakes of reclamation in the West when soil preparation came after settlement. Only land having good soil or soil whose fertility could be restored quickly should be included in the projects. In the new colonies settlers must also be eligible for credit large enough for cooperative purchasing and marketing, and the colonies should be located on good transportation routes, either roads or railroads. By imposing a system of national settlement, Reclamation could enrich the lives of poor people and break the cycle of monoculture (cotton or tobacco), illiteracy, and tenancy that retarded southern agriculture. But the problems could not be solved overnight. According to Branson "The way out lies in model settlements and the contagious example of such settlements throughout a generation or two." Mead's interest in model or demonstration communities in the South grew out of his belief that individual farmers must have "know how" and ambition before projects could succeed. Facilities alone could not insure success. Model communities that served as "contagious examples" must first be established before serious government reclamation could go forward in the South.[41]

The Way to Boulder Canyon

What now interested western reclamation supporters and particularly California advocates of the big dam on the Colorado was Congress's response to the crisis in the lower Mississippi Valley. So great was the crisis of the flood, and so great was the public outcry, that Congress moved rapidly to protect against future catastrophes. Western representatives, led by California, came willingly to the support of the South and the Midwest on the matter, and they demanded reciprocation when western proposals, especially the Boulder Canyon Project, reappeared on the floor of Congress.

Congress had a well-established history of appropriations for river and harbor improvements. Now it stood ready to make the federal government responsible for flood control in the large area of the lower Mississippi River Basin which covered multiple states. Some Congressmen in the upper reaches of the Mississippi Valley wanted more. "More" meant the extension of flood prevention projects to rivers reaching into the eastern and western portions of the Mississippi's watershed. Those outside the Mississippi Valley might have been expected to demand flood protection too, but they bided their time. Congressman Swing of California remarked on the floor of the House that, "Coming from the Imperial Valley, below the uncontrolled waters of the Colorado River, I have an appreciation of the menace of floods as great as anyone in Congress … [but] the Boulder Dam project is not going to be used to embarrass or harass you in the advancement of your legislation." However, he quickly added, "We expect to give to your problem of the Mississippi the same sympathetic and earnest and helpful consideration that we expect you of the Mississippi Valley to give to the problems of other parts of the country when they in turn are presented to Congress." The parsimonious President Coolidge threatened to veto any kind of broader and more expensive effort and Phil Swing knew this. For a time, there was grave doubt about whether he would sign a Mississippi flood control bill. Finally, after much hesitation, he signed the legislation on May 15, 1928, to provide large federal appropriations for the building of extensive flood control measures on the Mississippi. It committed the federal government, according to the *New York Times,* to the greatest expenditure of monies, aside from the Great War.[42]

After approval of the Mississippi Flood Control Act in 1928, California interests bided their time until after the summer congressional recesses to renew efforts for the Boulder Canyon Project. As in previous attempts, Representative Swing backed the bill in the House and Senator Hiram Johnson spurred it on in the Senate. Several factors now favored the bill's passage – namely an impending change in administration and a receptive outlook by representatives from the South and Midwest. In the upcoming presidential election in the fall of 1928, President Coolidge declined to run. The mantle fell to Herbert Hoover, who supported a Colorado River Project. After all, Hoover had negotiated the Colorado River Compact in 1922.

Hoover was a progressive Republican who embraced a positive role for the federal government in economic development and regulation. He accepted the public power provisions of the Swing-Johnson bill in contrast to his opposition to proposals for public power at the Muscle Shoals facility on the Tennessee River.[43] He differed noticeably from the doctrinaire Coolidge wing of the Republican Party that adhered to a separation of government and the economy wherever possible. Hoover's election as president in November added assurance that development on the Colorado River would occur. The new president, however, vainly hoped that interstate rivers could be developed without expanding federal authority. As California water historian Norris Hundley notes, this was not possible. The completion of Hoover Dam marked "the emergence of the government in Washington as the most powerful authority over the Colorado River and, by extension, over other interstate and navigable streams as well."[44]

In May 1928, the House passed the Boulder Canyon Project Act. In mid-December, just two months after the presidential election, the Senate approved its version of the bill and returned it to the House for approval. A major question revolved around the Senate's exclusion of the $25,000,000 needed to construct the All-American Canal for the Imperial Valley from the $165,000,000 earmarked for the project. The Senate's version directed that the $25,000,000[45] should be paid "out of the excess profits" that the water would bring to the agriculture of the valley. This meant reimbursement of the canal's construction costs from the benefited landholders in accordance with the Reclamation Act and permission to use hydroelectric power revenue derived from the waters of the canal to defer construction costs. It also meant that, in accordance with the Reclamation Act, no interest would be due on the repayments. The House wanted to charge off the cost of the canal to the project, "because flood control is now generally considered to be a Federal function," as noted by Congressman Swing during debate over the Senate's version of the bill. This would mean no reimbursement to the government by the farmers of the Imperial Valley.

While Swing preferred the House version, he was not overly averse to having the Imperial Valley pay for the canal. His primary concern was passage of the bill. By accepting the Senate's version of the funding distribution and payback requirement, he declared to critics that "the All-American Canal is now to be paid for wholly and distinctly by the lands which will be benefited … there is no way that it can be made a burden upon the revenues which are secured from the sale of power or power possibilities at the dam." The bill provided that the government would sell power under contract from the dam, and construction would not begin until contracts had been signed to cover the costs of the first phases of construction.[46]

Some House members wanted a provision to protect consumers against profiteering and exorbitant charges in the sale of project electricity. Congress had written a protective provision into the Muscle Shoals bill in the previous session of Congress. Swing assured members that such protection already existed in the wording of the act, which required that the Secretary of the Interior conform to the provisions of the Federal Water Power Act of 1920. In the absence of effective state regulation, the Federal Power Commission had the authority to regulate rates for the protection of the investor and consumer. Swing interpreted this as "a direct mandate in the law that the Secretary of the Interior in the making of his contracts for the sale of power shall protect the ultimate consumer." Congressman Fiorello LaGuardia of New York wanted to know if municipalities would receive preference in obtaining the power, and Swing responded that, "I am quite sure that is safeguarded in this bill." House members resented the fact that the Senate omitted wording in the House bill that declared, "Every contract for electrical energy shall provide that the holder of such contract shall guarantee that in any resale of such energy to the consumer thereof the rate shall not exceed what is fair, just and reasonable, as determined by the Federal Power Commission."[47]

While some House members suggested that the bill be sent back to the Senate with the insistence that these words be included, Swing argued vigorously against delay, saying that "This House and its committees have labored for the past eight years to try to bring into the Colorado River Basin some sort of river control for the safety of human life and property," only to meet frustration in the Senate, and if it were to be sent back to the Senate "for the purpose of changing some phraseology, which I think is sufficiently covered in this bill, then, of course we are simply deferring the matter of the safety of 70,000 people and their property to a time which may take us beyond the present session of Congress." Swing said that the people who lived in southern California, and for him that meant not only the Imperial Valley but the growing city of Los Angeles and its suburbs, were willing to take their chances on this bill in order to safeguard life and property. In effect, the protection of life and property took precedence over the power question.

Safeguarding life and property resonated with southern Californians in 1928. The failure of the St. Francis Dam northwest of Los Angeles on March 12, 1928, a few minutes before midnight, unleashed 12 billion gallons of water into the Santa Clara Valley, with the loss of at least 450 lives. The disaster ruined the career of William Mulholland, Chief Engineer for the Los Angeles Bureau of Water Works, casting a shadow over his achievement as the chief architect of the city's water system and the economic foundation for its phenomenal growth. Mulholland conceived of and engineered an aqueduct and tunnel system that brought water 242 miles from the Owens Valley on the eastern side of the Sierra Nevada to Los Angeles. The water reached southern California in 1913 despite ongoing battles with the residents of Owens Valley. Construction of the St. Francis Dam to store Owens Valley water was to be the crowning event of that project. In the larger picture of the future, not even this was enough for the water moguls of Los Angeles, who eventually looked to northern California and to the east toward the Colorado River for water supplies to sustain the seemingly limitless possibilities for growth in southern California. Under a new state law, Los Angeles and its suburbs formed the inclusive Metropolitan Water District in late 1928, thus authorizing a wider tax base for investment into municipal water development. Former Chief Engineer of the Bureau of Reclamation, Frank Weymouth, was its first Director. He was also the author of the Weymouth Report ("Report on the Problems of the Colorado Basin") from the Bureau of Reclamation to Congress in 1924 that endorsed and outlined plans for the development of the lower Colorado River featuring a strategically located big dam.[48]

The failure of St. Francis Dam, as Congress was about to consider once again the Swing-Johnson bill, raised several questions about large dams. The Secretary of the Interior immediately ordered an inspection of all Reclamation dams. Local project officials welcomed the inspections of dams and reservoirs, but they objected to the charges for the inspections. When the Project Manager of the Truckee Carson Irrigation District in Nevada protested the inspection to the Commissioner of the Bureau of Reclamation, he received a reply noting that the December 1928 contract ceding control of the project to local water users provided that dam safety inspections should be paid by the district.[49]

The collapse and failure of the St. Francis Dam also cast a shadow over the feasibility of building a large water storage project on the Colorado. Critics of the western reclamation program seized upon the collapse as further evidence of the dangers involved in large dams. They believed that the St. Francis catastrophe should persuade Congress to delay consideration of the Boulder Canyon Project. A respected engineering group, the New York-based American Engineering Council, announced its opposition to the Hoover Dam project in April 1928. The project presented too many "unanswered questions," the council concluded. Governor W. P. Hunt of Arizona, a fierce opponent of the Boulder Canyon Project, seized upon the St. Francis catastrophe as evidence that "the engineering world" could not be completely relied upon to guarantee the safety of its dams. The Bureau of Reclamation felt the disaster keenly because it had earlier called upon Mulholland to support the Boulder Canyon Project in testimony before

5.18. St. Francis Dam after failure, 1928.

Congress. It now eagerly volunteered its engineers to join a task force, including Commissioner Mead, to investigate the causes behind the failure of the St. Francis Dam. Congressman Swing immediately recognized the St. Francis disaster as a threat to the passage of the Boulder Canyon Project Act. He ignored it in congressional floor debate. He noted only that the people of southern California were keenly aware of the loss of life and property from floods and that the Colorado River project would offer protection now direly needed.[50]

As returned from the Senate, the Boulder Canyon Project Act offered three alternatives for building the project: (1) build the dam and then lease the power privileges, (2) build the dam and the powerplant and then lease the powerplant, and (3) build the dam and powerplant and operate the power-plant, selling the power at the switchboard. "It would be," Swing said, "engineeringly unthinkable, contrary to all good business judgment, to do otherwise than for whoever builds the dam to build the powerplant." He was certain that the administration would choose to build the powerplant at the same time as the dam because it would save millions of dollars. "It is my personal opinion that the Government will find it good policy to build the powerplant." Swing persuaded Congress that the decision to build and who should operate the powerplant could be left up to the Secretary of the Interior. When the question of negotiating on the waters of the Colorado with Mexico occurred, Swing admitted that a treaty with Mexico was desirable, but that could come after the dam was built.[51]

Differences between the two legislative bodies, in Phil Swing's judg-ment, were "of small importance compared with the beginning of this great project." He saw it as the culmination of his life's work and noted that, "The eight years that have been spent upon it generally might be construed as the best years of a man's life." Swing described the project as one that took water that ran to waste, menaced life and property and turned it, "to serve man's purposes, and all without the ultimate cost to the Government of a single dollar, because in the contracts which must be entered into before any appropriation shall be made provision must be made for repayment to the Government of every single dollar; in other words, this is a proposal for turning a natural menace to a national asset."[52]

Swing's words brought immediate applause from the House. Yet the St. Francis Dam's collapse strongly influenced Reclamation decisions. Reclamation brought in an outside review committee to examine the designs for Hoover Dam, resulting in a dam with a thicker cross section. Hoover Dam was one of the most conservative high dams ever built – a thick gravity dam, not a relatively thin arch dam as originally envisioned. The review committee also recommended that the dam be located in Black Canyon, with its more stable geology, 35 percent larger reservoir capacity, and less foundation excavation. The Bureau of Reclamation, of course, had no connection with the St. Francis Dam's design and failure, but it had to contend with now widespread suspicions about dam safety and assure the public that its newest proposal for a large dam met every specification for strength and integrity.[53]

5.19. June 23, 1929, the Board of Engineers for Hoover Dam posed on a viewpoint above the Black Canyon Damsite. Probably left to right: A. J. Wiley and Louis C. Hill, consulting engineers; Chief Designing Engineer J. L. Savage, Bureau of Reclamation; Chief Electrical Engineer L. H. McClellan, Bureau of Reclamation; Designing Engineer B. W. Steele, Bureau of Reclamation; and Project Construction Engineer Walker R. Young, Bureau of Reclamation.

Along with the pall cast over the project by the failure of the St. Francis Dam, political opposition from Arizona continued. Politicians in that state argued that the bill and the Colorado River Compact were part of a scheme to drain Arizona's water into California. Others, in the Midwest and East, feared that with the development of efficient internal combustion engines and coal and oil plants electrical energy could be produced cheaper by these means. After investing millions in building the dam and the powerplant, the Government "will be found to have purchased a lemon," they worried. Indeed, some historians concur that by the mid-1930s hydroelectric power production was technologically dated when compared to the increasingly efficient methods of producing electricity from fossil fuel (coal and oil) fired plants.[54]

But these voices were clearly in the minority. The states of the upper Colorado believed that agreements to limit California's claim to just 4.4 million acre-feet per year were welcome checks on that state's ambitions. And without the Colorado River Compact, California could preempt a much larger share of the Colorado by prior right use should it choose to move rapidly to build dams on the Colorado. The project bill also required ratification of the Colorado River Compact, which was done without Arizona's participation. Congress agreed the Arizona problem could be dealt with at a future time, as it was when Arizona ratified the Compact in 1944. With these general agreements, the Boulder Canyon Project Act passed both Houses of Congress overwhelmingly, with much support coming from flood-ravaged states in the Midwest and South: Senate, 54 to 11; House, 167 to 122.[55]

5.20. Interested parties at the signing of the Boulder Canyon Project Act on December 21, 1928, were, Commissioner Elwood Mead, Congressman Philip D. Swing of California, President Calvin Coolidge, Senator Hiram W. Johnson of California, Congressman Addison T. Smith of Idaho, and W. B. Mathews of Los Angeles.

The Bureau of Reclamation Wins with the Boulder Canyon Project

Reclamation was "ecstatic" over the task handed to it in the Boulder Canyon Project Act. Although Reclamation may not have fully grasped its multiple-purpose future at this point, a later historian saw the Boulder Canyon Project as a monumental break with the past: "For the Bureau of Reclamation, now emerged from the chrysalis of the Reclamation Service, it was the start of a redefined mission and identity."[56] From this project, Reclamation was to become a major player in the creation of the West's modern hydraulic society. This meant building gigantic high dams that surpassed anything previously built in Reclamation's history; providing hydroelectric power for huge urban conglomerates, while delivering water supplies to the same populations in urban metropolises. The Boulder Canyon Project Act may be seen as the gateway to Reclamation's industrialization and urbanization. By 1929, all the signatory states, except Arizona, had ratified the Colorado River Compact. Construction could not commence, however, before the U.S. Supreme Court rendered its decision in *Arizona v. California* (1931). It overruled Arizona's objections and declared that the federal government possessed the authority to build the large dam under Congress's powers over navigable interstate waters.[57]

In 1930, Secretary of the Interior Ray Lyman Wilbur negotiated over $327,000,000 in electrical contracts for the sale of Hoover Dam power. The figure confirmed what Congress had been promised: the big dam would pay for itself. For Reclamation, hydropower revenues would be a major source of repayment for its future projects. In July 1930, Congress released over $10,000,000 in funds for the beginning of the Boulder Canyon Project as part of yet another Deficiency Act for the Reclamation projects that were suddenly caught in the grips of plummeting agricultural prices and the Great Depression. Elwood Mead now had the authority to move ahead with the Boulder Canyon Project. Earlier investigations (1922-1923) revealed that Black Canyon possessed superior hard rock into which to anchor the big dam, whereas the originally proposed site farther upstream, Boulder Canyon, appeared less favorable. The Boulder Canyon Project Act became a misnomer early on that presaged a struggle over the naming of the dam – now, even the location could not add legitimacy to the name "Boulder Dam." The changed location was not the reason Secretary of the Interior Wilbur renamed it Hoover Dam, but rather to honor President Hoover. As noted earlier, Secretary of Interior Harold Ickes took pains to change the name to Boulder Dam soon after he became Secretary of the Interior. A Republican Congress in 1947, however, changed the name back to Hoover with the consent of Democratic President Harry Truman.[58]

Reclamation appointed Walker "Brig" Young construction engineer in charge of the Hoover Dam project. He had been a long-time engineer with the Bureau of Reclamation and the man who, beginning in 1921, headed investigations for the best site for the dam. Black Canyon's superior bedrock would firmly anchor the dam, and the narrower gorge would require less concrete for a dam calculated to be over 700 feet high – the highest in the world when constructed. The dam would far surpass the 389-foot-high

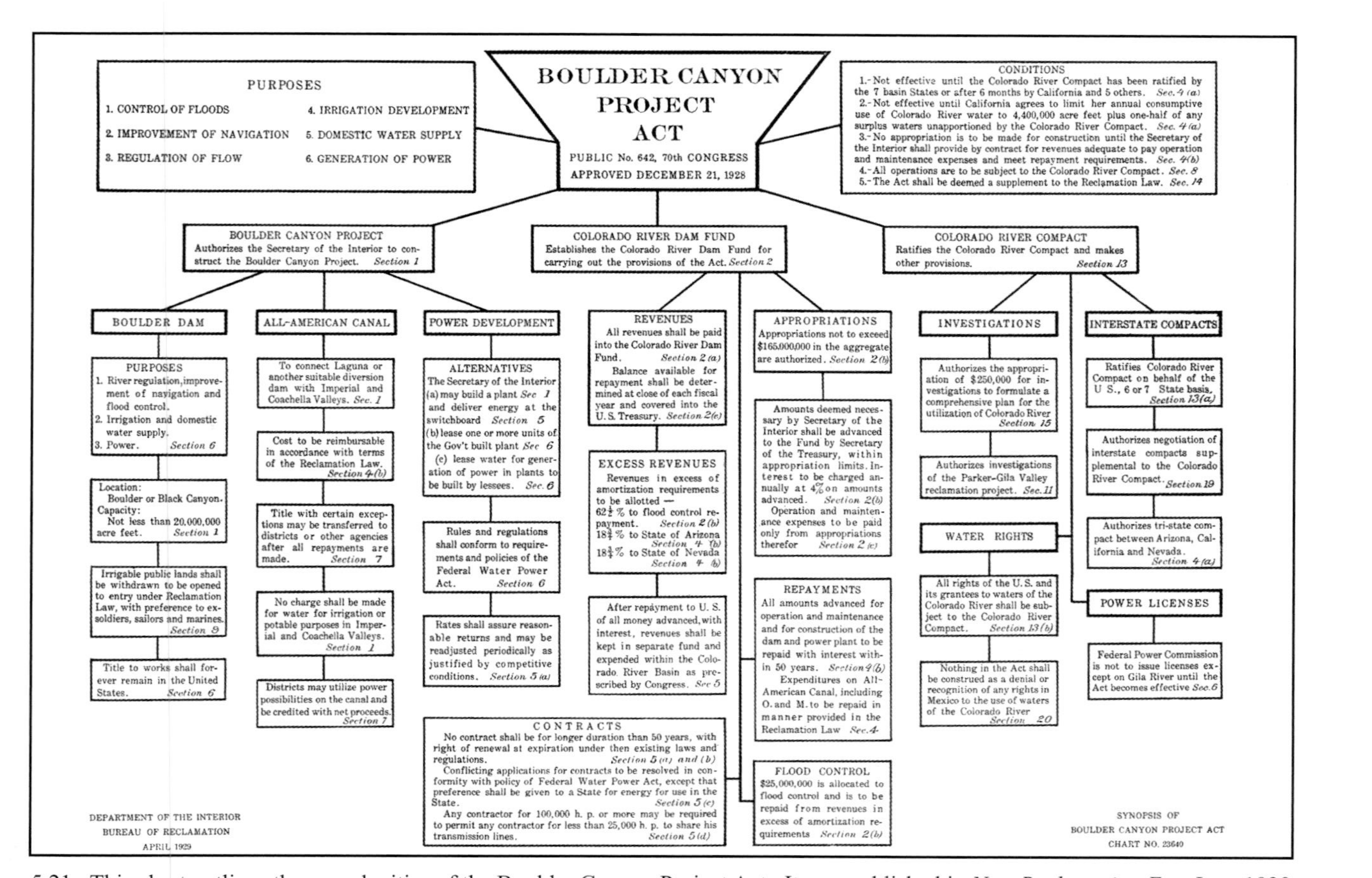

5.21. This chart outlines the complexities of the Boulder Canyon Project Act. It was published in *New Reclamation Era,* June 1929.

5.22. The official portrait of the principal engineers and officials of the Boulder Canyon Project. Left to right, standing: Chief Clerk Earle R. Mills; District Counsel James R. Alexander; City Manager Sims Ely, Boulder City. Seated: Office Engineer John C. Page; Construction Engineer Walker R. Young; Field Engineer Ralph Lowry.

5.23. September 17, 1930, Secretary of the Interior Ray Lyman Wilbur drove a silver spike at "Boulder Junction" on the construction railroad for Hoover Dam.

Diablo Dam, under construction on the Skagit River by the publicly owned Seattle City Light, which was the highest in the world when it was dedicated in August 1930. The first item on Young's construction agenda was the building of a railroad from a point on the Union Pacific line east of Las Vegas to the dam site. Sand and gravel deposits near Black Canyon would facilitate the production of the millions of yards of concrete needed for the dam.[59]

By the end of the 1920s, the Bureau of Reclamation took satisfaction in the revived strength of its program. Now, images of great multipurpose dams glimmered on the horizon and promised new constituencies in urban water and power consumers and downstream communities protected from floods. Recreation was not to be neglected. Commissioner Mead noted in the 1930 *Annual Report* of the Bureau of Reclamation that the plans for building Hoover Dam called for a permanent community to house dam workers. It would, in the future, grow into a popular resort in "this healthful region" on the edge of a great reservoir. He said, "The great lake will stretch away more than a hundred miles through a region of rare scenic beauty."[60] Reclamation now stood ready to offer the West not only water but power, flood control for the Imperial Valley, pleasurably navigable waterways that happily connected with recreation, and fine scenery on a vast body of water. To the north, however, on the sheep-grazing ranges of the Navajo Reservation in Arizona a later study observed that, "The dam [Hoover] was the catalyst that prompted drastic stock reduction." The outpouring of silt to Lake Mead from overgrazed ranges turned attention to grazing restrictions and a drive to protect the Navajo ranges in the interests of Lake Mead and the water power needs of southern California.[61] In addition to its good fortune in the passage of the Boulder Canyon legislation, the Bureau of Reclamation could claim that its western irrigation projects stood on a sounder footing by the end of the 1920s, paying back debt and enjoying attention to economic and community building tasks overlooked by the earlier leaders of reclamation.

Still, at the end of the decade, Commissioner Mead had to defend Reclamation from outright attacks by the Secretary of Agriculture, the National Chamber of Commerce, and the Association of Agricultural Colleges, all of which called for an end to western reclamation. President Hoover's Secretary of Agriculture, Arthur Hyde, argued that Reclamation should be transferred into the Department of Agriculture. Commissioner Mead, under mounting criticism, asked rural economist Dr. Alvin Johnson, editor of the *Encyclopedia of the Social Sciences* and founder of the

New School of Social Sciences in New York City, to study some of the less successful projects, where high rates of farm failures suggested that "economic conditions on these projects are not satisfactory." Having grown up on a Nebraska farm and been schooled in rural economics, Johnson was familiar with the challenges of western reclamation. He investigated the Lower Yellowstone Project in Montana and North Dakota, the Milk River and Sun River Projects in Montana, the Shoshone and Riverton Projects in Wyoming, and the proposed Casper-Alcova Project, also in Wyoming. Commissioner Mead described these projects as the most difficult to settle and to establish successful agriculture.[62]

Johnson's report lauded the goals of government reclamation, but it noted that in the projects he visited a large proportion of the farms were in the hands of tenants: 50 percent in Lower Yellowstone, over 47 percent in Shoshone, almost 46 percent in Milk River, and over 42 percent in Sun River. He denounced a system that might lend itself to putting "water on the land to grow thousand-acre fields of sugar beets with migratory Mexican labor." Johnson deplored any trend on the projects that might lead to land monopolization and factory-type migrant labor. If this were to be the end result of government sponsored reclamation, it would not be worth one dollar of subsidy. His remark was a prescient observation on the course of irrigated agriculture in the West. However, Johnson did not reject the legitimacy of government subsidy that enabled smaller farmers to acquire and stay on the land. The glaring problem was that, even with subsidies, the projects did not attract settlers. One of the reasons, he argued, was that the government had not done enough to prepare the land, which it had expected the settlers to do. That expectation was unrealistic in a time of urbanization, when settlers saw greater opportunities for their labor in non-farm industries.

In spite of the monumental mistakes and evidence that subsidized water contributed to land monopolies, Johnson accepted the arguments of Reclamation and its place in western agriculture. The Bureau of Reclamation, particularly Commissioner Mead, consistently argued that reclamation did not contribute measurably to agricultural overproduction and that it helped to sustain the economies of many hard-pressed western states. The total worth of crop and stock production from reclamation projects ($133,000,000 in 1927) was relatively insignificant in the total picture of American agricultural production, which ranged from twelve to fourteen billion dollars. Nonetheless, Reclamation provided major benefits to western states that would otherwise languish. In addition, Reclamation stood ready to develop the hydroelectric potential of the West's rivers. Johnson ultimately endorsed the reclamation program, but only "if it moves forward, as the times require, from its engineering achievement to equally distinguished achievement in the art of community building."[63]

Alvin Johnson's 1929 report dwelled too much on the problems of the past. By 1930, Reclamation's future was far from clear. The crushing realities of the Great Depression raised a specter of fear about the American economic and governmental structures. Those in the private sphere looked on in disbelief as the economy unraveled. Enormous wealth disappeared in the stock market, industrial production stalled in the face of inventories that did not move from salesrooms to buyers, and farm surpluses grew even as prices spiraled downward. With the collapse of private enterprise, the public sphere assumed unexpected importance in spite of the conventional wisdom that government should trim its sails in the face of economic downturns. Traditional economic wisdom held that without the tax generating revenues of the private economy, government public works must retreat in the face of slashed budget appropriations.

Yet, with passage of the Boulder Canyon Project Act, Reclamation found itself in a strong position regarding appropriations and future financing. Congress's more active role in river improvements also raised the possibility that the Corps of Engineers might play a larger part in western water development. The big question was whether the severe economic downturn would sabotage appropriations. While projected construction costs for the world's largest dam were astronomical and might seem foolhardy in the face of the national economic crisis, sound economics undergirded the project. The entire investment rested upon guarantees that sales of hydroelectric power would pay back the costs many times over. Given its passage in 1928, during the administration of President Coolidge, who once declared, "After all, the chief business of the American people is business," and ironclad provisions for its payoff in hydroelectric revenues, the project must not be classified amongst the subsidized public works projects of the Great Depression and the New Deal. It stood well within the traditional economics of conservative finance in the 1920s. In another sense, it continued a tradition of technical competency on the part of the American government, reaffirming its ability to undertake gigantic works such as the Panama Canal.

The timing of the construction, however, did coincide with the coming of the Great Depression. The rapid, successful construction of the dam in the early 1930s made Hoover Dam an icon in American history. It represented a beacon of hope and the promise of security amidst despair and fear in the depths of the Great Depression.[64] Thousands were employed building the dam (over 5,000 at one point, according to many well told accounts of its construction) who otherwise faced unemployment lines. While putting people to work in a time of need, the big dam on the Colorado River was not conceived to provide unemployment relief. Its engineering and political roots ran much deeper than the Great Depression. Still, its value as a model public works project was not lost, even on the Hoover Administration that generally opposed public works employment relief or even recovery projects. President Franklin D. Roosevelt's New Deal Administration in 1933 shared no such misgivings. The New Deal enthusiastically embraced the Boulder Canyon Project as a necessary work program that must be funded as part of an economic recovery agenda.

5.24. The employment office at Hoover Dam attracted many job seekers during the Great Depression.

Hoover Dam: Symbol of an Era

Astride the lower Colorado River, Hoover Dam stands as a monument to the ingenuity of engineers, the power of machines, and the strength of human muscles to impound the force of a mighty river. The dam was also a monument to political will and political power. Furthermore, it represented a major intervention upon the western landscape and waterscape by an urban-industrial-technical society to achieve what nineteenth-century forest hydrologists called "favorable" and "regulated" water flows. This is not to say that private enterprise had not already altered western landscape and rivers dramatically in developing and damming rivers for the production of, as one source called electricity, "white coal."[65] Over the years, Congress had struggled long and hard with the hydra-headed water question in the West. Now it dedicated one dam to multiple purposes to solve multiple water problems and meet multiple demands on the resource. As set forth in the Boulder Canyon Project Act of December 1928, these multiple purposes included: (1) flood control, (2) navigation improvement, (3) flow regulation, (4) irrigation, (5) domestic water supply, and (6) power generation. As a further benefit to irrigation downstream, the huge dam on the Colorado would free the downstream water of the heavy sediment burden or silt that the Colorado River carried in its natural state. While identified as a benefit, critics often point to the accumulation of sediment in reservoirs as ultimately destructive to the long term utility of dams.[66] However, on-the-other-side, basin-wide development of dams and reservoirs reduces flood-stage erosive power of fast-flowing streams, and that reduces the production of sediment to a limited degree.

Before Reclamation could fully embrace the multipurpose mission, the dam itself needed to be built. For contracts to be negotiated and construction to go forward, according to the Boulder Canyon Project Act, seven states of the Colorado River Basin must ratify the Colorado River Compact or, after six months, five states plus California must do so. In the process, California agreed to limit its annual consumption of Colorado River water to 4,400,000 acre feet plus one-half of any surplus waters unapportioned by the Colorado River Compact. This was accomplished by June 1929, with Arizona still refusing to come into the Compact. Also, no appropriation became available for construction until the Secretary of the Interior negotiated contracts for revenues sufficient to pay for maintenance and construction costs. By April 26-27, 1930, Secretary Ray Lyman Wilbur announced that he had in hand power contracts with Southern California Edison, the City of Los Angeles, and the Metropolitan Water District (MWD), with options presented to Nevada and Arizona to exercise when their consumers provided power markets, and, finally, various contracts with small companies in southern California. The MWD contract was principally for power to pump water from the Colorado to metropolitan Los Angeles. The contracts projected over $327,000,000 in revenues over fifty years. This figure, one of the largest power transactions ever negotiated, met the provision that the Secretary of the Interior "shall provide by contract for revenues adequate to pay operation and maintenance expenses and meet repayment requirements."[67]

The sale of power to private companies, however, in particular Southern California Edison, removed the threat of public power to the interests of the private sector. Almost from the beginning, the forces of private power worked to ensure that they would not have to face what they saw as "unfair competition" from government power. Public power supporters, of course, saw the arrangement as a compromise that yielded too much to private interests. Congress, with the Administration's assent, built into the Boulder Canyon Project Act provisions for private power to be a prominent player in the distribution and sale of the dam's hydroelectricity. In this light, Hoover Dam was not a victory for public power. Such assertions amounted to pure "chimera," according to one critical historian, who asserts: "The power company was in a 'no lose' situation. The genius, as well as the curse, of Hoover Dam was the way in which it blended private and public enterprise." The blend was necessary, given the political realities of the 1920s. Those who advocated a larger role for the federal government in the production and distribution of power did not hold sway in either the House or the Senate in this decade of pro-business Republican ascendancy. The contracts with both private and public agencies, announced in April of 1930, opened the way for the inclusion of an appropriation for $10,660,000 in the Second Deficiency Act of July 1930, and enabled work on the Boulder Canyon Project to begin.[68]

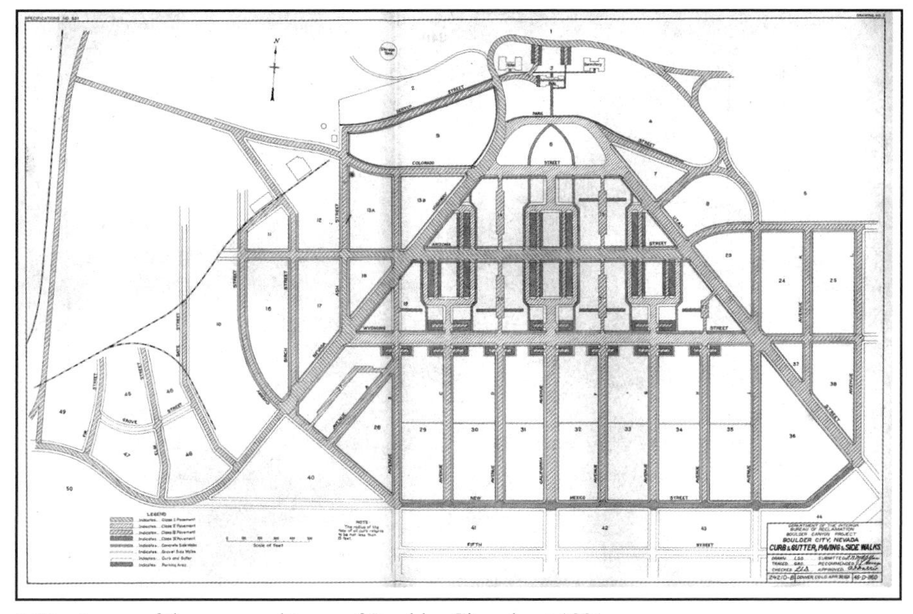

5.25. A map of the proposed town of Boulder City, about 1931.

The first money appropriated for construction was used: (1) $2,500,000 to build a railroad to the site from the Union Pacific line in Las Vegas, and (2) $525,000 to design and construct a new town to house workers, including waterworks, streets, sewers, and other conveniences;

5.26. Housing for Reclamation employees in Boulder City. December 1931.

construction of a main office building for government engineers and clerical staff; and twenty-five homes for permanent employees at the dam. Monies would also have to be spent on buying up small areas of privately owned land, especially in the Virgin River Valley, to clear the way for the 150,000 acres to be flooded by the dam's reservoir. Finally, at least five million dollars were to be used to plan and map construction strategies on the four tunnels to divert the river and begin building the coffer dams. All these tasks were undertaken while the Bureau of Reclamation invited bids for the building of a dam now planned for a height of 726 feet – recently revised upward from 557 feet by the Colorado River Board of Engineers at a meeting in Denver on April 12, 1930. The higher dam would provide for greater water storage for flood control and power production. Reclamation's dam design team, headed by Chief Designing Engineer John "Jack" L. Savage,

5.27. Married housing for Six Companies employees in Boulder City.

had been working on the dam's design for years. "Long before the Boulder Canyon Project bill received favorable action by Congress it had been decided definitely that the development was possible from an engineering point of view," wrote one Boulder Canyon Project historian. [69]

5.28. Dormitory housing for Six Companies employees, probably after the labor strike in 1931.

By December 1930, Reclamation officially called for bids to build Hoover Dam. Former Bureau of Reclamation engineer and construction engineer for many Reclamation dams, Frank Crowe, desperately wanted to build the big dam on the Colorado. He had left government service in 1924 to build dams for the Morrison-Knudsen Company and had prodded the company to join the Utah Construction Company to put together a consortium to bid on the Boulder Canyon Project. Later Crowe recounted, "I was wild to build this dam." [70] These two companies eventually brought in four more companies to form Six Companies, Inc. Their combined strength enabled them to raise the $5,000,000 bond required to bid on the project. Incorporated under the name "Six Companies," that name became inextricably linked to the construction of Hoover Dam. Six Companies, Inc., consisted of Morrison-Knudsen, Utah Construction Company, Pacific Bridge Company, Shea Company, MacDonald and Kahn, and Kaiser-Bechtel. Each of these companies had its own complicated history, and shrewd executives managed all of them. They had been building the physical infrastructure of the American West since the turn of the century: bridges, roads, pipelines, aqueducts, and dams. Utah Construction Company, for instance, had built O'Shaughnessy Dam on the Tuolumne River in Yosemite National Park, creating the much-contested Hetch Hetchy Reservoir.

For his part, Crowe had studied the Boulder Canyon Project for years. He had worked on some of the preliminary designs with Reclamation Service engineers Davis, Weymouth, and Savage. His work with Reclamation began under the supervision of Frank E. Weymouth, in the summer of 1904 on the Lower Yellowstone Project. He left to return to the University of Maine, where Weymouth also graduated, to complete studies for his degree in civil engineering. Crowe rejoined Weymouth's staff in late 1905 doing survey and canal construction work. He then spent two years working for a private contractor and rejoined Reclamation when Weymouth hired him to work on the Minidoka Project. During 1909, he identified a site for the permanent Jackson Lake Dam on the Snake River, and in 1910 he supervised construction of the new dam after the temporary rockfilled crib dam Reclamation had built there failed. He earned a reputation for efficiently utilizing labor and raw materials, transporting supplies, and in mechanizing dam building techniques, such as overhead cable delivery and excavation. One source sums up Crowe's special abilities: "Crowe was a gifted technician whose ability to interact effectively with the special breed of men who worked on the dams supplemented his talent for designing and putting in place the systems that harnessed their labor." With the six companies

5.29. Aerial views of Boulder City in 1936 and 1946 show the effects of landscaping by Wilbur W. Weed on the town. Reclamation brought Weed to Boulder City from the National Iris Garden in Beaverton, Oregon, in late 1931. By February of 1932, he was planting 9,000 trees in town.

incorporated and able to post the construction bond, Crowe's judgment was critical in preparing the Six Companies' final bid for work on the dam. On March 4, 1931, Reclamation officials opened the bids in Denver. The Six Companies bid of $48,890,955 was the winning figure. It was only $24,000 above the cost projected by Reclamation analysts. The figures were so close one might suspect collusion, but Crowe's experience and long familiarity with the challenges of building dams for Reclamation enabled this conglomerate to prevail. Six Companies applied its combined experience, gained over many years on large construction projects in the West, to the challenges of the big dam on the Colorado.[71]

5.30. The two primary managers stationed at Hoover Dam during construction: Project Construction Engineer Walker (Brig) R. Young, Bureau of Reclamation; and Superintendent of Construction Frank Crowe, Six Companies, Inc.

The construction site in Black Canyon on the lower Colorado presented a challenge in itself. Construction was to be concentrated and directed from the west bank of the Colorado in the southern portion of the most desolate and least populated state in the West and nation (Nevada's 1930 population: 84,515). The state was geographically huge but had a population the size of a small county. The nearest town, Las Vegas, with a population of only about 5,000 assumed great significance because the Los Angeles and Salt Lake City Railroad, of the Union Pacific system, ran through it. A spur from the railroad would serve to bring building materials to the river, and Las Vegas, a mere dot on the map, would become a boom town, providing temporary housing as well as notorious entertainment for the workers. Under Nevada's permissive gambling and prostitution laws, Las Vegas offered attractions for dam workers in a wide-open town with red light districts, liquor, and gambling.

While power to build the dam had to be brought all the way from Riverside, California, adequate gravel deposits existed nearby to supply the aggregate for the millions of tons of concrete needed. Reclamation engineers and private firms with years of experience building dams addressed the problems of raw materials, power, and transportation with energy and determination. One important ingredient remained to be found, labor.

By the spring and summer of 1931, unemployment had spread throughout America as the Great Depression deepened. The *New Reclamation Era* warned those seeking work to delay their departure for the work site until adequate housing and facilities had been constructed. The Bureau of Reclamation had designed a new town to accommodate over 5,000 workers. According to Walker Young, the town site for Boulder City was on a bluff high above the dam. He recalled, "It wasn't a very attractive site for us, but better than anyplace else in that particular area. We went there so many times, and all we found when we went was greasewood and sagebrush, a few burros, tarantulas, centipedes, snakes – not particularly attractive."[72] Reclamation laid out the town and its water supply, and it was also responsible for building quarters for Reclamation employees. Six Companies, whose onsite operations were under the direction of the "Big Boss," Frank Crowe, erected small houses of a temporary nature to house workers with families and also barracks for single men. The new community was christened Boulder City, not Hoover. President Hoover's name was in ill-favor. Throughout the country, under the deteriorating economic conditions of the Great Depression, ramshackle squatter towns of homeless people appeared on the edges of towns and cities. They became widely known as "Hoovervilles."

Although warned against striking out for "this deadly desert place," workers and even their families arrived before Boulder City had been completed. With nowhere to live, these hopefuls set up makeshift camps in tents and lean-to shelters made from cardboard and other materials at hand. Known as rag towns, the primitive accommodations posed a threat to health and even life as warm spring temperatures turned into an unbearably hot summer. Occasional fierce wind storms swept away makeshift shelters and tents. For the sake of jobs, the desperate willingly endured the heat, wind, and sand in the year 1931. The Hoover Administration, too, was desperate for jobs and immediately grasped the "public works" import of the Hoover Dam construction project, although it refused to endorse "public works" as a viable program to meet the problems of the Depression. In what was probably a political decision, it ordered construction on the dam to begin six months prior to the original date set for the fall of 1931. The order condemned the work force to labor through the hot summer months. They had to make-do until quarters became available in the promised "government town." Walker Young, Reclamation's project construction engineer, recalled, "It was anticipated that after the construction of the dam was completed and the powerhouse in operation, Boulder City would become simply a residential area for operators. That's why the permanent buildings were built for the government forces, whereas temporary buildings were built for the contractor." [73]

5.31. Two scenes from Ragtown at Hoover Dam in 1931.

The Bureau of Reclamation rushed to build accommodations for its workers at a site overlooking the reservoir. Built to last long after the construction phase ended, the permanent residences for dam employees were built of brick with a stucco exterior in a Spanish-style mission architecture popular in the Southwest. Six Companies built small, two- and three-room, wooden-framed houses with porches for married employees with families and large, two-story barracks for single men. Barracks near the work in Black Canyon at the "River Camp" were built with inadequate sanitation facilities, no cooling, and no running water. There was no way the company could provide adequate housing and begin work on the dam at the same time. The lack of housing and the order to start immediately on the dam contributed hugely to the outbreak of serious labor troubles that culminated in a strike on August 11-12, 1931. Untenable strains on the work force occurred from eight-hour shifts worked seven days a week and around the clock throughout the hottest days of summer. From late June to late July, a man died every other day, mostly from heat prostration and dehydration. The importance of drinking adequate water to prevent dehydration in heat that reached 120 degrees was not thoroughly understood on the project during the first year of construction.[74]

5.32. Six Companies' "River Camp" at Cape Horn in Black Canyon. June 2, 1931.

The voice of labor was not strong. Workers were thankful to have a job. Some skilled workers from the American Federation of Labor had made their way onto the project, but the vast majority of labor was without union representation. Given Frank Crowe's reputation as a hard driver, and the disdainful attitude of Six Companies toward organized labor, disputes over wages and working conditions reached a breaking point throughout the work site by mid-summer 1931. Nevada's labor laws protected the 8-hour day, as did Reclamation law, and Nevada required mine ventilation and banned the use of gasoline engines underground. The latter became an issue in the excavation of the diversion tunnels necessary to dewater the river bed for the construction of the dam. Working conditions as well as living conditions deteriorated during the first hot summer on the job. A move by management to reassign some workers to lower paying jobs out of the tunnels brought a crisis in labor relations. The lack of housing, injuries and death on the job, the insufferable heat, and a dispute over wages brought workers to the point of calling a strike. There was also a radical group of organizers from the Industrial Workers of the World (IWW) who saw opportunity in the almost unbearable working and living situations. In the years after World War I, state and federal authorities effectively suppressed this radical union movement and subjected it to prosecution under anti-syndicalism laws. [75]

Radical agitators seized upon the legitimate complaints of workers and became prominent voices in the crowd. They called for a strike for better working conditions, better housing, and a fair wage scale. When a strike was approved and the demands presented to construction boss Frank Crowe, he refused the demands and denounced the strike and the agitation as a plot on the part of the radical IWW. The company had no intention of permitting workers to dictate labor conditions in the building of Hoover Dam. The workers backed down under threat of the arrival of National Guard troops to protect government property. Not satisfied with the capitulation, Crowe fired nearly 1,200 dam workers and vowed to hire new people from the large pool of unemployed men in Las Vegas. Both Six Companies and Commissioner Mead insisted that the wages paid on the Hoover Dam project were higher than the prevailing wage for similar work in most of the southwestern states. The aggressive stance of the company indicated the desperate plight of labor during these years at the beginning of the Depression. Six Companies had every confidence that local, state, and federal government officials (including the Bureau of Reclamation and the Interior Department in the Hoover Administration) would support the company against labor. Only the most skilled workers involved in the strike were rehired. Those who promoted the strike found their names on a "no rehire" list, and certainly this included all known members of the IWW.

5.33. The concrete testing laboratory for Hoover Dam in Las Vegas, Nevada, was in this building, reconstructed from the old adobe"Mormon Fort" built in 1855-57. The Mormon Fort is now a historic site in Las Vegas.

The end of the strike spelled the end of the IWW in the work camps of the Boulder Canyon Project. Yet IWW influence should not be overestimated. Its organizers joined the labor unrest and walkout rather than instigated it. The "Big Boss" Frank Crowe, speaking for the Six Companies, was all too quick to paint the entire incident as a creation of radical labor. He seized upon underlying community fears of the IWW. Those fears had existed ever since it began campaigning for one big industrial union as far back as 1905 and subsequently opposed American participation in World War I. Tarring all labor with the radical brush of IWWism was a favorite anti-labor tactic of management.

The disruption of the work schedule lasted less than a week. But in response, the Six Companies removed the barracks in River Camp and housed single workers in Boulder City dorms with flush toilets, showers, heating-cooling systems, and safe drinking water. These amenities had not been available in the barracks at the bottom of the canyon. In the fall, cooler temperatures prevailed and many workers and their families began to move into newly framed houses in Boulder City. The major labor crisis in the construction of Hoover Dam had passed. It might not have occurred at all had the Hoover Administration not insisted, for what can only be seen as political motives, that the start-up date for construction occur before adequate living quarters and facilities for workers had been provided.

5.34. The north wing of the Six Companies mess hall for Hoover Dam workers in Boulder City seated 650, as did the south wing.

While Reclamation took pride in the construction of Hoover Dam, Six Companies managed the labor force. This meant that contract deadlines must be met and hard-driving foremen supervised workers who could be dismissed at will. The companies were required to honor only the

Reclamation labor law that included the 8-hour day (it made no mention of around-the-clock shifts) and prohibited the employment of "Mongolian" workers. Overall, the Hoover Dam project was a "white only" labor project and Congress further declared, in the legislation authorizing the project, that only American citizens should be hired and gave preference to veterans. Complaints by African-Americans that Six Companies would not hire blacks prompted a visit from NAACP leaders in the summer of 1932. They pushed Six Companies to hire blacks and publicized their efforts. By the end of the summer of 1932, some African-Americans appeared on the payrolls. Still, none were permitted to live in Boulder City, which was the original excuse the Six Companies offered in defense of their discriminatory hiring policies. They refused to provide living facilities for black workers and their families in the white-only community. After Harold Ickes became Secretary of the Interior, in March 1933, he ordered that blacks employed on the project had a right to live in Boulder City and freed them from a long ride each day between Las Vegas and the dam site. Yet, he could not directly interfere with the Six Companies' hiring policies that still admitted only the employment of a few dozen African-Americans.[76]

5.35. In December 1931 Six Companies jumbos were hard at work on the tunnels to divert the Colorado River from its course.

On the construction front, engineers and workers with heavy equipment tackled the building of the dam. Before dam construction began, three major tasks had to be accomplished: (1) to divert the river from its main channel, two tunnels on each side of the river, fifty-six feet in diameter and lined with three feet of concrete, needed to be drilled, blasted, and lined, (2) cofferdams had to be built above and below the portion of channel to be dewatered, where the dam was to stand, and (3) "dental work" had to be done on the sides of the canyon by high scalers to remove loose rock

5.36. By May of 1933 Six Companies had diverted the Colorado and built cofferdams to protect the foundation of Hoover Dam while it was cleaned. Here jackhammers are in use cleaning loose material away.

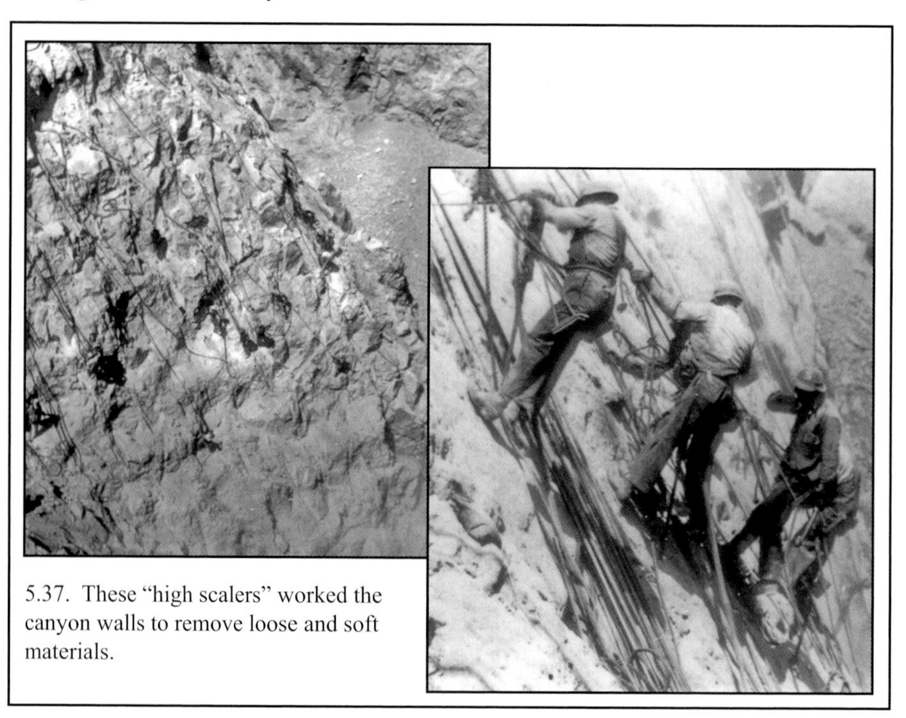

5.37. These "high scalers" worked the canyon walls to remove loose and soft materials.

down to bedrock. The "scalers" worked far above the river, suspended on bosun's chairs by ropes (their lifelines) over the walls of the canyon while they "cleared" the rock faces of the canyon with high pressure pneumatic drills. It was their job to plant charges to blast away loose rock so the dam's side wings could be fixed to solid rock in the canyon. Similar "dental work" proceeded on the floor of the canyon after the cofferdams and tunnels had diverted the river. The loose material in the channel had to be removed before the massive concrete structure could rise up on this anchorage. After building the diversion tunnels and cofferdams and blasting and excavating the sides and base, dam construction began. The river was successfully diverted in November 1932. Building the two cofferdams was a major task because the upstream dam was to be at least 100 feet high. Impressive for a temporary structure, Crowe wanted it strong enough and high enough to hold back anything the Colorado could throw at it during construction.

The first concrete was delivered via Crowe's famous overhead cable system on June 6, 1933. The aerial delivery system had first been employed in 1911 at Arrowrock Dam. Within each form, sixteen-ton buckets of concrete were dumped methodically around the clock, and one-inch pipe, five feet apart, circulated refrigerated water to cool the concrete. The pipe (approximately 662 miles of it) dissipated the chemical heat produced as the cement and water gave off heat in the dehydration or curing process. To

5.38. Eight-cubic-yard-capacity concrete bucket discharging into a dam column form in September of 1933. The bucket was delivered by the overhead cable system Frank Crowe used so effectively.

permit the heat to generate unchecked would invite expansion which, in a concrete mass of this size, would then result in serious cracking and flaws when the structure finally cooled. Reclamation successfully devised this cooling process in the construction of Owyhee Dam, completed in 1932.[77]

5.39. In March of 1933 this skipload of workers was on its way to work at Hoover Dam.

The dramatic story of Hoover Dam's construction has been told many times. Blocking the mighty Colorado with a gigantic dam has captured the imagination of journalists, publicists, writers of books, and film makers. The story validated the New Deal and its struggle to put America back to work in the Depression years, although it was conceived and launched before the Depression struck. That it captured American attention testified to the uplifting impact that the dam had upon the American public in a time of national crisis and disillusionment. After that difficult summer of 1931, construction continued ahead of schedule. Crowe's aerial cableway for delivering buckets of concrete and other construction materials to the dam proved efficient and reliable. By the fall of 1935, the dam was ready for dedication, a year ahead of schedule.[78]

5.40 One of the concrete mixing plants on the steep sides of Black Canyon. June 29, 1934.

The dedication ceremony on September 30, 1935, attracted not just the attention of the leadership of Reclamation and political leaders in the Southwest, but also President Franklin D. Roosevelt and his Secretary of the Interior. At the ceremony, Secretary of the Interior Harold Ickes dedicated "Boulder Dam." He said, "This great engineering achievement should not carry the name of any living man [i.e., former President Hoover who had not been invited to the occaion] but, on the contrary, should be baptized with a designation as bold and characteristic and imagination stirring as the dam itself." In response, one Las Vegas newspaper termed Ickes's action "a studied and intentional insult to a former President of the United States." But beyond the question of the dam's name, the Secretary of the Interior emphasized how the structure would provide an economic foundation for the future economic growth of the Southwest. His words spoke to the fear Americans felt about the future, but here was a concrete symbol that seemed to transcend the fears of the Great Depression. [79]

5.41. President Franklin Delano Roosevelt at the dedication of Hoover Dam. September 30, 1935.

When President Roosevelt rose to speak to the estimated 10,000 people at the ceremony, he praised the genius of the designers as well as the grit, determination, and skill of the workers who had built the engineering marvel of the age. Mindful, however, of strained relations between labor and the private construction

5.42. Secretary of the Interior Harold Ickes insisted Hoover Dam be known as Boulder Dam from 1933 to 1947.

company contracted to build the dam, Roosevelt made no mention of Six Companies. Most Americans were familiar with the President's voice. They heard it weekly over the radio in his "fireside chats" about the problems of everyday people caught in the grip of the Depression. Now his rich, confident voice lauded the completion of a desert monument dedicated to progress and economic expansion. At hand was access to vast, new amounts of electrical power and water for the growing urban populations of the American Southwest. Surely, this great structure was about to set in motion forces that would bring a brighter future. It meant the transformation of "an unpeopled, forbidding desert." The President's words sounded much like those of reclamation enthusiasts almost four decades earlier who had promised that the Reclamation Act and the Reclamation Service would transform arid America. For Reclamation, the Boulder Canyon Project committed it "irrevocably to hydroelectric development as an integral part of the federal reclamation program, opening the way to large multiple-purpose projects."[80]

Endnotes:

1 Elwood Mead, *Helping Men Own Farms: A Practical Discussion of Government Aid in Land Settlement* (New York: The Macmillan Company, 1920).

2 Paul K. Conkin, "The Vision of Elwood Mead," *Agricultural History* 34 (April 1960), 89.

3 Robert G. Dunbar, *Forging New Rights in Western Waters* (Lincoln: University of Nebraska Press, 1983), 99-132; I. G. Baker, "Elwood Mead in Australia: An Historical Survey," typescript, 1-19, Carton 2, Mead Papers; Kluger, *Turning on Water*, 6-11, 157-9; biographical description Packard File, Carton 2, Mead Papers.

4 Hundley, *Great Thirst*, 209,

5 The feasibility study was authorized by the Kincaid Act. Moses P. Kincaid, Chairman of the House Irrigation Committee from Nebraska, was noted for his promotion of enlarged homestead acts for the settlement of the high plains and the Sand Hills of Nebraska, 1906-1909.

6 First named Boulder Dam by Secretary of the Interior Hubert Work, soon after he took office. Secretary of the Interior Ray Lyman Wilbur named it Hoover Dam, after the then President Herbert Hoover, a key figure in achieving the Colorado River Compact of 1922. Secretary of the Interior Harold Ickes, however, named it Boulder Dam and reinforced that name at the dedication in September of 1935, claiming that such an enduring structure should not bear the name of one man.

7 "The Week," *New Republic*, 33 (January 17, 1923), 185.

8 George Kreutzer to Elwood Mead, April, 1922; Elwood Mead to Frank Mondell, May 29, 1922, Carton 2, Mead Papers.

9 A. P. Davis to Elwood Mead November 29, 1924, Davis Papers, Box 6, "Elwood Mead Folder."

10 A. P. Davis to Frank C. Wright, editor, *Engineering News-Record*, July 21, 1924, Box 4, Davis Papers.

11 *Denver Post*, September 20, 1924 announced Weymouth resignation; "F. E. Weymouth Leaves Bureau of Reclamation," *New Reclamation Era*, 15 (October 1924), 160.

12 Rowley, *Reclaiming the Arid West*, 128-9; Hundley, *Great Thirst*, 230-1.

[13] Tyler, *Silver Fox of the Rockies*, 124-6; Hundley, *Great Thirst*, 211-5; *Wyoming v. Colorado* 260 U.S. 1 (1922).

[14] For a sensitive treatment of this development see William deBuys with photographs by Joan Myers, *Salt Dreams: Land & Water in Low-Down California* (Albuquerque: University of New Mexico Press, 1999); William E. Smythe, "An International Wedding: The Tale of a Trip on the Borders of Two Republics," *Sunset* (October 1900), 286-300, as noted in Evan R. Ward, *Border Oasis: Water and the Political Ecology of the Colorado River Delta, 1940-1975* (Tucson: University of Arizona Press, 2003), 3-13.

[15] Mildred de Stanley, *The Salton Sea: Yesterday and Today* (Los Angeles: Triumph Press, Inc., 1966), 17-8; R. J. Glennon and P. W. Culp, "The Last Green Lagoon: How and Why the Bush Administration Should Save the Colorado River Delta," *Ecology Law Quarterly* 28 (October 2002), 903-90.

[16] deBuys, *Salt Dreams*, 64, 104-5; Orsi, *Sunset Limited*, 226-31.

[17] Stevens, *Hoover Dam,* 10-5; Michael P. Malone and Richard Etulain, *The American West: A Twentieth Century History* (Lincoln: University of Nebraska Press, 1989, second paperback edition, 1990), 122.

[18] "Definite Federal Policy on the Colorado River Urged," *New Reclamation Era*, 15 (April 1924), 51-4.

[19] A. P. Davis to Frank C. Wright, July 21, 1924, Davis Papers; Pisani, *Water and American Government*, 145.

[20] Brookings Institution, *U.S. Reclamation Service*, 35; "Six New Projects Are Recommended," *New Reclamation Era*, 15 (May 1924), 69.

[21] Pisani, *Water and American Government*, 145-6.

[22] "Recommendations for New Projects," *New Reclamation Era*, 15 (May 24, 1924), 77; this issue also contains an abridged version of the Fact Finders' Commission report and welcomes the recommendations under the headline: "Impending Reclamation Disaster May be Averted," *New Reclamation Era*, 15 (May 24, 1924), 67-9.

[23] Rexford G. Tugwell, "Reflections on Farm Relief," *Political Science Quarterly* 43 (December 1928), 481-97.

[24] R. P. Teele, "Land Reclamation Policies in the United States," USDA *Bulletin No. 1257*, August 23, 1924 (Washington, D.C.: Government Printing Office, 1924), 7; Elwood Mead, Address to Western Society of Civil Engineers, October 26, 1925, RG 115, Entry 7, Box 25, NARA, Denver.

[25] Teele, "Land Reclamation," 7, 8, 10.

[26] Teele, "Land Reclamation," 10, 31, 34-5.

[27] Teele, "Land Reclamation," 38, 40.

[28] Pisani, "Reclamation and Social Engineering in the Progressive Era," 46-83; Elwood Mead to H. H. Dare, Commissioner, Water Conservation and Irrigation Commission, Sydney, Australia, June 1, 1922, Mead Papers Box 1; Elwood Mead to Frank Mondell, February 7, 1922, Mead Papers, Box 2.

[29] Hubert Work to Congressman Addison T. Smith, November 10, 1927, RG 115, Entry 7, Box 198, NARA, Denver.

[30] Elwood Mead Memo, December 14, 1924, Box 2, Mead Papers; Elwood Mead, Address to Western Society of Civil Engineers, October 26, 1925, RG 115, Entry 7, Box 25, NARA Denver; Conkin, "The Vision of Elwood Mead," 88-97.

[31] "Operations on Indian Projects Transferred," *Reclamation Record*, 15 (February 1924), 31.

[32] Elwood Mead, "Construction of Irrigation Works and Selection of Settlers Directed by Bureau of Reclamation," *U.S. Daily,* September 17, 1928, clipping in RG 115, Entry 7, Box 31, NARA, Denver; Robinson, *Water for the West*, 46: Kluger, *Turning on Water*, 121.

[33] Kluger, *Turning on Water*, 124-5; "Pan-Pacific Conference Announcement," Memo from Elwood Mead, 1927, RG 115, Entry 7, Box 133, NARA, Denver.

[34] *Project Data*, 733-35; Jackson, *Ultimate Dam*, 188-92; George T. Wisner and Edgar T. Wheeler, "Investigations of Stresses in High Masonry Dams of Short Spans," *Engineering News* 54 (August 10, 1905), 141-4; David P. Billington, Donald C. Jackson, and Martin V. Melosi, *The History of Large Federal Dams: Planning, Design, and Construction in the Era of Big Dams* (Denver, Colorado: Bureau of Reclamation, 2005), 145-52 and chapter 2 "Theories and Competing Visions for Concrete Dams."
[35] Billington, Jackson, and Melosi, *Large Federal Dams,* 145-52.
[36] Swain, *Federal Conservation Policy, 1921-1933*, 85-6; Elwood Mead, Address to Western Society of Civil Engineers, October 26, 1925, RG 115, Entry 7, Box 25, NARA, Denver; Chief Engineer R. F. Walter to Elwood Mead May 12, 1928, and B. E. Stoutmyer to Elwood Mead, December 29, 1929, RG 115, Entry 7, Box 884, General Administrative Project Records, 1919-1945, Owyhee, 303. – 400.02, File: 320 Owyhee Project Power Development, NARA, Denver; "Owyhee Project History to January 1, 1928," RG 115, Entry 10, Box 428, Project Histories, Features, and Reports, 1902-1932, Owyhee, NARA, Denver.
[37] Elwood Mead, Address to WSCE, October 26, 1925, 3.
[38] Pisani, *Water and American Government*, 113-4; Pelz, *Federal Reclamation and Related Laws Annotated*, volume I, 111, 169.
[39] John M. Barry, *Rising Tide: The Great Mississippi Flood of 1927 and How It Changed America* (New York: Simon & Schuster, 1997), 240, 286, 417; Clements, *Hoover,* 111-2; Pete Daniel, *Deep'n as It Come: The 1927 Mississippi River Flood* (New York: Oxford University Press, 1977); Pisani, *Water and American Government,* 248-53.
[40] Kluger, *Turning on Water*, 123-4.
[41] U.S. Department of the Interior, Bureau of Reclamation, Elwood Mead, Commissioner, E. C. Branson, "Planned Colonies of Farm Owners" in *Economic Problems of Reclamation* (Washington, D.C.: Government Printing Office, 1929), 17-26; Kluger, *Turning on Water*, 123.
[42] U.S. Congress, House Committee on Flood Control, "Flood Control," 70th Congress, January 1928, 25, as quoted in Barry, *Rising Tide*, 405.
[43] "Without Hoover's personal guidance and constant pressure and without Swing-Johnson's threat to the recalcitrant States, there would have been no agreement at all." See Clements, *Hoover,* 85.
[44] Clements, *Hoover*, notes support of public power from the Colorado River by Hoover, 132-3; Norris Hundley, Jr., *Water and the West: The Colorado River Compact and the Politics of Water in the American West* (Berkeley: University of California Press, 1975), notes that the construction of Hoover Dam marked the expansion of federal authority over the Colorado and other navigable streams, 333.
[45] Paul L. Kleinsorge, *The Boulder Canyon Project: Historical and Economic Aspects* (Palo Alto, California: Stanford University Press, 1941), 88-9.
[46] *Congressional Record*, 70th Congress, 2nd Session, 70, part 1, (Dec. 3, 1928 to Jan. 4, 1929), 831.
[47] *Congressional Record*, (Dec. 3, 1928 to Jan. 4, 1929), 832.
[48] "Report on the Problems of the Colorado River, volume 5, Boulder Canyon: Investigations, Plans and Estimates," February, 1924, RG115, Entry 7, Box 477, Project Files, 1919-1945, NARA, Denver.
[49] Jared Orsi, *Hazardous Metropolis: Flooding and Urban Ecology in Los Angeles* (Berkeley: University of California, 2004), 68; J. W. Dent, Acting Commissioner to D. S. Struver, Project Manager TCID, July 15, 1929, RG 115, Entry 7, Box 139, NARA, Denver.
[50] Donald C. Jackson and Norris Hundley, Jr., "Privilege and Responsibility: William Mulholland and the St. Francis Dam Disaster," *California History* 82, 3 (2004), 28-9.
[51] *Congressional Record*, (Dec. 3, 1928 to Jan. 4, 1929), 831-2; Hundley, *Great Thirst*, 222.
[52] Beverley B. Moeller, *Phil Swing and Boulder Dam* (Berkeley: University of California Press, 1971).

[53] J. David Rogers, "Reassessment of the St. Francis Dam Failure," *Engineering Geology Practice in Southern California,* edited by B. W. Pipkin and R. J. Proctor, *Association of Engineering Geologists, Special Publication No. 4* (1992), 639-66; J. David Rogers, "A Man, a Dam and a Disaster: Mulholland and the St. Francis Dam," *Southern California Quarterly* 77 (Spring/Summer 1995), 100-9; Jackson and Hundley, "Privilege and Responsibility" refute the attempt by Rogers to exonerate Mulholland from responsibility for the failure of the St. Francis Dam, 8-47. Also of note, Charles Outland, *Man-Made Disaster: The Story of St. Francis Dam* (Glendale, California: Arthur H. Clark Co., 1977).

[54] *Congressional Record*, (December 3, 1928 to January 4, 1929), 834; Pisani, *Water and American Government*, 232; Louis C. Hunter and Lynwood Bryant, *History of Industrial Power in the United States, 1780-1930*: *The Transmission of Power,* volume 3 (Cambridge: Massachusetts Institute of Technology Press, 1991), 353, 360, 364; Thomas H. Gammack, "Hydroelectric Myths," *World's Work* 58 (May, 1929), 120.

[55] Hundley, *Great Thirst*, 222; Pisani, *Water and American Government*, 262 .

[56] deBuys, *Salt Dreams*, 162.

[57] Pisani, *Water and American Government*, 232, 235; Hundley, *Great Thirst*, 222; *Arizona v. California*, 283 U.S. 423, L.ed 1154, 51 S. Ct. 522, 1931.

[58] Stevens, *Hoover Dam*, 32.

[59] Stevens, *Hoover Dam*, 24-5, 32; Paul C. Pitzer, *Building the Skagit: A Century of Upper Skagit Valley History, 1870-1970* (Portland, Oregon: The Galley Press, 1978).

[60] U.S. Department of the Interior, Bureau of Reclamation, *Twenty-Ninth Annual Report of the Commissioner of Reclamation for the Fiscal Year Ended June 30, 1930* (Washington, D.C.: Government Printing Office, 1930), 3-4.

[61] Richard White, *The Roots of Dependency: Subsistence, Environment and Social Change among the Choctaws, Pawnees, and Navajos* (Lincoln: University of Nebraska Press, 1983), 251-2.

[62] Clements, *Hoover*, 133; Pisani, *Water and American Government*, 140; Alvin S. Johnson, *Pioneer's Progress: An Autobiography* (New York: Viking Press, 1952).

[63] U.S. Department of the Interior, Bureau of Reclamation Elwood Mead, Commissioner, Alvin Johnson, "Economic Aspects of Certain Reclamation Projects," *Economic Problems of Reclamation* (Washington, D.C.: Government Printing Office, 1929), vii, 15-6.

[64] Richard Lowitt, *The New Deal in the West* (Bloomington: Indiana University Press, 1984).

[65] Steen, *The U.S. Forest Service*; Jessica B. Teisch, "Great Western Power, 'White Coal,' and Industrial Capitalism in the West," *Pacific Historical Review* 70 (May 2001) credits Lewis Mumford for the term "White Coal," 222.

[66] Pelz, *Federal Reclamation and Related Laws Annotated*, volume I, 414; Golzé, *Reclamation in the United States,* 163-4; McCully, *Silenced Rivers*, 107-112; for a discussion of sediment and the Colorado Delta see, Glennon and Culp, "Save the Colorado River Delta," 903-90; Evan Ward, "The Twentieth-Century Ghosts of William Walker: Conquest of Land and Water as Central Themes in the History of Colorado River Delta," *Pacific Historical Review* 70 (August 2001), 359-85.

[67] Kleinsorge, *The Boulder Canyon Project*, 147-54; "Boulder Canyon Project Act," *New Reclamation Era*, 20 (May 1929), inside front cover.

[68] Pisani, *Water and American Government*, 234; Kleinsorge, *Boulder Canyon Project*, 185; "Construction of Boulder Dam Begun on July 7, 1930," *New Reclamation Era*, 21 (August 1930), 146; Hundley, *Great Thirst*, 222.

[69] Kleinsorge, *Boulder Canyon Project*, 191, 194-7.

[70] From "The Earth Movers I," *Fortune*, 28 (August 1943), 119-24, as quoted in Stevens, *Hoover Dam*, 38-9.

[71] Rocca, *America's Master Dam Builder*, 178; Wolf, *Big Dams and Other Dreams,* 33-5; Stevens, *Hoover Dam*, 46.

[72] Andrew J. Dunar and Dennis McBride, *Building Hoover Dam: An Oral History of the Great Depression* (Reno: University of Nevada Press, 1993), 59.
[73] Edmund Wilson, "Hoover Dam," *New Republic*, 68 (September 2, 1931), 66-69; Dunar and McBride, *Building Hoover Dam*, 59.
[74] Stevens, *Hoover Dam,* notes that physiologists from the Harvard University Fatigue Laboratory visited Black Canyon in the summer of 1932 and concluded that "heat prostration" was caused by dehydration and the remedy was to encourage the workers to drink more water, 60-1.
[75] Guy Louis Rocha, "The IWW and the Boulder Canyon Project: The Final Death Throes of American Syndicalism," *Nevada Historical Society Quarterly* 21 (Spring 1978), 3-24.
[76] Stevens, *Hoover Dam*, 176-7; Rocca, *America's Master Dam Builder*, 227; David Thomson, *In Nevada: The Land, the People, God, and Chance* (New York: Vintage Books, 1999), 183; Wendy Nelson Espeland, *The Struggles for Water: Politics, Rationality, and Identity in the American Southwest* (Chicago: University of Chicago, 1998); Jeanne Nienaber Clarke, *Roosevelt's Warrior: Harold L. Ickes and the New Deal* (Baltimore, Maryland.: Johns Hopkins University Press, 1996), "When he [Ickes] learned, for instance, that there wasn't a single black [sic.] employee in the Mormon-dominated and rapidly expanding Bureau of Reclamation, he promised that would change as soon as vacancies occurred," 180; Harvard Sitkoff, *A New Deal for Blacks: The Emergence of Civil Rights as a National Issue* (New York: Oxford University Press, 1978), 69-70.
[77] Rocca, *America's Master Dam Builder*, 257; Stevens, *Hoover Dam*, 196.
[78] Wolf, *Big Dams and Other Dreams*, 53.
[79] Lowitt, *New Deal in the West, 206*; Las Vegas *Age,* October 11, 1935; Tyler, *Silver Fox of the Rockies,* 241.
[80] Stevens, *Hoover Dam*, 246-7; Rocca, *America's Master Dam Builder*, 271; Las Vegas *Age*, October 4, 11, 1935; Las Vegas *Evening Review Journal*, October 1, 1935; "President's Talk at Boulder Dam," New York *Times*, October 1, 1935, 2, 2; Swain, *Federal Conservation*, 122.

Chapter 6

THE BUREAU OF RECLAMATION AND THE GREAT DEPRESSION

Introduction

By the beginning of World War II, the Bureau of Reclamation's multipurpose dams supported a world-renowned power and water infrastructure in the American West. From the maelstrom it faced less than two decades earlier, the Bureau of Reclamation had become "the mightiest federal agency in the American West." Hoover Dam and the high dams that followed provided jobs for the unemployed and concrete symbols of American industry, particularly its organizational capacity and engineering talent. Foremost, they demonstrated the courage, strength, and ability of American labor to accomplish great works. A prolonged western drought accompanied the Great Depression, adding urgency to irrigation and water storage issues. Finally, the Bureau of Reclamation's big

6.1. This farm showed the effects of the 1930s Dust Bowl.

dams of the 1930s gave the United States the industrial muscle to meet the challenges of a two ocean war after the Japanese attack on Pearl Harbor, December 7, 1941.[1]

When President Roosevelt dedicated Boulder Dam in the fall of 1935, the Las Vegas *Evening Review Journal* noted that the President, Secretary of the Interior Harold Ickes, and Works Progress Administration (WPA) Director Harry L. Hopkins could not help but understand the dreams and possibilities represented by this magnificent structure: "All, after seeing Boulder Dam actually completed and ready to function, could never doubt that the other projects we've visualized, can be realized. For after Boulder Dam, nothing, however fanciful, seems impossible." And other projects did go forward – the Columbia Basin Project including Grand Coulee Dam; Bonneville Dam (under the Corps of Engineers); the Colorado-Big Thompson Project; and the Central Valley Project, including Shasta and Friant Dams. The Roosevelt Administration spent huge sums on big dams, in part for their long term benefits, and in part to provide public works projects in the jobless years of the Great Depression.[2]

Facing the Crisis of the Great Depression

The election of a new president in 1932 with a mandate to confront the problems of a failing economy might very well mean severe cutbacks in all parts of government, if not the complete dismantling of the reclamation program. Faced with such uncertainty, Mead promoted the formation, in 1931, of the National Reclamation Association (after World War II named the National Water Resources Association). He was well aware of the key role the National Reclamation Congresses had played thirty years earlier in the promotion of national reclamation. In subsequent years, however, support for Reclamation declined among private organizations. As local irrigation projects became formally organized irrigation districts or water users associations, the early private sector support often transformed into an adversarial relationship. This had been disastrous for reclamation in the 1920s. Political scientists cite the phenomenon of the "iron triangle" that offers support to public agencies and guarantees their survival and support in Congress. The three sides of the triangle include private groups benefiting from the services of government agencies, Reclamation itself, and congressional members who respond to the lobbying of constituents voicing demands through private organizations. Mead particularly felt the need for a supportive organization as the Bureau of Reclamation faced the economic disaster of the Depression and a new national Administration bent upon change and reform. It was time to resurrect support from a local and regional base similar to early movements that had enthusiastically supported national reclamation. Since its early years, the Reclamation Service/Bureau of Reclamation had not enjoyed the support of an iron triangle and, in subsequent years, faced changing fortunes and faltering visions. The task at hand was to keep alive visions that included: water storage for an arid West; a self-sustaining Reclamation Fund based on project repayments to the government; and, finally, preventing the Bureau of Reclamation from being eliminated altogether or converted into a rural relief bureau.[3]

The Great Depression unexpectedly opened a wide range of opportunities for the Bureau of Reclamation. For once, it was the right bureau in the right time and place. Traditional economic doctrine suggested a thorough cutback of government services in line with its reduced and straitened circumstances. Unlike earlier hard times, expansion rather than retrenchment became the goal of depression-era planners. In contrast, during the relatively prosperous times of the 1910s and 1920s, the Bureau of Reclamation had barely survived congressional attacks and complaints from project settlers. As the New Deal Administration launched program after program to combat the Depression, Reclamation, especially after 1937, received ever larger mandates to build a water and hydroelectric power infrastructure for the western economy that exceeded, by far, any of its previous ambitions. In the meantime, before the transition to the New Deal Administration in 1933, retrenchment seemed the order of the day. Repayments from the projects showed a revival in 1929 but fell off sharply thereafter. Congress further reduced this source of income for Reclamation projects by placing various moratoria on repayments. During the days leading up to the reorganization of the program in the Fact Finders' Act, Congress passed moratoria for the period 1921-1924. As the Great Depression struck, it also passed moratoria for the years 1931 to 1936. Congress's leniency on the repayment issue doomed the Reclamation Fund as a source of funding to keep Reclamation alive in those hard times. This occurred while general tax revenues fell off at every governmental level. All this spelled disaster for any plans to maintain, let alone expand, the services of Reclamation.[4]

But the political will lay elsewhere. The depth of the economic crisis prompted the election of a new national administration under President Franklin D. Roosevelt and a mandate to act. New programs and experimentation became characteristic of New Deal initiatives that favored trial and error over inaction. All involved the expansion of federal power and the authorization of expenditures to meet the emergency, and some expanded state power by depending on state agencies to implement federal policies. All hoped to revive the economy, restore confidence, relieve unemployment, and return industrial production to pre-Depression levels. The Bureau of Reclamation suddenly became one of those programs. With the Hoover/Boulder Dam construction well

6.2. In 1937, the contractor's employment office at Grand Coulee Dam drew many applicants.

6.3. Payday at Grand Coulee was particularly popular during the Depression year of 1937.

underway by the spring of 1933, the Bureau of Reclamation set the example for what could be accomplished in public works. Its construction program promised to assist in recovery and its hydroelectric power offered a basis not only for recovery but for future growth. At the same time, dam-construction provided immediate work and jobs to relieve the crisis of unemployment. The President's speech at the Boulder Dam dedication emphasized the role of public works to uplift humanity and at the same time provide jobs that "created the necessary purchasing power to throw in the clutch to start the wheels of what we call private industry."

Building big dams and other public works projects with funds from the New Deal's Works Progress Administration (WPA) and the Public Works Administration (PWA) exemplified the economics of "pump priming." Government became the source of investment capital as it spent large sums for major projects — dams, highways, schools, libraries, and recreational facilities — to help restart the stalled engines of the private economy. Big dams, in particular, commanded special attention in this scheme. The finished product stood as a monument not only to the ingenuity of the present but also to a prosperous future. In economic terms, President Roosevelt's Boulder Dam speech extolled public works projects as economic multipliers: "Such expenditure on all of these works, great and small, flow out to many beneficiaries; they revive other and more remote industries and business, money is put in circulation, credit is expanded and the financial and industrial mechanism of America is stimulated to more and more activity."

The president noted that Reclamation was "the instrument" that finally realized the goals of the legislation first introduced in Congress by Senator Hiram Johnson and Congressman Phil Swing. While originally intended to regulate and control the flow of the Colorado River, the authorization of the big dam also brought "cheap power" to the Southwest. When Roosevelt emphasized "cheap power," he meant public power, which was not totally accurate with the sale of most of the dam's electricity to private power companies. Yet, the president insisted in his rhetoric that public power "will also prove useful yardsticks to measure the cost of power throughout the United States." The new sources of public hydropower, he was certain, would put a check on monopolistic pricing practices exercised by private power interests. These words sharpened the battle lines that had long been drawn between the forces of private and public power. On the crest of a great public works dam above the rising waters of its reservoir, the president proclaimed a victory for public power and its restraint upon the price of power everywhere.[5]

The Great Depression politicized the struggle between the public and private spheres of the economy as never before. The two camps stood divided between forces endorsing the New Deal's activist governmental effort to meet the problems of the Depression and those who objected to what they viewed as the unwarranted expansion of federal power into the private economy. They argued that if government remained passive, but benign, the private economy would recover from the Depression without the "pump priming" of the federal government that only discouraged initiative and enterprise. As the New Deal Administration continued in power from 1933 through World War II, Reclamation benefited from an activist government. Reclamation's track record in the construction of dams, tunnels, and other water projects made it the obvious choice to receive huge appropriations for structures that commanded public attention. Its new edifices displayed (often literally) concrete examples of the progress being made in the battle against the Depression. Expenditures of the Bureau of Reclamation averaged $52 million a year during the New Deal years, compared to $8.9 million prior to 1933. [6]

This good fortune for the Bureau of Reclamation was by no means predictable when the Roosevelt Administration moved into the White House. Although Commissioner Mead demonstrated an ability to survive through different Republican Administrations in the 1920s, his reappointment by

President Roosevelt in 1933 was an exception amidst the almost total dismissal of Republican appointees. His survival rested on his "reputation" and "positive image," according to his biographer.[7] But he now faced a new Secretary of the Interior, Harold Ickes, who was, at best, indifferent to the entire enterprise of Reclamation. In 1934, Ickes proposed a transfer of Reclamation to the Department of Agriculture in exchange for the Department of Agriculture transferring the Forest Service to the Department of Interior. At the last minute, Secretary of Agriculture Henry A. Wallace refused to consider it. Yet, Secretary of the Interior Ickes's indifference and suspicion gradually turned into admiration and respect as he grasped the Bureau of Reclamation's potential for producing hydroelectricity from large dam facilities. The turn of events was surprising, considering Ickes's reputation of irascibility. During the dedication at Boulder Dam, Ickes, for the first time, publicly embraced the grand designs and accomplishments of the Bureau of Reclamation. He said:

> Here behind this massive dam is slowly accumulating a rich deposit of greater wealth than all the mines of the West have ever produced – wealth to be drawn upon for all time to come for the renewed life and the continued benefit of generations of Americans.[8]

On the other hand, farmers on reclamation projects still struggled to escape their burden of debt and disastrously low agricultural prices. After Congress and the new Administration approved the National Industrial Recovery Act in June of 1933 and embraced the principle of funding public works, the Public Works Administration approved $103 million for Reclamation. Under the new appropriations, the Denver Office's work force increased from 200 to 750. New designs and new investigations were the order of the day for an organization that already had won a much-admired reputation for constructing over a hundred dams throughout the American West. By the beginning of 1934, Reclamation had some fifteen dams under construction, from small projects to large, multi-million dollar projects like Hoover and Grand Coulee Dams. Reclamation engineers had also been asked to design two Tennessee Valley Authority dams and the Caballo Dam on the Rio Grande, built under the authority of the International Boundary and Water Commission, but later operated as part of the Rio Grande Project. Suddenly, in the midst of depression gloom, the Bureau of Reclamation was abuzz with new activity as "the instrument of construction" for the development of multiple resources — rural and urban water, power, water storage, flood control, and navigation — in the river basins of the West.[9]

As the Bureau of Reclamation moved into a decade that reshaped its destiny, its old leadership was passing from the scene. From Oakland, California, Arthur Powell Davis took satisfaction in the progress underway on the big dam on the Colorado for which he was one of the earliest advocates and visionaries. He knew that, upon its completion, construction of the All-American Canal and the aqueduct to Los Angeles must soon get underway. Accordingly, Frank Weymouth, now Director of the Metropolitan Water District of Southern California, who had left his position as Chief Engineer with Reclamation in 1924, invited Davis to serve as a consultant on the construction of Los Angeles's aqueduct from the Colorado. Davis had just returned from a trip to Russia. As he noted in a letter to Frederick Newell, he was now giving speaking engagements to people who seemed intrigued with communism. Many, he said, thought it a remedy for the economic problems of the United States, but he strongly discouraged such ideas. Still, he admitted, "Until we can point a way out, it is pretty hard to make much progress against such propaganda." This exchange between old colleagues Davis and Newell occurred in 1931 Christmas messages. Newell died only seven months later, in July 1932, and Davis died in 1933. Morris Bien, Assistant to the Director of the early Reclamation Service, died at his home in Washington, D.C., at age 73 in 1932. Bien had developed a code of irrigation laws (the "Bien Code") to facilitate the operations of the Reclamation Service in western states that only Nevada fully adopted, while nine other states adopted aspects of it for their irrigation laws. [10]

6.4. The Boulder Dam commemorative stamp was ready for the dedication. A portion of the first-day commemorative covers are shown here on September 30, 1935.

When Boulder Dam was dedicated in September of 1935, Mae Schnurr, Assistant to the Commissioner of Reclamation and often Acting Commissioner when Mead was out-of-town, wrote Mrs. Davis, "With the dedication of Boulder Dam by President Roosevelt, I naturally would think of you and Mr. Davis." With her letter she enclosed a small booklet prepared for the dedication of the dam and a commemorative postage stamp issued for the occasion. She pointed out that the booklet cited A. P. Davis's early role in enabling the undertaking of the large dam project. Just four months later, shortly after the New Year, Commissioner Mead died at his post at age 76. After attending Mead's funeral on January 29, 1936, Secretary of the Interior Ickes noted in his diary that the Commissioner was an incorruptible, intelligent public servant who had few enemies. "He always seemed to be able to get along with Senators and Congressmen and others in the West where his work lay," wrote Ickes, "and yet he wasn't a yes man by any means." Later in the year, Ickes remarked that Reclamation, now led by Acting Commissioner Page, "was one of the best run and most efficient agencies in the whole Government." [11]

Both project settlers and Congress squeezed the Bureau of Reclamation in the early years of the Depression. Some congressmen wanted to use the Reclamation Fund to make unemployment payments in western public land states. Commissioner Mead responded that such a proposal was impossible because the money in the Reclamation Fund fell far short of the amount to meet even existing construction obligations. In fact, Congress made an emergency appropriation ("advance") to Reclamation of $5 million in March of 1931 to keep its operations afloat, including the preliminary work on the Boulder Canyon Project.

On the projects themselves, the Bureau of Reclamation resisted efforts by water users to further reclassify lands into nonproductive categories to avoid construction charges. The Adjustment Act of 1926 had established seven land classifications. Category six was "temporarily unproductive," and category seven was "permanently nonproductive." Chief Engineer Raymond F. Walter recommended against accepting any new classification. He predicted that this would mean large amounts of productive land would be reclassified in "an attempted evasion of the joint liability obligation which was [the] basis of all write-offs and concessions given them under the new contracts authorized by the Adjustment Act." He concluded that existing economic conditions were "abnormal" and that, "Advantage is being taken of the present unprecedented depression to use the argument for reclassification of lands as a pretext for writing off charges on all lands which cannot be profitably farmed under present conditions." Also, in Public Law 232, the 71st Congress, in 1930, authorized the sale "of vacant public lands designated under the Act of May 25, 1926, as temporarily unproductive or permanently unproductive" on federal irrigation projects to resident owners and resident entrymen under terms and conditions to be fixed by the Secretary of the Interior. The act hardly caused a rush on the part of impoverished farmers to buy up "worthless" lands, but it was an indication that the government was searching desperately for additional funds from the sale of public lands even on the projects themselves. Another way of interpreting this legislation was that Congress and Reclamation were abandoning large acreages of land on many irrigation projects. The Williston Project in North Dakota, for instance, was abandoned entirely in 1926.[12]

As Congress began to recognize the harsh realities of a prolonged Depression in both agriculture and industry after the fall of the stock market in 1929, some in and out of Congress saw the underpopulated reclamation projects with their vacant farm plots as homes for the resettlement of destitute farmers. Little mention was made of an earlier theme of settling the unemployed from the cities on reclamation projects, but echoes of the "Back-to-the-Land movement" emerged in some New Deal programs to meet the farm crisis and even urban unemployment. Experience, however, indicated that both farming skills and capital were a requisite for success on the projects, and the Bureau of Reclamation now required that new settlers have at least $2,000. Poor people would be hard put to meet the requirement. Still, within the New Deal leadership, the compelling idea of resettling people on the land persisted.

Resettlement Issues

By the autumn of 1933, Secretary of the Interior Ickes created a Subsistence Homestead Division within the PWA and made the New Deal agricultural economist M. L. Wilson its director. The Bureau of Reclamation immediately began receiving inquiries from western congressmen about "shifting farmers from sub-marginal lands to irrigated lands," and Reclamation saw the opportunity to tap resettlement resources available from the Subsistence Homestead Division. But in November 1933, Wilson had to explain to Acting Commissioner Mae Schnurr that his efforts were presently concentrated in the southern states. Still, he noted that he was giving some attention to the Milk River Project in northern Montana, with which he was familiar because of his previous career as a professor of agricultural economics at Montana State College in Bozeman, where he worked on the problems of sub-marginal lands in that state. By December 1933, Commissioner Mead agreed that help could be accepted from the Subsistence Homestead Division to place farmers on the Milk River Project, but Wilson said that the demands on his agency made it unlikely that assistance could be extended to other government irrigation projects. By 1935, the Resettlement Administration, under Rexford Tugwell in the Department of Agriculture, absorbed the Subsistence Homestead Division, first organized in the Department of Interior. By this time, the prolonged western drought and dust bowl conditions in parts of the Great Plains made irrigation projects with stored water an attractive resettlement option.[13]

Money soon became available for a "rehabilitation program" to move farmers from marginal land and place them on farms where, according to Mead, "rewards will be more certain." The Bureau of Reclamation notified the USDA's Resettlement Administration of its continued interest in working with the programs transferred to the Subsistence Homestead Division of the Department of the Interior. In May 1935, Mead requested permission from the Secretary of the Interior to open discussions with Tugwell about relocating farmers from marginal lands onto government reclamation projects that could provide settlers with an assured water supply. Mead mentioned that on the Belle Fourche Project in South Dakota, up to 20,000 acres of privately owned land could be made available on reasonable terms; on the Riverton Project in Wyoming, 40,000 acres of public land might be obtained by payment of $1.25 an acre to the Wind River Indian Reservation; and on the Owyhee Project in Idaho and Oregon, up to 30,000 acres of public land were available. Mead envisioned "teamwork" that, by 1937, could turn over to Tugwell's organization many acres of irrigated land. Accordingly, the Secretary of the Interior granted Mead permission to confer directly with Tugwell on the matter. By the summer of 1934, Tugwell was also Acting Secretary of Agriculture. In early November 1934, Mae Schnurr, in her capacity as Assistant to the Commissioner, visited the Resettlement Division to open more direct communication. Shortly thereafter, the Resettlement Administration decided to begin resettlement work on 3,000 acres in the Riverton Project in Wyoming; on 1,600 acres in the Uncompahgre Project in Montrose and Delta Counties, Colorado; and on 2,000 acres in the Grand Valley Project in Mesa County, Colorado. In addition, information was requested on projects in Montana.

Commissioner Mead soon became uneasy about the activities of the Resettlement Administration. He feared that its purchase of lands within reclamation projects might make the Bureau of Reclamation a bystander in the process. Mead wrote Ickes, "We have adopted the policy of welcoming the activities of the Resettlement Administration on all our projects," but "The fact that we are not buying [the land] for settlement ought not to lead to a policy where Reclamation would be prohibited from such settlement activity." Bureau of Reclamation economist, L. H. Mitchell, commented in a letter to Mead in August of 1935 that the proposals presented dangers to the entire Reclamation undertaking. This communication occurred after Mitchell attended the Regional Conference of the Rural Resettlement Division of the Resettlement Administration, held August 15-17, in Salt Lake City. He wrote, "It is my opinion [that] the placing of settlers on federal reclamation projects without the clear understanding every dollar must be repaid is dangerous. It will have a bad influence on those who are now trying to make good and repay their obligation to the United States and other creditors."[14]

Yet, Mead kept the lines of communication open to Tugwell because the Resettlement Administration promised funds that could be used to expand Reclamation's irrigation projects. Reclamation did not have the money to buy vacant lands within reclamation projects. For example, there was considerable unsettled area in the Belle Fourche Project, some of which was owned by the state of South Dakota, some by the county, and some by mortgage companies. Mead believed it should be "resettled" and welcomed help from the Resettlement Administration. He explained that the Belle Fourche Project was settled and developed at a time when no inquiries were made into the capital and experience of settlers. Miners took up homesteads with no agricultural knowledge. The result was farm failures and farm sales. Much of the land passed into the ownership of non-residents. The project desperately needed a resettlement program, and Mead informed Tugwell that, "The activity of your Administration in carrying it out is invited." Tugwell replied that he fully understood the limitations under which Reclamation worked in regard to land purchases and pledged the cooperation of the Resettlement Administration. He asked agricultural economist Dr. L. C. Gray to discuss the matter in detail with the Commissioner. Secretary Ickes was of the view that, "the Bureau of Reclamation ought to be made the agent of the Resettlement Administration for the purchase of lands in connection with reclamation projects."[15]

Still, the plans of the Resettlement Administration remained more of an ideal than a practical program. Land purchases occurred in many states, but the program became a target of critics in Congress. The Administration soon lost interest in resettlement, and only 4,441 families were resettled. Though he did not have the necessary budget, Mead advised purchase of lands in the Columbia Basin Project. He felt this was the best means to obtain the most economical development of the land, layout of canals and roads, and shaping of farms for the kinds of agriculture to be practiced. Moreover, during the summer of 1935, Tugwell became aware that the Bureau of Reclamation guarded its turf with respect to other agencies pursuing land purchases and settlement policies within its projects.

In 1935, Tugwell informed the Secretary of the Interior that the Division of Land Utilization of the Resettlement Administration had withdrawn from the purchase of lands in the Columbia Basin. Originally, the Division of Land Utilization had intended to purchase marginal lands within the project to remove them from farming. But the program on this project did not appear feasible when restricted to land not subject to irrigation because the two types of land could not be easily separated. Tugwell explained that, in order to eliminate any possibility of competition between the two bureaus involved, it was deemed advisable to withdraw from the project entirely. Tugwell probably foresaw that irrigation on the Columbia Basin Project would face years of delay and criticism even if the project was completed. Underlying Tugwell's suspicion of rural land purchases was his hesitancy to encourage uneconomical rural settlements and small producers who could not compete in the larger market or in economies of scale.[16]

In the fall of 1934, Commissioner Mead noted that the Columbia Basin Project, then under consideration, did not include irrigation. He told a member of the Wenatchee Chamber of Commerce that he could not predict when "the irrigation feature" would be added to the project. While he thought some wanted to test the ability of settlers on the project to make a living on small tracts of land, it was clear he believed the farms would be of the usual size with little hope for small subsistence plots. In loyalty, however, to the small farm idea, he pointed out that Reclamation had tried to develop the idea of farm laborers' allotments that ranged in size from five to twenty acres on the Riverton and Shoshone Projects in Wyoming. The small holdings could enable farm laborers to feed themselves, with a small surplus to market, but the major income would come from their work on neighboring farms. Mead was clearly trying to hold on to the ideal of the small farm, but his efforts remained more rhetoric than action. And despite the ambitious plans of Tugwell's Resettlement Administration for the western reclamation projects, it did little. Most of the 4,441 resettled families took up land in the South. Much of the Resettlement Administration's work became diverted to other endeavors: farm rehabilitations, camps for migratory workers, community cooperatives, greenbelt towns, and land utilization projects that converted submarginal farms into grazing lands.[17]

In 1937, the Water Facilities Act permitted the Department of Agriculture to develop water for small farm communities for the purpose of resettlement. Interior Secretary Ickes was unhappy over this development because he viewed it as duplicating the work of Reclamation in the USDA. By 1939, he succeeded in moving some of its functions to Reclamation, but had to agree to cooperate with the Department of Agriculture in a program to resettle 2,500 families on twelve irrigation projects in eight western states. While Secretary Ickes fought to direct resettlement funds into the Interior Department, Reclamation leadership expressed some hesitancy about resettling poor farmers on its projects. Also, economists expressed reservations about expanding irrigation acreage in order to provide new opportunities for farm families when only poor livings could be made from the land in those times of depressed prices.[18]

Farm and land questions did not command the major portion of Commissioner Mead's attention. While the retirement of lands played a role in the proposed movement of farmers to reclamation projects, in September of 1935 the Commissioner told the Secretary of the National Reclamation Association that he did not wish to see the question featured at the upcoming fourth annual meeting of the association at Salt Lake City in November. Land retirement suggested that reclamation imposed a financial penalty on the rest of the country. "In other words," he said, "if for every new acre of land ten submarginal acres are to be taken out, it puts a spotlight on the federal reclamation projects as contributing in a very material way to the surplus of agriculture." Reclamation adding to the farm surplus was an old question that would not go away for the Commissioner even in those drought years. For years, the Commissioner took criticism on this score from the Farm Bureau, the Grange, and, of course, the Department of Agriculture. Soon, however, such questions faded in comparison to the grander events on the Commissioner's agenda, such as the dedication of Boulder Dam.[19]

New Agendas

The new agenda of Reclamation was indeed grand. Millions of dollars came to it for the construction of high dams, tunnels, canals, and hydroelectricity. Multipurpose river development and vitally needed jobs underlay many of the schemes. On the already-existing projects, settlers also benefited directly from other New Deal programs. The Rural Electrification Administration (REA) financed the construction of electrical lines to rural areas where the expense of extending lines was too great for the small number of rural customers served.

Although the Minidoka Project prided itself on having provided electricity to farmers and nearby communities at an early date, only 50 percent of the farms on the project received electric power. When the Board of Directors of the Minidoka Irrigation District in Rupert, Idaho, learned of the REA program in early 1935, it was eager to make electricity available to all farms on the project. Electricity meant greater convenience to farmers and their wives. Supporters rarely mentioned electricity without noting how it eased the work burden for farm wives.[20]

The Civilian Conservation Corps (CCC) was another prominent New Deal program that assisted the reclamation program. It achieved high visibility in National Forests and National Parks, but also found assignments on reclamation projects building and renovating recreation facilities for boating, camping, and picnicking at new and old reservoirs as well as assuming some traditional operation and maintenance tasks. For example, on the Vale Project in Oregon, CCC enrollees helped repair canals, laterals, and gates. Elsewhere, the CCC built dams, extended canals, eradicated rodents and weeds, and performed operations and maintenance duties. According to *Reclamation Era*, "By their work … they are raising to themselves a mighty monument whose component parts will be green fields, trees, and comfortable farm homes." [21]

6.5. Civilian Conservation Corps (CCC) enrollees laying concrete pipe on the Vale Project, Oregon, in May of 1937.

6.6. A Reclamation photographer labeled this November 16, 1935, CCC image "Uncompahgre Project. Truck drivers and their trucks in background after their Saturday bath."

The CCC workers, made up of single men age 18 to 25, showed up with picks, shovels, and hammers to lend a hand in construction and maintenance tasks on the projects. At its height, the nation's CCC enrollees totaled 505,782 men in 2,652 camps in the summer of 1935. In 1934, Reclamation received authorization for its first nine CCC camps. The first was assigned to the North Platte Project. CCC operations and camps occurred at over eighty different locations on forty-five Reclamation projects in fifteen western states. Some camps were financed under drought relief appropriations. Bureau of Reclamation personnel cooperated with the Army in administering CCC camps and work.

6.7. The semi-military character of the CCC program is illustrated by this March 4, 1937, assembly of enrollees at Camp BR-44 on the Umatilla Project.

Beginning in June 1937, the CCC program required that ten hours each week, out of the 40 hour/5 day week, to be devoted to job training. Supervisors on Reclamation projects recognized the need for vocational training and welcomed this additional requirement for CCC workers. Christine Pfaff's careful and important study of the CCC work on Reclamation projects noted the importance of the program for western irrigation farmers: "By 1934, it had become critical for the federal government to address the plight of western farmers and to protect its hefty investment in irrigation projects." Pfaff concluded that the CCC provided the perfect program to improve Reclamation projects during hard times and drought, and it helped put unemployed men into productive work under competent supervision.[22]

By the mid-1930s, the CCC's work became a regular topic of discussion in the pages of the *New Reclamation Era,* a topic often explored by historians attempting to understand the local impacts of the New Deal programs. From the new Vale Project in Oregon came the report that CCC enrollees were helping to transform a "sagebrush waste" into a prosperous farming community. Fending off the ever-present criticism of New Deal work programs, one article asserted

6.8. The ECW tent camp at Belle Fourche Dam, Belle Fourche Project, South Dakota, in August 1934. ECW, Emergency Conservation Work, was the early name of the CCC.

that the CCC accomplished its work "with great pride" that reflected favorably on the ability and efficiency of the supervisory personnel. "With competent supervision, the CCC man will turn out as much and as good work per hour as the average man working for a contractor or on force account."[23]

In 1938, the CCC contributed almost a million man days to the reclamation projects. CCC operations included: moving 1,165,600 cubic yards of earth; moving 40,300 cubic yards of rock; placing 161,200 square yards of concrete lining; building 3,000 water-control structures; constructing 590 miles of operating roads; building 82 vehicle bridges, mainly over canals; cleaning and clearing 7,463,300 square yards of canals; laying 92,600 feet of pipe and tile lines; riprapping 234,000 square yards of canal banks; and moving 382,800 cubic yards in leveling work. Some of this activity brought protests on the part of private businesses, which charged that the CCC diverted business away from contractors and that CCC work on reclamation projects benefited private individuals rather than the general public. In response, Reclamation drew up new guidelines that confined CCC camps and their work to lands in government ownership. For fiscal year 1939,[24] the Director of the Civilian Conservation Corps authorized ten additional CCC camps to the Bureau of Reclamation, bringing the total number of camps operating on the projects to forty-four.[25]

6.9. CCC enrollees lining a ditch on the Truckee River Storage Project near Reno, Nevada. December 1935.

On the Newlands Project in Fallon, Nevada, CCC enlistees arrived in June 1935 to set up Camp Fallon on the edge of town. The camps were not under the direct supervision of Reclamation. They were commanded by Army officers and adhered to a military regimen. The CCC men received salaries of from $30 to $45 a month, plus room and board. A portion of their salaries had to be sent home to families. Two camps of about 225 men took up bivouacs that soon became permanent barracks near Fallon. There, local farmers provided the CCC with jobs that taught them building, masonry, and various construction skills. The Truckee-Carson Irrigation District (TCID) provided teams and other equipment as the CCC rebuilt diversion gates, cleaned canals, opened new ditches, and renovated the Truckee Canal from the Derby Diversion Dam, on the Truckee River, to Lahontan reservoir. These CCC camps published a newspaper,

6.10. CCC enrollees from BR-34 gathering rocks for riprap and an enrollee from BR-35 sealing a crack in a structure at Lahontan Dam. Newlands Project, Nevada.

6.11. Moon Lake Dam, Utah, while under construction by the CCC in 1940.

6.12. Basin No. 3 of the All-American Canal desilting works unwatered for reconditioning and major repairs in 1946.

the *Carson River Mustang* that described social activities and reflected the good relations between the CCC "boys" and the town that enjoyed their presence and the extra cash they spent on main street. By 1938-1939, some CCC men left the Fallon camps to work on Boca Dam, a Reclamation reservoir on the Little Truckee River in the Sierra Nevada west of Fallon and Reno. In 1942, World War II brought the closure of the CCC camps in Fallon and nationwide.[26]

By January 1934, Reclamation announced that the costs of its current and planned projects amounted to at least $218,440,000. The money encompassed a long list of items. These included ongoing appropriations for completion of Boulder Dam, including building the powerhouse, and equipping it with turbines and generators. PWA Funds supported many other projects:

- The All-American Canal
- The Verde Project, located near Prescott, Arizona [never completed]
- The Humboldt Project, in north central Nevada, to store waters for irrigation of hay lands near Lovelock, Nevada
- Truckee Storage Project for Boca Dam on the Little Truckee River in California before it flows into the Truckee River upstream from Reno, Nevada

- Provo River Project, above the town of Provo, Utah, to supply a canal running north to the suburban areas southeast of Salt Lake City for municipal, domestic, and irrigation purposes
- Moon Lake Project, in northeastern Utah
- Ogden River Project, in Utah, for building the Pine View Dam, on the Ogden river a few miles above the city

6.13. The All-American canal.

- Ephraim Project, in southwest Utah near Ephraim and Spring City
- Hyrum Project for the construction of a dam on Little Bear River near Hyrum, Utah
- Uncompahgre Project in Colorado, one of the older Bureau of Reclamation projects, for the construction of a dam and reservoir near Gunnison, Colorado, and for the completion and lining of the Gunnison Tunnel near Montrose, Colorado, and other betterment works
- Casper-Alcova Project, near Casper, Wyoming, for the construction of dams and hydroelectric power facilities to irrigate 66,000 acres to be paid back mostly from power revenues
- Milk River Project, between Havre, Montana, and the Canadian boundary, for flood control and more stable water supplies for irrigation

6.14. In August 1935 irrigation tunnel No. 2 on the Casper-Alcova (Kendrick) Project was under construction.

- Sun River Project, in Montana, to build laterals from canals already completed and improve drainage on lands already under irrigation in this older Reclamation project
- Upper Snake River Division of the Minidoka Project, in Idaho, for the improvement of Jackson Lake and American Falls Reservoirs, on the Snake River, to provide adequate water storage for the lands served by this stream
- Owyhee Project, on the Oregon-Idaho border, for the completion of irrigation works and water delivery systems made possible by the already-completed Owyhee Dam

6.15. August 11, 1930, Reclamation's Owyhee Dam was still at its foundations.

- Vale Project to provide for the construction of the Agency Valley Dam that would create a reservoir for the storage of 60,000 acre feet of additional water for the project near Vale, Oregon
- Stanfield Project near Stanfield, Oregon, that became part of the Umatilla Project, for the rehabilitation of irrigation facilities under the direction of the Stanfield Irrigation District

6.16. Owyhee Dam. 1935.

The Bureau of Reclamation also entered into a contract with the Metropolitan Water District of Southern California to design and construct Parker Dam on the Colorado River at the head of the Colorado River Aqueduct. The PWA also permitted the Bureau of Reclamation to design two major dams for the Tennessee Valley Authority: the 268-foot-high Norris Dam and 76-foot-high Wheeler Dam. Finally, the largest appropriation from the PWA to the Bureau of Reclamation ($15,000,000 of a total of $63,000,000 projected requirement for the project) came for the newly authorized Columbia Basin Project in Washington State.[27]

6.17. Parker Dam. 1946.

The Bureau of Reclamation on the Columbia River

Aside from Hoover Dam, Grand Coulee Dam was the largest undertaking of Reclamation in the 1930s. While Reclamation showed an early interest in the Columbia, it "eschewed involvement." The Boulder Canyon Project overshadowed everything else including the Columbia. Meanwhile, in 1925 Congress ordered the Corps of Engineers to estimate the cost to survey major river systems (except the Colorado) and formulate integrated plans regarding navigation, power, flood control, and irrigation. In 1926, the Corps presented the result of these studies, House Document 308, to the First Session of the 69th Congress. House Document 308 estimated that the "308 Studies," as they were to be called, would cost $7,322,400. All major United States rivers and streams would be studied for the purpose of determining their potential for multiple-use development. The Rivers and Harbors Act of January 21, 1927, directed the Chief of Engineers to undertake the studies, and the first funds became available to the War Department in March of 1928. The disastrous floods on the Mississippi in the spring of 1927 placed the Mississippi and Missouri Rivers high on the Corps' priority list, but Senator Wesley L. Jones of Washington State insisted that the Corps also give special attention to the Columbia River. The task fell to Major John S. Butler, District Engineer in the Corps' Seattle office.[28]

In Washington State, controversy had raged for nearly a decade over two plans to irrigate the dry lands of the Columbian Plateau in eastern Washington. Support for a high dam at Grand Coulee came from an enthusiastic triumvirate, the "Heroes of Columbia Basin," including newspaper editor Rufus Woods of the *Wenatchee Daily World,* Ephrata lawyer Billy Clapp, and contractor-lawyer James Edward O'Sullivan. Their approach proposed pumps powered by electricity generated at the dam to lift water from the reservoir behind the dam into the natural reservoir of the Grand Coulee some miles away. From this natural reservoir in the plateau lands of the Columbia River, water could be directed into the rolling lands of its Big Bend country. The campaign for a huge, publicly-built dam with tremendous hydroelectric potential received broad support among business and professional people in the central part of the state and appealed to advocates of public power at the state and national levels. William Warne's[29] history of the Bureau of Reclamation called local efforts for the high dam at Grand Coulee by Woods, Clapp, and O'Sullivan "the most remarkable promotion in the development of the West." The unrelenting, almost obsessive work on the part of Woods

6.18. Rufus Woods with Billy Clapp, two tireless promoters of a high dam at Grand Coulee. Courtesy of *The Wenatchee World.*

6.19. Rufus and Mary Woods with pipe-smoking James O'Sullivan. O'Sullivan was a third tireless promoter of the Columbia Basin Project. Courtesy of *The Wenatchee World.*

and O'Sullivan for the dam gained them the reputation of "amiable nuts, bitten by a promotion bug."[30]

The dam they envisioned would produce more power than Hoover Dam. A small portion of it would be used to pump water up from the reservoir behind the dam into the "Grand Coulee." The river eroded this huge gorge during a monstrous flood in the geologic past, after which the river fell back to the course it followed before that flood. The Grand Coulee, a steep-walled valley or ravine, provided an enormous potential reservoir from which water could be directed to lands of the Columbian Plateau that lay in a rough triangle to the southwest and north of where the Snake River joins the Columbia at Pasco, Washington. It was an ingenious plan to build a dam and a reservoir with the skills of advanced technology and industry and at the same time utilize an ancient structure of nature – the Grand Coulee.[31]

This plan, known as the "the pumpers' plan," competed with "the gravity plan," advocated by the Spokane Chamber of Commerce and Washington Power and Light Company. The gravity plan projected a diversion of the Pend Oreille River in Idaho into the Spokane River with water eventually reaching the Columbian Plateau (the lands of the Big Bend Country of the Columbia River) through aqueducts and tunnels by gravity. The plan also called for a series of small dams with a fraction of the power production that a large dam at Grand Coulee might produce. Private power interests argued that using public funds to build a gigantic dam on the Columbia would produce a surplus of power, that would drive the price down at a time when there was little market for power in a Pacific Northwest not yet industrialized. In the 1920s, in the wake of the post-World War I "Red Scare," private power interests vilified government projects as steps toward socialism and bolshevism. This charge "became the bugaboo" in the arguments manifested against "the pumpers' plan." Public power and big dam advocates countered that cheap power would facilitate industrialization as well as provide a new irrigated empire on the Columbian Plateau. They argued that the dam should be built to stimulate economic growth rather than simply wait for it.[32]

Democratic presidential candidate Franklin D. Roosevelt agreed. In his presidential campaign swing through the Pacific Northwest in 1932, Roosevelt promised to support the development of hydroelectric power on the Columbia. This meant a big dam that would guarantee abundant, cheap power for the region. Not to be outdone in the campaign for Pacific Northwest votes, in late October of 1932, President Hoover, too, declared his support for construction of a Grand Coulee Dam. By this time, the verdict was in from the U.S. Army Corps of Engineers. Major Butler's report favored the pumpers' plan and the construction of the Grand Coulee Dam on the Columbia. As early as 1931, it was rumored, and in some circles well-known, that Major Butler's 308 study probably would recommend the big dam over the gravity plan. Commissioner Mead, in 1931, finally broke the near-silence of Reclamation not only on the question of Columbia River development but also on the controversy over the two plans by declaring Reclamation's interest in building the big dam at the Grand Coulee. Reclamation had, since World War I, resisted local appeals and campaigns for a large dam. Its silence, and even diffidence toward a Columbia River project, contributed to delays in answering questions on what to do about the Columbia. The big dam backers probably would have met defeat in the 1920s if a decision about the water development of the Columbia River country had been forced upon the Pacific Northwest in that decade.

The advent of the Great Depression changed everything. As events moved toward the realization of a project on the Columbia, Reclamation favored the big dam with its pumping plan and revealed a support that had, "rested secretly within the Bureau of Reclamation, whose engineers saw it as preferable to the gravity plan." Not all engineers were of one voice on the issue. George W. Goethals, the Supervising Engineer who oversaw the completion of the Panama Canal, supported the gravity plan as did the later Commissioner of the Bureau of Reclamation, Harry W. Bashore.[33]

6.20. Harry W. Bashore, Commissioner of the Bureau of Reclamation from August of 1943 to December of 1945.

The Corps of Engineers' report ultimately favored the big dam with its pumping capacity rather than the gravity plan. That it did so can be seen as a testimony to the impartiality of Major Butler's investigations and report. The recommendation for a large dam seemed to play to the interests of both Reclamation and the Corps. On the one hand, Reclamation clearly was in the business of building large dams. But, on the other hand, the Columbia was definitely a navigable river, and the Corps had some expectation of building it for that reason. In any event, Butler seemed to ignore the pressure from the powerful private power interests in Washington State and delivered a report that was an "honest, professional, detailed assessment" of the situation and "laid the foundation for the Grand Coulee." While Woods, Clapp, and O'Sullivan stood out as the local heroes of the Columbia Basin Project, supporters of public hydroelectric power at the state and national level played crucial roles that all came together under the enthusiasm for public works projects during the New Deal. Washington State Governor Clarence D. Martin, as well as the state's U.S. Senators, Homer T. Bone and Clarence C. Dill, cooperated with Harold Ickes and the president in passing a 1933 appropriation bill to begin construction at the site of the Grand Coulee.[34]

6.21. Governor Clarence D. Martin releasing concrete during the first official placement at Grand Coulee Dam in November 1935.

In response to President Roosevelt's appeal for emergency funds to put America back to work in 1933, Congress appropriated $500,000,000, which included funds to begin work on Grand Coulee Dam and also for the U.S. Army Corps of Engineers to build Bonneville Dam, upriver from Portland. The two bureaus would share in the Columbia's development, but their competition militated against any future consolidation of the Columbia River under one single development administration or authority. The Columbia, Missouri, and Tennessee River Valleys were the major river valleys in which the New Deal sought to develop public works programs related to water, hydroelectricity, flood control, and navigation.[35]

For its part, the Bureau of Reclamation moved ahead with preliminary work on the Grand Coulee, soon issued specifications for construction of the dam, and invited bids. On June 18, 1934, Reclamation opened the bids. Present were Elwood Mead; Washington's Governor Martin; Project Engineer Frank Banks; Bernard Stoutemyer, Chief Legal Counsel for Reclamation; and

Reclamation's Chief Engineer Raymond F. Walter. Thousands jammed Spokane, where the decision was made, to learn who would build the largest structure in the world and bring progress and industry to the Pacific Northwest. Most expected the Six Companies to walk away with the prize, given their success in the competition for the Boulder Canyon Project. To the surprise of almost everyone, another consortium made up of the Silas Mason Company from Kentucky, Atkinson-Kier Company of San Francisco, and the Walsh Construction Company of Davenport, Iowa, incorporated under the name MWAK, submitted the low bid of $29,339,301.50. (The Six Companies' bid came in at $34,555,582.) Six Companies representative, Henry Kaiser, tried to conceal his disappointment. Later, Six Companies, Inc., would obtain contracts for some of the work on the project, but Kaiser was clearly frustrated by the results of the original bidding.

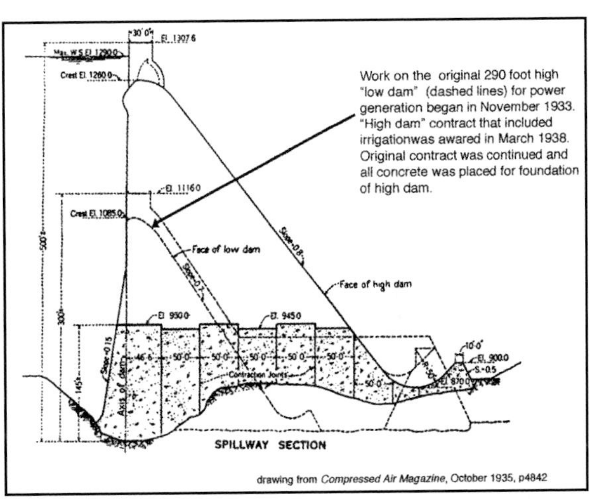

6.22. This spillway section of Grand Coulee Dam shows both the original low dam plan and the final high dam as built. With permission from *Compressed Air Magazine,* a publication of Ingersoll-Rand Company.

Less than two months later, President Roosevelt and his entourage made a well-publicized visit to the Pacific Northwest. Major stopovers included both the Bureau of Reclamation's construction site at Grand Coulee and the Corps of Engineers' Bonneville Dam site some thirty miles upstream from Portland, Oregon, on the Columbia. The two projects, assigned to different bureaus, signaled that development on the Columbia would not be unified or under the direction of one administrative unit like the TVA. The president appeared at Bonneville on August 3, and then

6.23. Project Construction Engineer Frank A. Banks with President Franklin Delano Roosevelt at Grand Coulee Dam in October of 1937.

journeyed by train north to Grand Coulee. Secretary of the Interior Ickes estimated that 20,000 people drove long hours across the hot, dry plateau of eastern Washington to see and hear the president. According to Paul C. Pitzer, in his history, *Grand Coulee: Harnessing a Dream*, after looking over the desolate country in the middle of nowhere, First Lady Eleanor Roosevelt commented that, "It was a good salesman who sold this to Franklin." Good salesmen or not, President Roosevelt staked much on the success of the New Deal's public work projects, and Grand Coulee Dam would be the crown jewel of his efforts in the Pacific Northwest. It was a monument to his Administration's attempt to industrialize the Pacific Northwest and demonstrate what the federal government could do to promote the general welfare through water projects.[36]

Reclamation's original plan was for a low dam that could be raised in height later, but in August 1935, after heavy lobbying by the original local backers of the Grand Coulee site, Rufus Woods, James O'Sullivan, and Billy Clapp, Congress authorized the full high dam along with the irrigation project. MWAK proceeded to build the foundation of the dam quickly, and, from 1938 to 1941, Consolidated Builders, Inc., a combination of MWAK and the Six Companies under Henry Kaiser, completed the high dam. This included the left powerhouse along with a foundation for the right powerhouse and pumping facilities.[37] To meet wartime demands for power in the Pacific Northwest for the production of aluminum, airplanes, and ships, six 108,000-kilowatt Grand Coulee generators and two 75,000-kilowatt generators were installed. These latter two had been scheduled for Shasta Dam in California. As the United States rushed to complete the world's first nuclear bombs for use at the end of World War II, the available power and water from the Columbia made Hanford, Washington, downstream from

6.24. Cleanup of loose material at Grand Coulee required many drill bits. Here they are being tempered. 1936

6.25. Cleanup of bedrock at Grand Coulee Dam. 1937.

6.26. Excavating the foundation behind the west coffer dam at Grand Coulee. 1936.

6.27. The west concrete plant at Grand Coulee and the beginnings of the trestle system used to place concrete at Grand Coulee.

Grand Coulee, an ideal site for the production of plutonium for the Manhattan Project. After the two Shasta generators had served the war effort at Grand Coulee Dam, they were moved to the completed Shasta Dam in California. [38]

The question of whether power or irrigation would be the focal point of the Columbia Basin Project was somewhat confused, but most conceded that enormous production of power would be the immediate benefit. Secretary of the Interior Ickes was thoroughly enamored with the electrical potential of the Grand Coulee Dam. He saw electricity as the key to unlocking natural and human resources yet unimagined. The Bureau of Reclamation, once dismissed by him as "expendable," made the Department of the Interior a major player in the modern world of kilowatts and comprehensive river basin planning. The Columbia Basin Project exemplified both. The journalist, and later U.S. Senator from Oregon, Richard L. Neuberger expressed faith in both irrigation and electrical power from the Columbia River in his book, *Our Promised Land* (New York: The Macmillan Company, 1938).[39]

Still, it was difficult for the Bureau of Reclamation to concede that its traditional responsibility of irrigating lands would be subordinate to the production of power. F. A. Banks, the Columbia Basin Project engineer, for whom Banks Lake is named, declared in an address at Washington State College in 1936, after Congress approved the irrigation aspect of the project, "[it] was essentially an irrigation project with power as an incidental product." He predicted, "Out of the now semi-arid tract, it creates an empire equivalent in area and productivity to three Yakima Valleys." He envisioned as many as 200,000 people living on the land in eastern Washington using water supplied to them from Lake Roosevelt, the reservoir behind Grand Coulee Dam. From Lake Roosevelt, water is pumped up about 300 feet into the Feeder Canal which flows into the Banks Lake reservoir in the Grand Coulee. From there,

6.28. These workers were vibrating concrete at Grand Coulee Dam to remove air bubbles and improve the quality of the final product. 1936.

water is distributed throughout the project. Although efforts to create new farms did not occur until after World War II, the words of Banks suggest that, in its rhetoric and dogma, the Bureau of Reclamation still defended its mission to build homes in the desert. And, Banks pointed out, the 1937 Columbia Basin Project Act gave the Secretary of the Interior the authority to reduce the maximum size of farms to as little as forty acres.[40]

6.29. A 1937 view of the high trestle and the uneven bedrock at Grand Coulee Dam.

Some have argued that the irrigation aspect of the project was not needed and should never have been developed. And later generations saw Grand Coulee Dam as the major event that opened the way for multiple dams on the Columbia and its destruction as a living organic river. But Grand Coulee Dam historian Paul C. Pitzer writes: "The opposition to Grand Coulee then [in the 1930s and 1940s] was political and economic, not environmental. And it must be repeated that then most people saw the dam and irrigation as conservation measures that enhanced and protected the environment, a view that has been altered over the ensuing decades."[41] Progressive conservation in the early twentieth century abhorred waste and espoused efficient resource use for the benefit of the people. It denounced both the misuse and the waste of natural resources, including rivers that ran wildly to the sea. This contrasted sharply with the values of environmentalists later in the twentieth century who deplored the destruction of natural river systems.

6.30. The contractor's dining room in Mason City at Grand Coulee Dam in 1935.

6.31. Grand Coulee Dam during construction in 1937.

In the 1930s, *Reclamation Era* did express concern over the fate of Columbia River salmon. The adverse effects of dams upon fish had long been a concern throughout the United States. Ivan Bloch, in a 1938 article, assured readers that efforts were underway to ensure the survival of the salmon industry in the Pacific Northwest. He noted that, besides the dams, there were other longstanding threats to the Columbia River salmon runs: overfishing, pollution, and silting caused by soil erosion. Dams on the scale of Grand Coulee, however, called for extraordinary measures to protect the fishery. Congress responded in 1938 with legislation to provide for the conservation of fishery resources on the Columbia River. It appropriated $500,000 to carry out investigations and

6.32. The L-3 rotor during installation at Grand Coulee Dam. July 5, 1941.

construct devices for the protection and improvement of feeding and spawning conditions for fish and to facilitate "free migration of fish over obstructions."

This act paved the way, in 1939, for the Bureau of Fisheries, then under the Department of Commerce, to be transferred to the Department of the Interior as the Fish and Wildlife Service (FWS). The Biological Survey, in the Department of Agriculture, was also consolidated into the FWS. All were part of an effort to protect the fishery that, according to Bloch's article in *Reclamation Era*, must lay "the foundation for an intelligent approach to the use of abundant natural resources of the northwest corner of our nation." The success of this effort would be watched by the world, he concluded, as an example that, "Man can manage his planet by making use of his intelligence and knowledge when it subordinates his otherwise reckless desire to ravish his natural resources." While these words expressed an optimism about the abilities of science and technical efforts "to save the salmon," decades of experience since the damming of the Columbia have revealed the loss of the salmon fishery. Reclamation now expends considerable money and effort trying to deal with this issue.[42]

Still, the power delivered by Grand Coulee Dam during the war emergency overcame earlier arguments against its construction. The output of its power facilities exceeded that of Hoover Dam by 1947. In hindsight, events seemed to confirm the argument for the big dam that, "If we build it, they will come." When the war came, the power of the dam helped draw population and industry to the Pacific Northwest. To the building of these two major dams in the 1930s — Hoover and Grand Coulee — the United States owed much of its security and ability to successfully conduct the two-ocean war that followed in the 1940s.

Reshaping California's Water Landscape

In the midst of the Great Depression, California launched the state's ambitious Central Valley Project (CVP). It had long since abandoned the experiments to build irrigation settlements in the two small communities at Durham and Delhi. In the early 1920s, Mead oversaw these efforts as Chairman of the California Land Settlement Board while he served as professor of rural institutions at the University of California. The CVP was designed to serve extensive farm holdings in the Central Valley. The delivery of government water to private owners, often possessing land far in excess of the 160 acre limitation specified in the national Reclamation Act, posed several problems but did not deter

Reclamation's enthusiasm for the project after it was initiated by the state. The more exciting challenge for Reclamation engineers came with the restructuring of the state's waterscape. Lawyers, judges, or Congress could resolve the 160 acre rule at a later date. Along with farms, the CVP aimed to serve urban centers and replenish ground water in the San Joaquin-Sacramento River Delta. As industrial and agricultural uses reduced freshwater supplies, the encroachment of salt water from San Francisco Bay posed a serious threat to the waters of the rich delta lands.[43]

In the Central Valley lay the bulk of California's agricultural lands. From west to east, the valley lies between a low coastal mountain range and the Sierra Nevada and Cascade ranges whose eastern slopes reach into Nevada. At the south end of the valley, where the Tehachapi Mountains separate the San Joaquin Valley from the Los Angeles Basin, the average rainfall is five inches per year. In the north, where the Sierra Nevada converges with the Cascade mountain range, the rainfall is nearly thirty inches a year. The mountainous northern part of California generally experiences abundant rainfall that turns to snow as storms sweep in from the Pacific during the winter months. Californians early grasped the possibilities of connecting the thirsty Central Valley with the water-rich mountains that occasionally sent gushing torrents over valley farmlands and cities.

6.33. Two monitors (nozzles or water cannons) mining gold by "hydraulicking" in the California gold fields. Courtesy of the California Historical Society.

During the Gold Rush, early miners moved water from place to place on a large scale. They diverted, rerouted, and dammed streams – all in the attempt to separate sand, rocks, and debris from alluvial, free standing, or placer gold. Gravity carried water from high elevation reservoirs into flumes, ditches, and ultimately the water cannons [monitors] that washed away entire hillsides with hydraulic mining techniques in search of ancient placer deposits. The message was clear: if water could be diverted, ditched, flumed, or siphoned for mining purposes, the same could occur for agriculture and over greater distances. Californians understood the need to move and store water to realize the potential of their geography, topography, and climate.[44]

The intensification of agriculture in the Central Valley as it evolved from stock raising, to wheat, to specialty crops, increased demands upon the groundwater. State officials and citizens increasingly looked to schemes to import surface water from mountain streams. Agricultural, urban, and industrial consumers stood to benefit. As early as 1873, at the instigation of both California and Nevada representatives, Congress authorized a commission to study irrigation in California. Headed by B. S. Alexander of the U.S. Army Corps of Engineers, the five member commission published its report in 1874. On the western side of the valley, it considered a north-south canal from Red Bluff to the Sacramento Delta and, on the east, from the San Joaquin Delta to Bakersfield. The eastern side of the valley was to be fed by the multiple streams coming from the Sierra Nevada. While estimated costs exceeded the ability of landowners to pay, making the project impractical, the report established the fact that the plan was feasible from an engineering standpoint. The great water transfers awaited further economic development and a favorable political climate for approval and implementation.[45]

6.34. B. S. Alexander, U.S. Army Corps of Engineers. Courtesy of the U.S. Army Corps of Engineers.

By the 1920s, those events seemed at hand. The success of San Francisco's Hetch Hetchy and Los Angeles's Owens Valley projects, both transporting water over great distances, augured well for plans to irrigate the Central Valley with a complicated, extended system of canals and reservoirs. In the 1850s, the state's first Surveyor General noted the need for a comprehensive state water plan, as did California's first State Engineer (1878-1888), William Hammond Hall, a champion of California water development. But only in 1915 did the legislature begin to consider a state water plan. It authorized a water conference to develop a plan that would address irrigation, reclamation, water storage, flood control, municipalities, drainage, and the needs of navigation and water power development. Still, the legislature did not act on the basis of a report delivered from the conference in 1916. This opened an opportunity for Robert Bradford Marshall to present what became known as the "Marshall Plan" to the governor of California in 1919. Marshall had been a prominent geographer for the U.S. Geological Survey in California and, since the 1890s, a longtime advocate for tapping the waters of the Sierra Nevada for the irrigation of the Central Valley. His plan touched off a campaign that lasted over a decade.

Complicated and extensive, the plan rested upon a key feature, the construction of a large dam on the Sacramento River at Kennett, which was later named Shasta Dam. During the 1920s, the plan was an issue in the California legislature (also on the ballot in 1922, 1924, and 1926) until interest declined in the latter part of the decade amidst battles between private and public power and concerns about the daunting costs. In 1929, however, President Hoover suggested that Reclamation make common cause with the states, local communities, or even private individuals to build irrigation projects that could ultimately be administered by the state. Commissioner Mead, who had long experience with California water questions, represented Reclamation on yet another commission to study the Central Valley. This study echoed earlier calls for construction of Kennett Dam north of Redding on the Sacramento River and Friant Dam on the San Joaquin River near Fresno, in the south, to provide large reservoirs on both ends of the valley. The plan also involved a freshwater barrier against saltwater encroachment from San Francisco Bay and provided a basis for an alliance between the bay area and delta farm interests in any upcoming battle for acceptance in the legislature. In 1933, the legislature passed a comprehensive Central Valley Project Act to be funded by state bonds and, given the involvement of Reclamation in the planning of the project, possibly by federal dollars. Few dreamed that the Bureau of Reclamation would eventually finance and build the project.[46]

Governor James "Sunny Jim" Rolph signed the act, but it was challenged in a referendum sponsored largely by the Pacific Gas and Electric

Company. The campaign against the Central Valley Project trotted out a litany of evils that the project would produce, but the real, though faintly veiled, opposition sprang from the company's fear of public power. Voters north of the Tehachapi Mountains overwhelmingly supported the state's project, but the vote was close — 459,712 to 426,109. The margin of victory was probably fueled as much by the fear of persistent drought in California from 1929 to 1935 as by any arguments about the project's advantages for the future growth of California or the possibility of federal funding. In the midst of the Depression, the construction bonds could not be sold, and California appealed to the federal government for help just at the time the Roosevelt Administration made federal relief and recovery dollars available. In 1935, the Emergency Relief Appropriation Act authorized funds to Reclamation to build the project with the consent of California. In the years after World War II, efforts to build out the Central Valley Project continued. Eventually the CVP became one of the largest accomplishments of Reclamation. In October 1937, work began on the Contra Costa Canal, and the contract for the construction of Shasta Dam, "the keystone of the Central Valley Project," was awarded in 1938. With delays occasioned by World War II, the dam was not completed until 1945. Nevertheless, water storage began in 1944, and Shasta Dam delivered electric power just in time to contribute to war production.[47]

Backed by generous emergency funds from the New Deal, the principle of multiple-use came to dominate the Bureau of Reclamation's large projects in the 1930s, and officials made less and less mention of the ability of agricultural lands to pay back costs. Multiple-use offered a rationale to build projects without repayment guarantees, while agricultural development was expected to return costs to the Treasury. Emphasis upon this hard-nosed economic principle got lost in the fray of multiple-use appropriations. Much of

6.35. A vertical section construction joint with keyways and galleries in Shasta Dam. March 7, 1942.

6.36. Shasta Dam under construction.

the cost of a project could be charged off to flood control and navigation, while revenue to pay for construction could be expected from the sale of hydroelectric power and water to urban users. Professor Donald Pisani notes in his early work on California:

> Just as the Reclamation Service had abandoned the principle that federal reclamation should benefit primarily public lands early in its existence, in the late 1920s and 1930s it discarded the requirement that the land benefited should bear the cost of reclamation. By the 1930s the Reclamation Bureau fully grasped the implications of the doctrine of multiple use which freed it from an economic straightjacket [that confined its role to developing and serving the interests of irrigated lands]. The answer to the increasing cost of arid land reclamation was to build bigger projects to serve a variety of needs, urban as well as rural... .[48]

Emphasis upon multiple-use marked a change that ranked in importance with earlier decisions by Congress to permit delivery of federal water to private lands. Water service to private lands, often called "supplemental water," permitted land owned in excess of 160 acres to benefit from federal water. The 1902 national reclamation law prohibited individuals from receiving water from

6.37. Shasta Dam, Central Valley Project, California.

Reclamation projects for lands owned in excess of 160 acres. Wide distribution of the public domain became a goal, a principle established earlier by the Homestead Act of 1862, which offered virtually free land to heads of families in parcels of 160 acres. The Bureau of Reclamation did not always hold to the letter of national reclamation law. When Secretary of the Interior Ickes raised the issue of acreage limitation in 1934, the Bureau of Reclamation leaders indicated that it had been "a dead letter for years" and it was best "to let sleeping dogs lie." By the 1930s, various methods had been applied to circumvent the law and allow water to flow to larger holdings, especially private lands. A man and a wife could jointly receive water for 320 acres, and land could be parceled out to children and other parties, but essentially kept under a corporate ownership. In 1914, Reclamation Chief Counsel Will R. King ruled (some said dubiously so) that landowners who paid back the construction charges in full on the excess lands could be entitled to water over the 160 acre limit. While these circumventions played a role, primarily lax enforcement by Reclamation did the most to undermine acreage limitation. When enforcement did take place, as on the Vale, Owyhee, and Deschutes Projects under the 1926 Omnibus Adjustment Act, it experienced some success. Excess landholders had to sign "recordable contracts" to sell off their excess land at prices set by the Secretary of the Interior to avoid the practice of speculation or profiting unduly from the delivery of government water to the lands.[49]

The excess lands issue continued to be the "elephant in the room" when it came to making federal water available to private lands. Reclamation's assumption of the CVP and the Central Valley Project Act of 1937 carried no exemption from the 160 acre rule. However, these lands had been in private ownership for generations and in holdings much larger than the maximum 160 acre limitation. In the Colorado-Big Thompson Project, where larger holdings were also the rule, Congress lifted the 160 acre limitation in 1938 on old farms but imposed the requirement on newly irrigated land within the project. On the Newlands and Humboldt Projects in Nevada, where the production of forage crops for livestock prevailed, Congress, much to chagrin of Secretary of the Interior Ickes, lifted the limitation in 1940 in recognition of the large amount of acreage needed to raise these crops profitably. Contention over the 160 acre rule reflected the continued commercialization of American farming and the break-down of an earlier idealism that sought to protect and encourage only moderate sized landholdings for families and homemakers. [50]

Restructuring Nature: The Colorado-Big Thompson Project

In Colorado, the Colorado-Big Thompson Project became the Bureau of Reclamation's major undertaking of the 1930s. It went forward with the assumption that the excess lands provision would not apply and that the project would enjoy the congressional appropriations as well as revenue from sales of power and water to industries and urban residents. Approved by Congress and President Roosevelt in 1937, Coloradoans had long dreamed of bringing water from the western slope of the Rockies to the east side of the Continental Divide to serve both a rural and urban population. The interruption of World War II, delayed completion of the Alva B. Adams Tunnel through the Rockies until 1947. While hydropower was not as significant as in California or Washington, power from the Green Mountain Dam and from various plants along the water's route down the East Slope of Colorado's Front Range paid back more than half the project's costs. This emphasized the thoroughly multiple-use character of the project's conception and financing. Needs of the population on the Front Range of the Rockies and in the agricultural lands of northern Colorado varied. The transfer from one watershed to another did not come without a price. It took most of the 1930s to negotiate agreements to meet West Slope demands that adequate "replacement" storage capacity be built to protect the West Slope.[51]

This was no ordinary transfer from one basin to another. The project breached the Continental Divide, the most striking watershed division in North

6.38. Working on the gate chamber of the Alva B. Adams Tunnel, Colorado-Big Thompson Project, Colorado, in 1939.

America. Bringing water from the relatively water rich western slopes of the Rocky Mountains through a thirteen-mile-long tunnel beneath Rocky Mountain National Park was a major engineering feat. The project symbolized a willingness to use federal dollars "in the massive reordering of nature." If the largest rivers of the West could be dammed, diverted, and harnessed for hydroelectric power, so could water be transferred through that great cordillera that forms the backbone of the continent.[52]

The Colorado-Big Thompson Project's thirteen-mile-long tunnel was characteristic of the audacious spirit of the Bureau of Reclamation during the 1930s. The multiple-use nature of the project freed it and others undertaken in this decade from the single purpose of irrigation agriculture. Multiple purposes opened new avenues of financing for projects, freeing them from the limited capability of irrigated farms to repay project costs. Urban users as well as irrigators combined in the Northern Colorado Water Conservancy District to pay back part of the costs. Without the conservancy district and its power to levy taxes to support repayment charges, the project might not have occurred. This was one of the acknowledged advantages of reclamation projects throughout the West, for they helped create valuable and taxable property not only for the benefit of reclamation projects, but for local and state governments as well.

Flood control and navigation fell comfortably within "the general welfare" and "promotion of commerce" functions of the federal government that could justify outlays by Congress without requirements for reimbursement. Financing this multiple-use project presented few problems. The delay came in the complicated maneuverings of local interest groups and the inter-bureau competition between the National Park Service and Bureau of Reclamation. The West Slope-East Slope negotiations within Colorado and Secretary of the Interior Ickes's hesitancy about allowing a tunnel beneath Rocky Mountain National Park in the face of opposition by the National Park Service delayed serious work on the project until 1939. When World War II caused serious shortages in labor and building materials, construction proceeded at a slow pace.[53]

During the New Deal, Reclamation's expanded activities brought it into conflict with the National Park Service, but ultimately Reclamation won its point. Commissioner Page acknowledged long-standing quarrels between the two agencies when, on occasion, the National Park Service opposed water projects that impinged on free flowing streams within the national parks. Arno B. Cammerer, the Director of the National Park Service in the 1930s, suggested that the Park Service be included in the planning stages of projects to avoid misunderstandings. While Page acknowledged that he was probably being "presumptuous," he offered Cammerer his thoughts on new directions for the Park Service. He believed that few people in the West, or what he termed "local people," approved of the National Park Service's policy of strict "segregation" of its useful resources. The Park Service, he suggested, should create two classes or areas of land under its jurisdiction. The first would adhere strictly to National Park preservation policy, "with no invasion of any sort." The second would remain dedicated to public use but with less rigid restrictions than "purely park areas." On the less than "purely park areas," the administration of natural resources would be much like that followed by the Forest Service in national forests.

6.39. Arno B. Cammerer, Director of the National Park Service, opposed construction of the Alva B. Adams tunnel under Rocky Mountain National Park. Courtesy of the National Park Service.

Commissioner Page was convinced that these lands of "lesser majesty" could be opened to planned development with leased cabin sites and the construction of irrigation, reservoir, and power facilities – a policy that he predicted would meet with great approval in western states. More contact between the two agencies was important because, "We dislike very much to be held up as an agency forever at conflict with the National Park Service, when, as a matter of fact, no such attitude of antagonism exists in the thoughts of the Bureau." Page said he disliked seeing Reclamation forced into a position, "which appears to be inimical to the National Park Service." He believed this could be avoided, "if some such plan as I have outlined were in effect." Conciliation aside, Page's suggestions, at best, were "presumptuous" and, at worst, destructive to the National Park Service's preservation principles. Little wonder that Secretary of the Interior Ickes struggled to accommodate the new Bureau of Reclamation projects and the strikingly different view of resource management that prevailed within the National Park Service.[54]

Transitions and Ambivalence

The Great Depression and the New Deal enabled the Bureau of Reclamation to cast off some of the limitations imposed by earlier legislation, especially by the 1902 national Reclamation Act. The broader purposes of reclamation projects relieved Reclamation from relying on agricultural lands alone to pay back construction costs. Funds now came readily to Reclamation independent of the revolving Reclamation Fund that had ceased to revolve even before the Reclamation Service finished its first decade of work. In other respects, too, the decade of the 1930s permitted Reclamation to break free of the past. Although Congress increasingly ignored the 160 acre excess lands rule in the original Reclamation Act, the restriction still had its backers, as can be seen in the acreage limitations asserted by the 1937 Prevention of Land Speculation Act which pertained to the Columbia Basin Project. On the Colorado-Big Thompson Project, Congress granted an exemption, as it did for private lands in the Imperial Valley served by the All-American Canal under the Boulder Canyon Project. In 1936, the *Reclamation Record* justified construction of the All-American Canal under the auspices of Reclamation and its delivery of water to lands in the hands of private landowners in the valley. In part, it was responding to inquiries about whether "new lands will be available for settlement" in the Imperial Valley. Reclamation revealed that all lands were "taken up" (not open to homestead entry), and it described the All-American Canal as a rescue operation to save privately-owned irrigated land in a spirit of cooperation between

Reclamation and private landowners in the West. Some called this "the Betrayal of Reclamation Law" and the ultimate abandonment of the small-farm ethic that had first inspired the reclamation movement.[55]

The Bureau of Reclamation still clung to original ideals in spite of the plunge into building large projects under the banner of multiple-use and work relief. In 1936, as noted earlier, Frank Banks of the Columbia Basin Project looked forward to the development of irrigation agriculture on the project and the resettlement of displaced farmers on the land: "Isn't the rehabilitation of those tillers of the soil, who move from marginal lands to the irrigated farm home all worth to the Nation, the subsidy that it grants developing the irrigated features of this project?" The nation and the West must answer this question with a resounding affirmation, he said, "If reclamation as a national policy is to endure." Moreover, Banks said, in words written to appeal to an agricultural audience at Washington State College in Pullman, that "such a project is essentially an irrigation project with power as an incidental product and stream control as a substantial contribution toward development of the rest of the river for power and navigation purposes." Banks's statement suggested that irrigation was still central to the vision of Reclamation — even in the face of one the most "technologically sublime" dam and power production complexes ever constructed. His ambivalent rhetoric presented a confused message about Reclamation's mission. Identifying hydroelectric power as a "by-product," he overlooked the eagerness of Reclamation to sell power and use its revenues to develop the urban, industrial West. Yet, while touting agriculture, Banks pinned the entire viability of the project upon electricity because it would "insure the return of the Government's investment in the project."[56]

Commissioner Mead's sudden death came in 1936, when the Bureau of Reclamation was in full transition to its new status as the major water resource bureau in the West. From 1924 to 1936, his tenure as commissioner accomplished not only the survival of Reclamation, but also a new vitality for it in public works and big dam construction. Eulogies to Mead emphasized how he had constantly justified the economics of reclamation to the nation and to Congress and how every dollar invested in reclamation projects returned four dollars in benefits to the regional economy. Mead maintained that the $280,000,000 invested in reclamation had produced two billion dollars in property values. As Commissioner, Mead often sounded the theme that reclamation contributed to a happy balance between agriculture and industry in the nation.[57]

6.40. Grand Coulee Dam and its pumping plant were the centerpieces of the Columbia Basin Project's irrigation plans.

In 1935, Commissioner Mead issued a directive to restructure Reclamation. The directive, addressed to "All Field Offices," acknowledged the expanded activities of Reclamation since 1933. Mead first focused upon Reclamation's mission "to conserve and use to the best advantage the land and water resources of the arid states." To him that meant building irrigation works to help people "improve conditions under which they live." In times of industrial depression and farm crisis, the Reclamation bureau provided new opportunities to "farmers who now live on marginal land and for the unemployed who are attracted by rural life." While Mead emphasized "cooperation and teamwork," his directive separated the engineers who planned and built water projects and those who administered projects in conjunction with water users' associations. He now wanted project administrators and water users' associations to report directly to the Washington, D.C., office rather than the Denver Office to relieve it of these administrative details. In this letter, Mead continued to endorse the family farm as the basic mission of Reclamation, but he clearly recognized that the Denver Office should concentrate only on the planning and engineering tasks, given the increased demands upon it. [58]

Mead thought that the basic job of project superintendents was to improve the social and economic well-being of water users. Other concerns should include: (1) problems connected with water supply, storage, carriage, and delivery of water; (2) improvement of irrigation methods to secure a more efficient and economic use of water; (3) examination of seepage or drainage conditions; (4) revisions and improvements in the crop census; (5) the status of class five lands; and (6) the status of excess holdings. Many of these items represented operation and maintenance (O&M) issues that Reclamation and its projects confronted over many decades. Mead's directive reflected a recurring concern that the Denver Office of Chief Engineer paid insufficient attention to the economic needs and concerns of settlers on the projects.[59]

The following projects were ordered to report directly to Washington. The list indicates that Reclamation still administered the operations and maintenance on many projects, while an equal number had been turned over to the operation of water users' associations. On some projects, both Reclamation and water users performed operation and maintenance functions on different divisions.

O&M by Bureau	O&M by Water Users
Belle Fourche	Baker
Boise	Bitter Root
Carlsbad	Boise
Grand Valley	Huntley
Klamath	Lower Yellowstone
Milk River	Minidoka
Minidoka	Newlands
North Platte	North Platte
Orland	Okanogan
Owyhee	Salt Lake Basin (Echo)
Rio Grande	Salt River
Riverton	Shoshone
Shoshone	Stanfield (now part of the Umatilla Project)
Umatilla (McKay)	Strawberry Valley
Vale	Sun River
Yakima	Umatilla
Yuma	Uncompahgre
	Yakima-Kittitas

Engineering, construction, powerplant operation, purchasing, operations and maintenance, and personnel records remained with the Denver office as did, of course, the Office of Chief Engineer. O&M changes, along with public relations, were to be handled in the Washington office. George O. Sanford took charge of operation and maintenance oversight. Mead also created five districts "To further unify policies in operation and maintenance and to advance efficiency," and named five supervisors of the districts. District 1 was headed by H. H. Johnson (Montana and North Dakota); District 2 by J. S. Moore (Idaho,

Oregon, Washington); District 3 by H. D. Comstock (Wyoming, South Dakota, Nebraska); District 4 by L. H. Mitchell (Northern California, Nevada, Utah, Colorado); District 5 by R. C. E. Weber (Southern California, Arizona, New Mexico).[60]

A restructured Bureau of Reclamation that attempted to both centralize and localize authority awaited John C. Page when he assumed the job as Acting Commissioner of Reclamation following Mead's death in early 1936. Mead's death curtailed the reorganization, which accomplished little. When another reorganization occurred in 1943, the results were more substantial. A 1908 graduate of the University of Nebraska in civil engineering, Page worked his way up through the ranks of the Reclamation Service from topographical surveyor in Colorado to superintendent of the Grand Valley Project in 1925. He became chief administrative assistant for the Boulder Canyon Project in 1930, and, by 1935, he was head of the Office Engineering Division. He assumed the title of Commissioner on January 25, 1937, almost a year to the day after Mead's death.[61]

A Land to be Redeemed

In July 1936, the president appointed Acting Commissioner Page to the Great Plains Drought Area Committee. He, along with Morris L. Cooke, Administrator of the Rural Electrification Administration (REA) and chair of the committee; Colonel Richard C. Moore, Corps of Engineers; Frederick H. Fowler, Director of the Drainage Basin Study, National Resources Committee; Rexford G. Tugwell, Administrator of the Resettlement Administration; and Harry L. Hopkins, Administrator of the Works

6.41. Commissioner John C. Page, Jane Ickes, Secretary of the Interior Harold Ickes, and I. C. Harris, in the Nevada wing of the Hoover Dam Powerplant. February 23, 1939.

6.42. Notables in the construction of Hoover Dam posed in Boulder City in July of 1931. W.A. Bechtel, first vice president, Six Companies, Inc.; Walker R. Young, project construction engineer, Bureau of Reclamation; Commissioner Elwood Mead; Frank T. Crowe, general superintendent, Six Companies, Inc.; and Chief Engineer Raymond F. Walter of the Bureau of Reclamation.

Progress Administration, made up the government representatives on the drought committee. Page understood that Reclamation could play a major role in drought relief and recovery efforts and, like Mead, he was quick to grasp opportunities when they appeared. These included large as well as small projects. In discussions with Chief Engineer R. F. Walter and John "Jack" L. Savage, chief designing engineer in the Denver office, he was eager that Reclamation receive the bulk of emergency funds to construct small storage dams for rural and urban communities. Furthermore, he found it encouraging that the Great Plains Drought Area Committee favored upstream storage to protect against soil erosion and floods as well as to combat drought. Tugwell was so discouraged by the long drought

6.43. John ("Jack") L. Savage, chief designing engineer, Bureau of Reclamation.

that he advocated irrigation throughout the entire region. Page noted, with some satisfaction, "This is a radical departure from the previous policy of the Department of Agriculture."[62]

Page wasted no time in outlining his "vision of the land redeemed" to the committee chair, who also was Administrator of the REA. His plan for Great Plains recovery featured dam building, water storage, and hydroelectricity production. Page saw a common cause between Reclamation and the goals of REA. Dam and water storage facilities produced power that the REA could distribute for the benefit of sustainable communities on the Great Plains. Page's explanation for conditions on the Great Plains, i.e., Dust Bowl conditions, followed the general New Deal "gospel" of how the crisis came about in the controversial Farm Security Administration's 1937 film, *The Plow that Broke the Plains*. Farmers had plowed up marginal land in response to the higher prices offered for crops during World War I. They did this at a time of above average precipitation on the plains, with the wartime slogan "Food Will Win the War" still ringing in their ears. Too many farmers came to the Great Plains, and they plowed up too many acres that should have remained grazing land.[63] The water supply of the Great Plains declined not only because of drought, according to Page, "but more seriously because of loss of vegetative cover and the resulting increased run-off and evaporation with decreased absorption and percolation." His words suggested that the Dust Bowl was largely a man-made phenomenon, as did the film. Answers to the problem lay in the explanation of its origins: what man had taken apart, man could put back together again with, of course, the aid of Reclamation and others. Page believed that long-range planning and action could reverse the course of recent history and "redeem" the Great Plains. This meant, in his view, to restore the region "to conditions of farm ownership and tenure to the status of 1890-1900, which probably approached the maximum beneficial use of this general area by its combination of crop farming and livestock operations."[64]

All of this could be brought about through the cooperation of state agencies and federal bureaus. On the federal level, Page saw contributions by the Department of Agriculture through knowledge of soil tillage, forestry, and animal husbandry; the Rural Resettlement Administration through planning, land purchases, and encouraging population to shift away from marginal lands; and REA through production and distribution of electric power for domestic uses, pumping for irrigation, and multiple other purposes. Reclamation could expand the storage of surface water by building reservoirs and increasing irrigation and power generation. Finally, the Corps of Engineers, with its knowledge of flood control and river regulation, stood to play a part. "My vision of the redeemed land can materialize," Page wrote, if a national effort uniting private, state, and federal agencies occurred.[65]

Many believed that the future of the Great Plains was at stake, and Acting Commissioner Page was determined that Reclamation should have a role in that future. The Division of Great Plains Projects within the Chief Engineer's Office in Denver coordinated Reclamation activities with the Civilian Conservation Corps, Farm Security Administration, and drought abatement projects. The White House saw in Reclamation projects a stabilizing influence upon western agriculture. In August 1936, a letter from President Roosevelt to the Spokane Chamber of Commerce asserted that, "Federal reclamation projects served to mitigate the effects of the drought of 1934 and are this year in many localities giving another demonstration of the wisdom and usefulness of our national reclamation policy." On federal reclamation projects, the president cited four large "storage dams" completed recently and scores more that were in various stages of construction. He hoped they would be completed soon, to contribute to "the conservation of the West's primary resources, its water." He considered water as a plural resource that served many masters. But he took special interest in those "dislodged by droughts … who will desire to seek new homes on the West's last frontier, its irrigated lands." The statement recalled old themes in reclamation history.[66]

The crisis on the Great Plains generated plenty of rhetoric. By the fall of 1936, Commissioner Page understood that irrigation was not a panacea for drought problems nor could it bring about his "vision of a redeemed land." He now wrote that less than 1 percent of the drought areas of 1934 and 1936 could be reclaimed by irrigation. He thought that the primary need was for a survey of the water resources to determine the feasibility of new projects. While opportunities for project development existed, they were few in relationship to the large area under consideration. Page did praise the existing projects within the drought area and their ability to produce crops and feed livestock from drought stricken ranges. These included the Belle Fourche Project in South Dakota, the North Platte Project in Wyoming and Nebraska, and the Lower Yellowstone Project in Montana. These projects had prospered during the drought, he insisted, but they were the exceptions in the large picture and indicated that, "Under these circumstances [widespread, pervasive drought] irrigation can only assist in the solution of the problems of the drought area."

6.44. A Nebraska Dust Bowl farm.

By the end of 1936, Page's enthusiasm for measures to counter the drought appeared chastened. In 1940, he wrote to Senator Carl Hayden of Arizona that, "It is possible and it is desirable, through a program of small water conservation projects combining reclamation and relief in the Great Plains, to anchor in or near the areas they now occupy many of the people who otherwise will find it necessary to migrate." The undertakings, however, were small compared to the Boulder Canyon, Columbia Basin, and Central Valley Projects, and, at best, they could support only a type of stock-grazing agriculture rather than intensive crop production. Page remained concerned about how to best deal with water supply issues for communities on the Great Plains, although the drought lifted after 1938.[67]

Farther to the West, the Resettlement Administration continued its plans to place down-and-out farmers from the Great Plains on reclamation projects. While the Bureau of Reclamation saw some opportunity in the suggestion, the Department of Agriculture was more cautious. Under Secretary of Agriculture M. L. Wilson noted that destitute farmers were in no position to establish themselves on reclamation projects. Reclamation's requirement that they have at least $2,000 in capital to qualify for a reclamation project farm eliminated most. Migrants from the Great Plains could not be accommodated even by the new Columbia Basin Project. Perhaps 43,000 families could be provided with land eventually, but only 13,000 could be settled within five years. Wilson doubted that it would "be possible or desirable" to resettle all families from the Great Plains on land within federal reclamation projects. And he concluded that, "few of them [migrants from the Great Plains] are experienced in irrigation farming or in the production of specialty crops, and, consequently, they require training and supervision if they are to have a fair chance for success."

Wilson's letter to the White House echoed longstanding Department of Agriculture criticisms of the Bureau of Reclamation. Not even the Depression and drought could put these to rest. Still, Wilson said that, although "the problems to be met are knotty, they are not insolvable." He suggested changes in the procedure relating to the settlement of reclamation projects with particular attention to the Columbia Basin Project. These included a waiver of the $2,000 asset requirement for families; adequate training in irrigation farming and the production of specialty crops; assessment of water charges based upon the capacity of the farm to produce; market studies that advised farmers to plant crops that would not contribute to surpluses and undermine their profits; and the allocation of construction costs, especially at the Grand Coulee, amongst power, navigation, flood control, and irrigation. The "allocation," of course, should result in charging off most costs to purposes other than irrigation, thereby keeping annual charges on the land low enough for farmers to make a profit. All of this would require more study, but, in the meantime, Wilson concluded, "We shall be glad to cooperate as fully as our resources may permit … in the fields of farm management, land use planning, and land settlement."[68]

Behind the arguments and facts in Wilson's long letter was the mind of an economist at work. It was neither practical nor economical for all or most refugees from Great Plains farms to remain farmers. And it was neither necessary nor desirable for existing reclamation projects to accommodate emigration from the Plains. Many families should seek work outside of agriculture. Even the Columbia Basin Project posed problems. The Department of the Interior and the Bureau of Reclamation reacted politely to Wilson's letter saying, "It is heartening to learn that the Department of Agriculture likewise recognized the value of this irrigation project as a means of stabilizing the western states and of providing a refuge for those displaced in other areas by adversity." The Interior Department appreciated "recognition of the federal irrigation policy, but it is unfortunate that larger areas are not now available for the resettlement of the homeless." The Bureau of Reclamation recognized its responsibilities for "the development of such an empire as that within the boundaries of the Columbia Basin Project." Yet, despite all the urgent talk about how the Columbia Basin

Project should be settled, the agricultural phase of the project did not begin until after World War II and then never on the scale envisioned.[69]

President Roosevelt also backed a plan to settle Indian farmers on the projects. Already, Indian and white farmers worked lands close to reservations that had been dispersed to both under Indian allotment policies. While signing a series of "allotments" for additional Indian reclamation in early 1935, the president decided that the Bureau of Indian Affairs should work out a schedule to settle "Indians of the agricultural type" on a portion of the new irrigation farms in their vicinity. The Department of the Interior believed successful "colonization" would require skilled leadership. Indian colonies should be in compact groups, but at least three years would be needed to plan for their arrival. Only the Riverton and Shoshone Projects in Wyoming and the Gila Project in Arizona appeared to have sufficient public land to host such colonies.

If Indian settlement occurred, the Department of Interior believed it should be in set-aside colonies and not randomly distributed throughout the projects. Neither should those Indians willing to settle on government projects be saddled with debt. They should be exempted from construction and land improvement costs, as they were on Indian irrigation projects, and excused from maintenance and operation charges for at least five years. Thereafter, they should be allowed to work in lieu of paying cash for O&M charges. As much as $4,000 per family should be provided for housing, water, equipment, livestock, and other improvements, and the land should be tax exempt. Commissioner Mead responded negatively to the president's suggestion: "Generally speaking, Reclamation projects do not lend themselves to Indian settlement, because the vacant public lands are scattered." In addition, he noted that existing regulations required entrymen to qualify with farming experience, capital, character, and industry. Few Indian families, he implied, could meet the criteria. Another official within the Bureau of Reclamation objected to exempting Indian colonists from repayment or operation and maintenance charges. This would place an unfair burden on other settlers and create resentment. Most important, "If one group of settlers is not required to pay operation and maintenance charges, it will be increasingly difficult to enforce such collection from others." In the face of these criticisms, the president's plans for Indian settlement on project lands went nowhere.[70]

Altogether an Ambitious Decade

Reclamation prided itself on the important role it played in the New Deal's struggle against the Great Depression. Paraphrasing a theme in President Roosevelt's 1937 inaugural address in which the president said that he saw, "one third of the Nation, ill fed, ill clad, and ill-housed," Commissioner Page, in 1938, said he saw "one-third [of the Nation] ill-watered." Reclamation now possessed a well-funded dam building program to bring water and other benefits to the nation. At the same time, it put men to work completing structures that inspired a world wracked by depression, want, and desperation. The flurry of construction activity said that soon the world would be a better place. Dams, reservoirs, aqueducts, tunnels, and hydroelectric facilities marked a path to permanent recovery. Some derided the public works aspects of projects, which seemed blatantly political in their motivation, but few denied the sober realities that the taming of rivers and the development of hydroelectric plants promised a bright economic future. Others noted that the monumental structures that blocked rivers and reshaped river valleys represented the aggressive power of science, technology, and industry to which all in the society and in nature must conform, especially land and water. These achievements, however, contained a "sinister as well as utopian message."[71]

By the fall of 1938, Commissioner Page reported to the Secretary of the Interior that fourteen water "conservation and storage dams" had been completed in the previous five years under the construction program that had begun in 1933 with appropriations from the Public Works Administration. Page's report pointed out that the most recently completed dam near Austin, Colorado, Fruit Growers Dam, standing fifty-three feet high and 1,500 feet long, was an earth and rock-fill structure constructed at a cost of $200,000. It replaced a structure built by local irrigators in 1898 that had been destroyed during a flood in June 1937. Reconstruction of the dam was necessary to keep 2,050 acres of land productive and to save the farmers who lived on that land. These small dams illustrate the continued interest of the Bureau of Reclamation in family farms. Reclamation's program was not all large projects like Boulder Canyon, Grand Coulee, Central Valley, and the Colorado-Big Thompson Projects nor even the Norris and Wheeler Dams built with design assistance from the Bureau of Reclamation as a part of the Tennessee Valley Authority.

The other dam and storage projects completed in this five year period (1933-1938) included:

- Boulder Canyon Project in Arizona and California: Imperial Dam below Hoover Dam diverted water to the All-American Canal; and into the Imperial and Coachella Valleys.
- Parker-Davis Project in Arizona and California: Parker Dam, also below Hoover Dam on the Colorado River, was the Metropolitan Water District of Southern California's diversion point for the Colorado River Aqueduct and generates power for Gila Project pumps.
- Uncompahgre Project in Colorado, the Taylor Park Dam on the Taylor River above Altmont created a storage reservoir.

6.45. Taylor Park Dam during installation of the parapet wall. August 23, 1939, on the Uncompahgre Project.

- Humboldt Project in Nevada: Rye Patch Dam provided water storage.
- Carlsbad Project in New Mexico: Sumner Dam on the Pecos River near Fort Sumner to provide a reliable supply of water.
- Rio Grande Project in New Mexico and Texas: Caballo Dam on the Rio Grande below Elephant Butte Dam served to "re-regulate" the river below Elephant Butte and provided power and water for the project.
- Vale Project in Oregon: Agency Valley Dam on the Malheur River provided irrigation water.
- Burnt River Project in Oregon: Unity Dam provided water for the project.
- Hyrum Project in Utah: Hyrum Dam on the Little Bear River provided irrigation water.

6.46. The Agency Valley Dam on the Vale Project.

- Moon Lake Project in Utah: Moon Lake Dam, provided irrigation water.
- Ogden River Project in Utah: Pine View Dam to create a reservoir for irrigation water.
- Riverton Project in Wyoming: Bull Lake Dam on Bull Lake Creek to store water.
- Kendrick Project in Wyoming: Alcova Dam, to provide irrigation water.

6.47. Alcova Dam, Powerplant, and Reservoir on the Kendrick Project.

All the dams except Imperial Dam and Parker Dam, are earth and rockfill structures; Alcova Dam in Wyoming is the highest at 265 feet.[72]

This impressive list revealed Reclamation still engaged in smaller projects, not just large, multiple-use projects. There were funds available for both, mainly "emergency" Public Works Administration (PWA) and Works Project Administration (WPA) monies and direct appropriations from the general fund. Repayments for construction costs came primarily from power sales, and after a project was paid off, power revenues went to the U.S. Treasury. As noted earlier, from 1934 to 1940, annual Reclamation expenditures averaged near $52,000,000 while annual expenditures in the years before 1933 equaled approximately $8,900,000. With these increased funds, Reclamation tripled its staff during these years of feverish activity.[73]

There were also organizational changes. Reclamation sought an organizational structure that accommodated both local decision making and centralized authority. Complicating the attempt was the longstanding division of duties between the Denver Office and the Washington, D.C., office. Commissioner Mead's 1935 memo attempted to assign policy functions and relations with Congress to Washington and to assign engineering, personnel matters, and project problems to the chief engineer, who headed the Denver Office. Mead's designation of five districts in the West, under supervisors or directors, was a sure indication that growing federal bureaus often sought to decentralize power, not just consolidate it.

In 1939, Congress passed a major piece of legislation (Reclamation Project Act) to guide the future of the reclamation program. The act thoroughly legitimized the power production and municipal water purposes of Reclamation dams and projects. This meant that, in planning projects, Reclamation was to consider the multiple-uses to be achieved: irrigation, power, municipal water, navigation, and flood control. Most important, the act departed from the principle that all costs must be repaid by a reclamation project. It revised repayment requirements, pegging them to crop income. And it acknowledged that certain project benefits, such as flood control and navigation, were national in character or fell under "the general welfare," and therefore the costs of these benefits were not reimbursable by the water users. [74]

Under the 1939 Reclamation Project Act, repayment could be forty years, with an additional ten years added at the beginning. This granted irrigators a repayment moratorium to permit them to bring their lands under effective cultivation. Still, operation and maintenance payments must be made. This provision addressed one of the longstanding problems of reclamation. From the beginning of the Reclamation program, farmers complained that repayments interfered with bringing lands into production and curtailed the capital they could invest. Farmers could also now obtain loans from the Farm Security Administration under this act. [75]

Although the 1939 act offered loans and generous repayment terms for water users, it did not address immediate problems on the projects. The Great Depression's low agricultural prices and the almost decade-long drought in the West tested the will of project settlers to struggle on against ever increasing odds that they would fail. The projects proved not immune to drought and experienced shortfalls in water delivery for their crops. From Director Newell onward, Reclamation looked askance when Congress seemed too willing to grant extensions and moratoriums on repayments. To Reclamation, Congress's failure to pursue a tough policy on repayments had undermined the moral obligation of settlers to repay their contracts and, most importantly, drained the resources of the revolving Reclamation Fund — from which Reclamation was supposed to finance its operations and future projects. The crisis of the Great Depression, however, softened the attitude of Reclamation officials on this issue, particularly in light of New Deal emergency appropriations that enabled it to launch massive new big dam projects, as well as expand and renovate old projects.

Still, Reclamation saw dangers in the cancellations and deferments of charges. The old objections surfaced when it came to Indian reclamation projects. Leniencies on Indian projects carried implications for the Reclamation's projects. This, of course, was for the Bureau of Indian Affairs to decide. But Reclamation informed the Secretary of the Interior that more generous policies on Indian projects would generate pressures to relax terms of repayment on Reclamation's projects. "Any such pressure would be dangerous to the repayment policy of the reclamation law and would endanger the reclamation fund and the future program for reclamation," declared Acting Bureau of Reclamation Commissioner R. B. Williams, in the spring of 1939. He further noted that some Indian projects were located close to Reclamation projects. This, alone, suggested that the Interior Department should keep the repayment requirements on the Indian projects, where some white farmers also homesteaded, and Reclamation projects relatively in line with one another.[76]

These delicate considerations prompted Congress to turn over repayment questions largely to the discretion of the Secretary of the Interior. In the 1939 Relief to Water Users Act,[77] Congress authorized the Secretary of the Interior to determine whether construction repayments on reclamation and Indian reclamation projects should be delayed because of undue hardships that prevented repayment, such as crop failure, water shortage, or other causes beyond the control of the water users. The time periods specified by the Secretary of the Interior were to fall "within the probable ability of the water users to repay." In addition, the Secretary of the Interior must extend "preference" to municipalities, cooperatives, and nonprofit power distributors financed by the REA Act of 1936 in the sale of power and water from Reclamation projects. The 1939 legislation represented a high point of Congress's generosity to water users. It also made the determination that products of government water projects — power and water supplies — benefit public, nonprofit agencies rather than private corporations.[78]

The latter provision took on special meaning for Secretary of the Interior Ickes in the case of San Francisco's Hetch Hetchy project. While not a

Reclamation project, San Francisco's successful struggle to dam the Tuolumne River within Yosemite National Park dated from the beginning of the century. The controversy over "desecrating" a national park to fuel the growth of a city gained national notoriety and pitted municipal public power advocates against preservationists; private power interests (not always champions of conservation) rallied around the Sierra Club and John Muir. Congress ultimately approved the city's request with the passage of the Raker Act of 1913, but with the proviso, in Section 16, that the water and power from the dam and reservoir not be sold "to any corporation or individual, except a municipality or a municipal water district or irrigation district ." When the city completed O'Shaughnessy Dam in 1923, the city complied with the Raker Act by obtaining a bond issue from the voters to purchase the Spring Valley Water Company's delivery system in the city. The attempt to do the same with Pacific Gas and Electric's (PG&E) power grid did not work out. The city spent many years obtaining permission to dam Hetch Hetchy Valley and in subsequent construction. With the dam completed, PG&E offered to buy the power and deliver it.[79]

Although the Raker Act clearly prohibited it, the city had little choice other than to cooperate with PG&E, otherwise power from the municipal project would be unused and wasted. What it hoped was a temporary agreement — until voters approved purchase of the PG&E facilities — turned out to be a long-term expedient. When "public power ideologue" Harold Ickes became Secretary of the Interior, this arrangement infuriated him. He pushed and badgered San Francisco to assume ownership and initiated court cases in the cause, but still he was unable to persuade the city and its voters to purchase PG&E's electrical lines. Neither were the voters ready to pay for building an alternate system. By the 1940s, voters preferred to purchase their power from a regulated private power company. Ickes left office in 1945, having failed to carry out the mandate of the Raker Act. He considered it his special obligation to see that the public not only produced the power but delivered it, without a private corporation profiting from public investments. As the controversy between Ickes and the city of San Francisco ran its course, Reclamation faced the same issues. Despite the intent of the Raker Act and the Reclamation Project Act of 1939 that declared in favor of publicly owned utilities in water and power, Americans seemed to accept a partnership between public and private interests and Reclamation followed suit by acquiescing in the sale of public power to private distributors. Most important, the cooperation reflected the realities of a political world which was congenial to Reclamation at this juncture of its expansion. After Mead, the engineering leadership of Reclamation thrived on building great works and river basin planning. But when it came to making special provisions or even sacrifices to settle small farmers on the land in the cause of some abstract social policy demanding acreage limitations, Reclamation was no zealot.[80]

The 1939 legislation confirmed many changes in the Reclamation program since passage of the Reclamation Act. Some of the major changes included: (1) surplus waters in reclamation projects were permitted to serve private lands outside the projects already under irrigation that lacked an adequate water supply; (2) on matters of repayment of construction costs and operation and maintenance charges, the government was not to deal with individual water users but with irrigation districts or other forms of water users' organizations; (3) excess power revenues could be credited to the cost of the irrigation projects, "beyond the ability of the water users to pay;" (4) the reclamation program expanded from simply creating small irrigation projects to building large multiple-purpose projects; (5) construction costs in multiple-purpose projects for navigation and flood control must be omitted from repayment contracts with water users; and, (6) finally, at the discretion of the Secretary of the Interior, extension of up to fifty years for repayment on new projects and the development of new repayment formulas based on crop income and quality of lands.[81]

On the Great Plains, the 1930s drought compounded the misery of the Great Depression. The Great Plains Drought Area Committee was one of many efforts to study the situation and make recommendations. For its part, Reclamation moved in the direction of building small water conservation and utilization projects in the region. The passage of the Water Conservation and Utilization Act of August 11, 1939, authorized small water storage projects "to rehabilitate farmers on the land." As extensively amended on October 14, 1940, this act, also known as the Case-Wheeler Act, declared that the excess land provisions of federal reclamation laws did not apply to projects established under the Water Conservation and Utilization Act. Under this act, farm units served by dams and reservoirs could be larger than 160 acres. The provision was a clear indication that the Bureau of Reclamation saw advantages to getting rid of the 160 acre rule (or 320 acres for a married couple). In the big picture, western agriculture seemed to demand larger farm units.[82]

Removal of the 160 acre rule opened possibilities that more private lands could receive government water. However, title to dams, reservoirs, and irrigation works remained with Reclamation. None of the projects were to cost more than $2,000,000, up to $500,000 of which could be written off to flood control, which was not subject to repayment. Commissioner Page believed that the small water conservation projects combining reclamation with relief on the

Great Plains would help the victims, especially on the southern Great Plains, who suffered from what he termed drought, mechanization, and soil exhaustion, and "who would otherwise be forced to indefinite enlistments in the 'Grapes of Wrath' army in California." He also praised the Farm Security Administration's attempts to finance indigent and qualified prospective homesteaders on public land within existing reclamation projects.[83]

In the view of some Reclamation officials, the provision in the 1939 Reclamation Project Act, which allowed for up to a ten year period before repayments began, should have been very attractive to the Farm Security Administration (FSA). The FSA had expressed reservations about loaning money to settlers on projects where Reclamation held a prior lien on the land until construction costs were paid. Without the threat of impending construction cost repayments and with up to forty years to repay, in addition to the ten year moratorium, the FSA could have "a more aggressive" lending program on the projects. A. R. Golzé believed help from the FSA to struggling and promising young farmers on the projects would go a long way toward confronting what he called "the tenant problem on our projects." The tenant problem was but one of the chronic problems on the projects. Another was the extended periods for repayment and congressional moratoriums. Commissioner Page noted that under the August 4, 1939, Water Utilization and Conservation Act, which was an addendum to the May 1939 Reclamation Project Act, the Secretary of the Interior could waive construction charges for the period 1939 through 1943. While operation and maintenance charges must still be met by the water user associations or districts, the ready availability of moratoriums on repayments indicated a continued need to prop up many ailing projects.[84]

Shrinking Vistas on the Land

Facing an ongoing crisis over the viability of the farms in its projects, the Bureau of Reclamation had its hands full in the West. Since the 1920s, Reclamation had toyed with the idea of swampland reclamation in the South, and now some in the Pacific Northwest suggested stump land (cutover lands) reclamation and projects to reclaim the tide flats around Puget Sound and the Columbia River towns. Congressman Compton I. White, of Idaho, urged that Reclamation take "a great forward step" by seizing the opportunities presented to it "for what we may call conservation of the potential agricultural lands by reclaiming the cutover areas of the Northwest."

John W. Haw, Director of the Northern Pacific Railway Company's Agriculture and Immigration Department, opposed enlarging the scope of Reclamation. "Once the Bureau departs from arid land reclamation," he wrote Commissioner Page, "there is no good stopping place." If Reclamation is "importuned" to clear logged-off lands or drain swamps, it might find itself removing stones from the farms of New England. He cautioned against any support for the Arkansas Valley Authority proposal in Congress because this would require altering the national Reclamation Act to include Louisiana, Mississippi, Arkansas, and Missouri. There were already too many problems associated with arid land settlement, many of which confounded the railroads in their efforts to colonize people on railroad lands. A commissioner of the Chicago, Milwaukee, St. Paul and Pacific Railroad expressed doubts about government agencies trying to resettle people on the projects, "particularly as to the extent to which subsidization of busted or distressed or 'grapes of wrath' types is justified."[85]

Those conducting the Columbia Basin Plan of Investigations faced questions about what aid should be extended to prospective settlers whenever the irrigation phase of the project was launched. Extending aid to settlers on new projects might help meet problems of interstate migration. Reclamation claimed that its projects, including the Columbia Basin Project, offered economic and social stability to the nation. With this goal in mind, Reclamation prepared many river basin-wide plans, not only for the Columbia but for California's Central Valley and the upper Colorado River, while at the same time collecting information for similar plans on the Rio Grande, Arkansas, Nueces, and other rivers. All looked forward to providing homes and opportunities for an American population on the move in search of security, employment, and new communities.[86]

Congress recognized this migratory quest when it created a "Special Committee of the House of Representatives Investigating Interstate Migration of Destitute Citizens." Testifying before that committee, Reclamation officials continued to invoke the old idea that unused waters were wasted waters: "unused waters can be conserved to give an assured supply to the present lands and to reclaim an additional area slightly larger than the present area." But the focus was not entirely upon the 383,000 new farms Reclamation officials estimated could be provided, but also upon the towns that would grow up where 6,000,000 persons might make their homes. The figures presented a balanced picture of Reclamation serving both its traditional vision of an agricultural West while it provided power and water to a new urban West.[87]

Meanwhile, the estimated kilowatt power of Reclamation's Grand Coulee Dam and the Corps of Engineers's Bonneville Dam on the Columbia commanded the attention of the Pacific Northwest. Under Executive Order 8526, the Bonneville Power Administration came into existence in August 1940, as the marketing agency for the world's largest complex of hydroelectric power production. The government marketing agency aimed to integrate the power supplies of the two facilities for the development of the defense industries of the Pacific Northwest. Under this order, the Bonneville Administrator also set about to construct, operate, and maintain transmission lines necessary for the marketing of power. The measure, along with others in 1940, reflected a nation moving toward a war posture. Secretary of the Interior Ickes created a Defense Resource Committee within the Department of the Interior to coordinate the utilization of resources, should the nation suddenly find itself at war.[88]

The New Deal placed a great deal of stock in "planning" as it tried to cope with the economic crisis of the Great Depression. The Administration's National Resources Planning Board, by the 1940s, tried to develop broad plans for the future economic directions of the nation. Opponents in Congress and private industry feared that "planning" by national boards meant too much centralized federal control and succeeded in abolishing the National Resources Planning Board in 1943. Federal infringement on that most guarded western resource – water – was an issue that had involved the Bureau of Reclamation since the beginning of the twentieth century. As one concerned westerner put it, "We of the West have created our own appropriation system of water law. It differs from that of the East. We shall not permit, if we can help it, the extension of Federal control." Secretary of the Interior Ickes would have none of this. The new dominance of Reclamation over major western rivers resounded to the benefit of the region's people and the nation. While Ickes once viewed the Bureau of Reclamation as expendable, by the close of the depression decade, he regarded it as "a veritable Aladdin's lamp" in the far reaches of the West.[89]

Commissioners Mead and Page had guided Reclamation successfully through a decade that presented some of the same vexing problems as the 1920s. In large part, the Bureau of Reclamation achieved success, despite these problems, because it served as "an instrument" to realize the public works goals of the New Deal's economic strategy — investing in natural resource development even during the hard times of the 1930s. Hoover Dam demonstrated what the Bureau of Reclamation could accomplish (thereafter "nothing, however fanciful, seems impossible"), [90] but there was nothing inevitable about the following large dam era. It was the crisis of the Great Depression that opened the way for

Reclamation to become the dominant player in multiple-use water development. While some grew to fear the expanded power of government under the New Deal, others welcomed it, not only to meet the problems of the Great Depression, but now the looming dangers of another world war.

Endnotes

[1] Hundley, *Great Thirst*, 223; Lowitt, *New Deal in the West*, 81.

[2] "An Important Occasion," (editorial) Las Vegas *Evening Review Journal,* October 1, 1935.

[3] Robinson, *Water for the West*, 71; Grant McConnell, *Private Power & American Democracy*, (New York: Alfred Knopf, 1966), 199-200; Grant McConnell, *The Decline of Agrarian Democracy* (New York, Atheneum, 1969), 173-81; McCool, *Command of the Waters*, 5, defines "iron triangle," but in terms of efforts to persuade Congress to curb Indian water rights for the benefit of non-Indian constituencies in the West; Donald C. Swain, "The Bureau of Reclamation and the New Deal, 1933-1940," *Pacific Northwest Quarterly* 61 (July 1970), 140.

[4] Robinson, *Water for the West*, 58; Swain, "The Bureau of Reclamation and the New Deal, 1933-1940," 137, 141.

[5] Las Vegas *Evening Review Journal*, October 1, 1935. The reservoir was named Lake Mead by Secretary of the Interior Harold Ickes upon the creation of Lake Mead National Recreation Area in 1936.

[6] Lowitt, *New Deal in the West*, 89.

[7] Kluger, *Turn on Water*, 128-29.

[8] Swain, "The Bureau of Reclamation and the New Deal, 1933-1940," 138; Harold L. Ickes, *The Secret Diary of Harold L. Ickes*, volume I (New York: Simon and Schuster, 1953), 37-8; Harold L. Ickes' speech in *Reclamation Era*, 25 (November 1935), 209-10.

[9] Robinson, *Water for the West*, 56; "Eighteen U.S. Bureau of Reclamation Projects to Cost $218,440,000," *Western Construction News and Highways Builder* (January 1934), 5.

[10] F. H. Newell to A. P. Davis, December 20, 1931, A. P. Davis to F. H. Newell, December 30, 1931, Davis Papers, Box 6, Newell File; "Morris Bien Dies in Takoma Park," Washington *Evening Star*, July 29, 1932; "Report of the State Engineer," *Appendix to Journals of Senate and Assembly of the Twenty-Fourth Session of the Legislature of the State of Nevada, 1909*, volume 2 (Carson City, Nevada: State Printing Office, 1909), 41-2; "Report of the State Engineer," *Appendix to Journals of Senate and Assembly of Twenty-Second Session, 1905* (Carson City, Nevada: State Printing Office, 1905), 14-5; Dunbar, *Forging New Rights in Western Waters*, 120-2.

[11] Mae A. Schnurr to Mrs. Arthur Powell Davis, October 9, 1935, Miscellaneous Corres., Davis Papers; Ickes, *Diaries*, 528-9, 584.

[12] Elwood Mead in Memorandum for the Secretary (Secretary of the Interior, Ray Lyman Wilbur), January 23, 1932, RG 115, Entry 7, Box 461, NARA, Denver; R. F. Walter to Elwood Mead, December 31, 1932, RG 115, Entry 7, Box 781, NARA, Denver; Elwood Mead to Secretary of the Interior, October 9, 1930, RG 115, Entry 7, Box 781, NARA, Denver; Pelz, *Federal Reclamation and Related Laws Annotated*, volume 1, 487.

[13] M. L. Wilson to Mae Schnurr, November 21, 1933; M. L. Wilson to Elwood Mead, December 26, 1933, RG 115, Entry 7, Box 847, NARA, Denver; James S. Olson, editor, *Historical Dictionary of the New Deal: From Inauguration to Preparation for War* (Westport, Connecticut: Greeenwood Press, 1985), 419-20, 477-8.

[14] Elwood Mead to Secretary of the Interior, May 14, 1935, RG 115, Entry 7, Box 848, NARA, Denver; Elwood Mead to Secretary of the Interior, July 9, 1935, RG 115, Entry 7, Box 845, NARA, Denver; L. H. Mitchell to Elwood Mead, August 21, 1935, RG 115, Entry 7, Box 848, NARA, Denver; J. H. Jenkins, Assistant Director Resettlement Division, to Mae Schnurr, November 9, 1935, RG 115, Entry 7, Box 848, NARA, Denver.

[15] R. G. Tugwell, Administrator, Resettlement Administration, to Elwood Mead, July 5, 1935; Elwood Mead to R.G. Tugwell, July 8, 1935; R. G. Tugwell to Elwood Mead, September 11, 1935, RG 115, Entry 7, Box 845, Denver NARA; As quoted from Harold L. Ickes to Roosevelt, July 13, 1935, Box 258, Harold L. Ickes Papers, Library of Congress, Washington, D.C. in Lowitt, *New Deal in the West*, 245, n. 22.

[16] Tugwell to Secretary of the Interior, July 9, 1935, RG 115, Entry 7, Box 845, NARA, Denver.

[17] Olson, *Dictionary of New Deal*, "Resettlement Administration," 419-20; Elwood Mead to W. T. Clark, Jr., Wenatchee Chamber of Commerce, September 27, 1934, RG 115, Entry 7, Box 851, NARA, Denver; William G. Robbins, editor, *The Great Northwest: The Search for Regional Identity* (Corvallis: Oregon State University Press, 2001), 66-79; Paul K. Conkin, *Tomorrow a New World: The New Deal Community Program* (Ithaca, New York: Cornell University Press, 1959).

[18] Lowitt, *New Deal in the West*, 91-92; John C. Page, "Water Conservation and Control," *Reclamation Era*, 27 (March 1937), 48.

[19] Elwood Mead to Kenneth C. Miller, Secretary, National Reclamation Association, September 26, 1935, RG 115, Entry 7, Box 716, NARA, Denver.

[20] B. E. Stoutemyer, District Counsel, Portland, Oregon, to Commissioner Elwood Mead, April 27, 1935, RG 115, Entry 7, Box 657 and Elwood Mead to Secretary of the Interior, January 16, 1935, RG 115, Entry 7, Box 781, NARA, Denver.

[21] Henry L. Lumpee, "The Civilian Conservation Corps: Irrigators," *Reclamation Era*, 26 (November 1936), 268-9.

[22] Christine Pfaff, "The Bureau of Reclamation and the Civilian Conservation Corps, 1933-1942," (Denver: Cultural Resources Management, Bureau of Reclamation, 2000), 12, 17, 24, 11.

[23] Kenneth W. Baldridge, "Reclamation Work of the Civilian Conservation Corps, 1933-1942," *Utah Historical Quarterly* 39, 3 (1971), 265-85; Louis S. Davis, "E.C.W. Activities on Reclamation Projects," *Reclamation Era*, 26 (December 1936), 298-9. Note: "Force account' is an administrative term denoting occasions when Reclamation used its own employees for construction labor.

[24] Fiscal Year 1939 ran from July 1, 1938, to June 30, 1939.

[25] Alfred R. Golzé, "Civilian Conservation Corps Accomplishments on Federal Reclamation Projects: Fiscal Year 1938," *Reclamation Era*, 28 (September 1938), 190-2; "CCC Program for Federal Reclamation, Fiscal Year 1939," *Reclamation Era*, 28 (August 1938), 163; "CCC Work to Continue on Reclamation Projects," *Reclamation Era*, 28 (April 1938), 74.

[26] John M. Townley, *Turn this Water into Gold: The Story of the Newlands Project* (Reno: Nevada Historical Society, 1977), 125.

[27] "Eighteen U.S. Bureau of Reclamation Projects to Cost $218,440,000," *Western Construction News and Highways Builder* 9 (January 1934) 5-9.

[28] J. M. Young, "River Planning in the Missouri Basin," *The Military Engineer* 22 (March-April 1930), 152-3; Pitzer, *Grand Coulee*, 23, 43.

[29] Warne worked at Reclamation, where he served in various capacities, including Assistant Commissioner, and he was acting Commissioner on occasion. From 1947-1951, he was First Assistant Secretary of the Interior and later served as Water Resources Director of the State of California.

[30] Warne, *Bureau of Reclamation*, 145-7.

[31] Pitzer, *Grand Coulee*, 114-5; Michael Parfit, "The Floods that Carved the West," *Smithsonian*, 26 (April 1995), 48-75.

[32] Pitzer, *Grand Coulee*, 22.

[33] Pitzer, *Grand Coulee*, 44, 58; Warne *Bureau of Reclamation*, 148.

[34] Pitzer, *Grand Coulee*, 58; Warne, *Bureau of Reclamation*, 148.

[35] William Lang, "1949: Year of Decision on the Columbia River," *Columbia: The Magazine of Northwest History,* 19 (Spring 2005), 8-17.

[36] Pitzer, *Grand Coulee,* 101.

[37] *Project Data*, 382.

[38] Golzé, *Reclamation in the United States*, 178.

[39] Swain, "The Bureau of Reclamation and the New Deal, 1933-1940," 138; see also introduction to Richard L. Neuberger, *Our Promised Land,* edited by David L. Nicandri (Moscow: University of Idaho Press, 1989).

[40] F. A. Banks, "Significance of Grand Coulee Dam," *Reclamation Era*, 26 (December 1936), 278-80; Clayton R. Koppes, "Public Water, Private Land: Origins of the Acreage Limitation Controversy, 1933-1953," *Pacific Historical Review,* 47 (November 1978), 612.

[41] Pitzer, *Grand Coulee*, 362; White, *The Organic Machine*.

[42] Richard White, "The Altered Landscape: Social Change and the Land in the Pacific Northwest," in *Regionalism and the Pacific Northwest*, edited by William G. Robbins, Robert J. Frank, Richard E. Ross (Corvallis: Oregon State University Press, 1983), 120; Ivan Bloch, "The Columbia River Salmon Industry," *Reclamation Era*, 28 (February 1938), 26-30; "Columbia River Fishery Development," in Pelz, *Federal Reclamation and Related Laws Annotated*, volume 1, 602-3; see also Brown, *Mountain in the Clouds;* Jim Lichatowich, *Salmon without Rivers: A History of the Pacific Salmon Crisis* (Washington, D.C.: Island Press, 1999); Joseph E. Taylor III, *Making Salmon: An Environmental History of the Northwest Fisheries Crisis* (Seattle: University of Washington Press, 1999); Matthew D. Eveden, *Fish Versus Power: An Environmental History of the Fraser River* (Cambridge, U.K.: Cambridge University Press, 2004).

[43] Robert L. Kelley, *Battling the Inland Sea: American Political Culture, Public Policy, & the Sacramento Valley, 1850-1986* (Berkeley: University of California Press, 1989).

[44] Powell Greenland, *Hydraulic Mining in California: A Tarnished Legacy* (Spokane, Washington: The Arthur Clark Company, 2001), 71-112.

[45] Pisani, *Family Farm to Agribusiness*, 110-7.

[46] Pisani, *Family Farm to Agribusiness*, 425-37, see especially fn. 91, 437.

[47] Pisani, *Family Farm to Agribusiness*, 436; *Project Data*, 168.

[48] Pisani, *Family Farm to Agribusiness*, 437.

[49] Koppes, "Public Water, Private Land," 609-13.

[50] Golzé, *Reclamation*, 196-7, 65-9; see also Gerald R. Ogden, compiler, *The Excess Lands Provisions of Federal Reclamation Law: A Bibliography and Chronology* (Davis: Agricultural History Center, University of California, Davis, July 1980, typescript).

[51] Norcross, "Genesis of the Colorado-Big Thompson Project," 34; Tyler, *Last Water Hole in the West*, 41.

[52] White, *"It's Your Misfortune and None of My Own,"* 523.

[53] Tyler, *Last Water Hole in the West*, 58-80; "Correcting Nature's Error: The Colorado-Big Thompson Project," *Reclamation Era*, 30 (September 1940), 267-69; Donald B. Cole, "Transmountain Water Diversion in Colorado," *Colorado Magazine,* 25 (March 1948), 49-64; James F. Wickens, *Colorado in the Great Depression* (New York: Garland Publishing, Inc., 1979).

[54] John Page Memorandum for Arno B. Cammerer, July 25, 1938, RG 115, Entry 7, Box 781, NARA, Denver.

[55] "Imperial Valley, California," *Reclamation Era*, 26 (February 1936), 45-6; Hundley, *Great Thirst*, 223.

[56] "F. A. Banks Discusses Columbia Development," *Reclamation Era*, 26 (January 1936), 3-5; "Significance of Grand Coulee Dam," *Reclamation Era*, 26 (December 1936), 280.

[57] Marshall N. Dana, "Dr. Elwood Mead," *Reclamation Era*, 26 (May 1936), 115.

[58] Commissioner Elwood Mead to All Field Offices, May 2, 1935, RG 115, Entry 7, Box 639, NARA, Denver.

[59] U.S. Department of the Interior, U.S. Bureau of Reclamation, "Bureau of Reclamation Historical Organizational Structure," (Denver: Bureau of Reclamation History Program, 1997, typescript), 66.
[60] Elwood Mead to All Field Offices, May 2, 1935, RG 115, Entry 7, Box 639, NARA, Denver.
[61] Robinson, *Water for the West*, 76.
[62] F. D. Roosevelt to John Page, July 18, 1936; John Page to R. F. Walter, July 21, 1936, RG 115, Entry 7, Box 725, NARA, Denver.
[63] Lowitt, *New Deal in the West*, 46.
[64] John Page to Morris L. Cooke, Chairman Great Plains Drought Area Committee, July 30, 1936, RG 115, Entry 7, Box 725, NARA, Denver; three books on the Dust Bowl offer varying opinions on its origins: Donald Worster, *Dust Bowl: The Southern Plains in the 1930s* (New York: Oxford University Press, 1979); Mathew P. Bonnifield, *The Dust Bowl: Men, Dirt, and Depression* (Albuquerque: University of New Mexico Press, 1979); Douglas R. Hurt, *The Dust Bowl: An Agricultural and Social History* (Chicago, Illinois: Nelson-Hall, 1981).
[65] John Page to Morris L. Cooke, Chairman Great Plains Drought Area Committee, July 30, 1936, RG 115, Entry 7, Box 725, NARA, Denver
[66] "Bureau of Reclamation Historical Organizational Structure," 71; F. D. Roosevelt to J. A. Ford, Managing Secretary of the Spokane Chamber of Commerce, August 29, 1936, RG 115, Entry 7, Boxes 725, NARA, Denver.
[67] John Page to Cook, REA Administrator, November 18, 1936, RG 115, Entry 7, Box 725, NARA, Denver; John Page to Carl Hayden, April 12, 1940, RG 115, Entry 7, Box 843, NARA, Denver; Pisani, *Water and American Government*, 271; John C. Page, "Don't Forget the Drought," *Reclamation Era*, 31 (November 1941), 282.
[68] M. L. Wilson to James Roosevelt, Secretary to the President, January 24, 1938, RG 115, Entry 7, Box 843, NARA, Denver.
[69] B. K. Burlew, Acting Secretary of the Interior, to James Roosevelt, Secretary to the President, May 23, 1938, RG 115, Entry 7, Box 843, NARA, Denver.
[70] Franklin D. Roosevelt to the Secretary of the Interior, November 9, 1935; Charles West, Acting Secretary of the Interior to Roosevelt, January 13, 1936; Elwood Mead to the Secretary of the Interior, November 14, 1935; H. D. Comstock to Commissioner Elwood Mead, c. January 15, 1936, RG 115, Entry 7, Box 844, NARA, Denver.
[71] B. D. Zevin, editor, *Nothing to Fear: The Selected Addresses of Franklin Delano Roosevelt, 1932-1945* (Boston, Massachusetts: Houghton Mifflin Company, 1946), 91; John C. Page, "Also One-third is Ill-Watered," *Reclamation Era*, 28 (January 1938), 2; Carroll Pursell, *The Machine in America: A Social History of Technology* (Baltimore: Johns Hopkins University Press, 1995), 267.
[72] "Dams Completed During Past Five Years," *Reclamation Era*, 29 (February 1939), 38.
[73] "Bureau of Reclamation Historical Organizational Structure," 73.
[74] "The Reclamation Project Act of 1939," August 4, 1939; ch. 418, 53 stat. 1187, in Pelz, *Federal Reclamation and Related Laws Annotated*, volume 1, 634-44.
[75] Golzé, *Reclamation*, 106-7.
[76] R. B. Williams to Secretary of the Interior, April 10, 1939, RG 115, Entry 7, Box 781, NARA, Denver.
[77] "Relief to Water Users Act, an act of May 31, 1939," ch. 156, 53 Stat. 792, in Pelz, *Federal Reclamation and Related Laws Annotated*, volume 1, 632-3.
[78] "Reclamation Project Act of 1939," in Pelz, *Federal Reclamation and Related Laws Annotated*, volume 1, 663.
[79] Righter, *The Battle Over Hetch Hetchy*, 169-72.
[80] Righter, *The Battle Over Hetch Hetchy*, 176, 180, 185-6.
[81] Golzé, *Reclamation*, 108-9.
[82] "Water Conservation and Utilization Act, an act of August 11, 1939," ch. 717, 53 Stat. 1418, in Pelz, *Federal Reclamation and Related Laws Annotated*, volume 1, 668-79; Lowitt, *New Deal in the West*, 90.

[83] "Water Conservation and Utilization Act," in Pelz, *Federal Reclamation and Related Laws Annotated*, volume 1, 668; John C. Page to Senator Carl Hayden, April 12, 1940, RG 115, Entry 7, Box 843, NARA, Denver.

[84] A. R. Golzé, Memorandum to Commissioner, August 30, 1941, RG 115, Entry 7, Box 847, NARA, Denver; Commissioner John Page to Senator Rufus Holman of Oregon, February 26, 1940, RG 115, Entry 7, Box 639, NARA, Denver .

[85] Compton I. White to E. K. Burlew, First Assistant Secretary of the Interior, March 7, 1941, RG 115, Entry 7, Box 618; John W. Haw to John Page, February 10, 1941, RG 115, Entry 7, Box 618; R. W. Reynolds to H. W. Bashore, Acting Commissioner of Reclamation, November 14, 1941, RG 115, Entry 7, Box 845, NARA, Denver; Haw had been an active observer of the progress of western reclamation in the 1930s as indicated by the following report: F.E. Schmitt and John W. Haw, "Report on Survey of Federal Reclamation in the West," *Reclamation Era*, 25 (February 1935), 29.

[86] Patricia Kelly Hall, and Steven Ruggles, "'Restless in the Midst of Their Prosperity': New Evidence on the Internal Migration of Americans, 1850-2000," *Journal of American History,* 91 (December 2004), see graph, 831.

[87] Robinson, *Water for the West*, 77; E. B. Debler, "Stabilization by Irrigation," *Reclamation Era*, 30 (November 1940), 309-11.

[88] Executive Order 8526 "Coordinating the Electrical Facilities of Grande Coulee Dam Project and Bonneville Project," *Federal Register*, 3390, volume 5, August 29, 1940.

[89] Lowitt, *New Deal in the West*, 79-80; Marion Clawson, *New Deal Planning: The National Resources Planning Board* (Washington, D.C.: Resources for the Future, 1981); Olson, *Historical Dictionary of the New Deal*, 363-66; Philip W. Warken, *A History of the National Resources Planning Board, 1933-1943* (New York: Garland Publishing, Inc., 1979), 1-3; "Bonneville to Market Grand Coulee Power," *Reclamation Era,* 30 (November 1940), 322; "Defense Resources Committee Appointed," *Reclamation Era,* 30 (September 1940), 258; L. Ward Bannister, "National Resources Planning Board: How It should be Constituted," *Reclamation Era,* 30 (January 1940), 7; "Aladdin's lamp," as quoted in Swain, "The Bureau of Reclamation and the New Deal, 1933-1940," 146, but originally found in Harold Ickes, "Friant Dam Construction Started, Central Valley Project, California," *Reclamation Era,* 29 (November 1939), 291; Ickes, "Our Right to Power," *Colliers,* 102 (November 12, 1938).

[90] Las Vegas *Evening Review Journal,* October 1, 1935.

CHAPTER 7

THE MANY FACES OF RECLAMATION

Introduction

Japan's attack on Pearl Harbor, December 7, 1941, thrust the United States into full mobilization for a world-wide war to defeat enemies in Asia and Europe. On the Pacific Coast, from San Diego to Puget Sound, power from big western dams enhanced the ability of the nation to build airplanes and warships. Large dam projects, undertaken for the internal development of the Far West during the hard years of the Great Depression, now became important to the security of the nation. As war approached, Reclamation took steps to guard against sabotage at dams, particularly at Hoover Dam, where it was joined by the Army after the outbreak of war. Hoover Dam sent water and power to California's agriculture and the state's burgeoning cities and industries. Grand Coulee Dam generated power for the industrialization of the Pacific Northwest and, not incidentally, the production of plutonium at Hanford, Washington, for the construction of the atomic bombs that ended World War II and ushered in the nuclear age, with all its fears and promises.[1]

In 1941, completion of the powerhouse at Grand Coulee Dam guaranteed a supply of hydroelectric power from Reclamation's great dam on the Columbia River for war industries and quieted critics who once labeled the dam as the "White Elephant" of eastern Washington. But Grand Coulee Dam historian, Pitzer, cautions that, while Grand Coulee contributed to satisfying wartime power demands, it was not as critical to victory as some sources extravagantly asserted.[2] In northern California, Shasta Dam came online late in June of 1944 as part of the Central Valley Project. Along with Hoover Dam, these dams provided the enormous electrical power demanded for the production of aluminum aircraft, ships, and other implements of war without seriously inconveniencing civilian consumers and homes. These huge facilities in the Far West, which controlled floods as well as generated power, represented the hard side of Reclamation's achievements. The other side of Reclamation operations was a softer side. From its origins, rooted in the ideals of the Jeffersonian family farm, Reclamation remained involved in project settlement questions and the welfare of settlers. Reclamation negotiated agreements with water users, helped to adjudicate water conflicts, and distributed water to farmers. During the war emergency, the hard tasks of kilowatt production drew the attention of Reclamation away from social and economic questions on the projects, but, with little notice, it became engaged with the program to relocate the West Coast Japanese population, to house war resisters, and even prisoners of war — all of which drew upon Reclamation's experience in managing project settlements. While efforts in the previous decade to rehabilitate and resettle farm families involved far more talk than action, the Bureau of Reclamation showed a willingness to participate should programs go forward. Like other governmental service bureaus in the Great Depression, the Bureau of Reclamation took part in the social and economic relief programs of the New Deal. The flexibility demonstrated in that decade prepared Reclamation for new roles that it would assume in World War II.

Relocation of West Coast Japanese

With the war emergency came demands that the Bureau of Reclamation make room on its projects for the internment of West Coast

7.1. Part of the Hunt Relocation Center for Japanese internees, Minidoka Project, Idaho.

Japanese-Americans (and Japanese nationals). Just as it had initiated a social experiment in the construction and settlement of irrigation projects earlier in the century, Reclamation, in the words of one correspondent, participated in one of the "most significant social experiments arising out of the war." Author-journalist Carey McWilliams, a critic of the plight of Mexican labor in California, saw a positive side to the evacuation and relocation of the Japanese. In the summer of 1942, during the forced Japanese evacuation along the Pacific Coast, McWilliams's assessment was that the experiment in inland internment stood to achieve "excellent social objectives," if it democratized and Americanized the immigrant Japanese population.[3] And yet, two years later, McWilliams reversed himself arguing against Japanese internment. He now saw the forced evacuation as the culmination of a long history of anti-Japanese prejudice in California.

In Executive Order 9066, dated February 19, 1942, President Roosevelt authorized the Secretary of War to exclude, among others, all persons of Japanese ancestry from military areas designated by the Secretary or his designated commanders. In the end, Japanese-Americans and Japanese nationals were excluded from the western half of Washington, Oregon,

7.2. Chinese cabbage on the Tule Lake Japanese internment farm near Newell, California, on the Klamath Project.

California, southern Arizona, and other areas. Approximately 110,000 people fell under the exclusion order in one of the most controversial and regretted, if not shameful, moves by the United States on the West Coast in the anxiety-ridden months after the attack on Pearl Harbor. At first, the Army handled the assignment and made no distinction between non-citizens and citizens, loyal or disloyal, or among men, women, children, and the elderly. Once assembled at appointed "terminals" on the West Coast, the Japanese faced relocation to inland destinations. The Department of the Interior became immediately involved. Inquiries went out from the Secretary of the Interior's office to Reclamation concerning abandoned CCC camps on its projects. With the return of full employment, the CCC program was winding down and vacant CCC camps might provide housing for the Japanese internment. Reclamation responded cautiously, showing little enthusiasm for the relocation of a Japanese population near or on the irrigation projects.[4]

7.3. Sorting picked tomatoes at the Hunt Relocation Center on the Minidoka Project.

Information returned to the Secretary of Interior's office showed approximately fifteen CCC camps that were available to house evacuees on reclamation projects. Reclamation officials also quickly noted that labor from the camps could be used in the maintenance and rehabilitation of existing irrigation structures and for farming operations. New camps could be constructed on ten or more of the reclamation projects if necessary. But the message conveyed back to the Secretary of the Interior by Reclamation made clear that the recommendations were only possibilities and concluded, "You will recognize that there are many difficulties encompassed in such a program and that the Bureau is not seeking a part of it." Still, Commissioner Page recognized the probable participation of Reclamation in what he called, "the location of concentration camps for the Japanese who are being moved from the West Coast area." It was unclear what agency would handle moving the thousands of people whom he thought might be quartered in vacant CCC camps. The formation of the War Relocation Authority came on March 18, 1942, shortly after Page speculated about what agencies would be involved in the "concentration camps." The correspondence reflected a casual use of the term prior to the horrendous meaning it soon assumed in the course of World War II and after. [5]

At a meeting with Milton Eisenhower, head of the War Relocation Authority (WRA), Acting Commissioner Bashore learned that the plan was to resettle the Japanese into small towns or villages with a population of from three thousand to ten thousand. Articles in the *Fallon Standard* (Nevada) for May 13, 1942, expressed ambivalence about the prospect of a Japanese internment camp there. An editorial welcomed the prospect of using Japanese farmers to bring 20,000 additional acres of land under irrigation on the Newlands Project, but only if the land "is settled by white farmers after the war or after the development is completed." The editorial cautioned people not to be blinded by their prejudices: "People who would not have the Japanese in this district under any circumstances, it appears to us, are motivated by emotion rather than reason." By August 1942, hastily constructed relocation camps began receiving evacuees throughout the West. The Newlands Project, at Fallon, was not among them.

All the refugee settlements were to have, so far as possible, their own social services, such as hospitals, churches, schools, and public utilities. During the summer of 1942, the federal government designated the Minidoka, Klamath, and Shoshone projects, each "undeveloped" and "at a safe distance" from strategic locations, as eligible for relocation camps. On the Minidoka Project, the camp and barracks would be built by the Morrison-Knudsen Company, which had a long-time working relationship with Reclamation. The Tule Lake area received the relocation site on the Klamath Project, and Heart Mountain on the Shoshone Project, in northern Wyoming, became the third site on Reclamation lands. While remote, the availability of electricity from the recently completed Heart Mountain Power House on the

7.4. Cucumbers under hot caps on the Heart Mountain Relocation Center.

7.5. Harvesting potatoes on the Heart Mountain Relocation Center in Wyoming.

7.6. Processing table beets at the Tule Lake Relocation Center for shipment to other centers.

project in 1941 recommended this site. Heart Mountain was one of the few rural communities in Wyoming served by electricity. This is not to say that northern Wyoming was an ideal place for the internees to take up residence. Their barracks were flimsy, the winters cold, and the summers very windy. [6]

A significant issue emerged regarding the employment of Japanese labor when "evacuees" were sent from the Hunt Relocation Center to work on the Boise Project's Anderson Ranch Dam. In 1941, Morrison-Knudsen Company and other companies contracted to build the dam. The contract contained a clause that read: "no Mongolian labor shall be employed in constructing the Anderson Ranch dam." (Section 4 of the 1902 Reclamation Act stipulated: "That in all construction work eight hours shall constitute a day's work, and no Mongolian labor shall be employed thereon.") Reclamation held fast to this racial exclusion labor policy throughout the early decades of the twentieth century. It was an iron-fast rule until labor shortages during World War II forced a change. The Interior Department Appropriations Act for 1943 (Public Law No. 645 – 77th Congress, 2nd Session) authorized Japanese under control of the War Department or any other Federal agency to work on federal reclamation projects. It read:

> Labor and services of enemy aliens and of all American-born Japanese who are under the control of the War Department or other Federal agency may be utilized by the Secretary of the Interior, under such regulations and conditions as the War Department or other Federal agency exercising said control may prescribe in the construction of Federal reclamation projects and water conservation and agreement between said Secretary and said War Department or other Federal agency.

Acting Commissioner Bashore requested that authority to waive the provision be extended to cover all construction work of Reclamation. But only those Japanese under, "the control of the War Department or other Federal agency" could be so employed. The new law fell short of opening construction jobs to all races, but it recognized the difficulties of maintaining race-based restrictions upon labor during the desperate labor shortages created by the war. Carey McWilliams might have hailed it as achieving some positive goals for the Japanese in their War Relocation experience.[7]

Three of the ten relocation centers were placed on Reclamation projects. At first, the Bureau of Reclamation saw these sites as merely expanded CCC camps, but it soon became clear that the relocation centers demanded far more facilities to handle people ranging in age from the very young to the very old. Financing for the hastily built camps came from the WRA and not from Bureau of Reclamation funds. On the Minidoka Project, or Hunt Center, the population reached a high point of 9,397; on the Heart Mountain Center, 10,767; and at the Tule Lake Center, associated with the Klamath Project, over 18,000.[8]

At first, these projects welcomed the prospect of thousands of workers available to do reclamation work. When it became apparent that entire communities were to be established, however, resistance and protests began. Project settlers feared that Japanese families would take up permanent residence after the war, even though the vast majority of evacuees were not farmers but rather merchants, urban workers, and professional people who had little experience with farming. Ironically, young men in the relocation camps became subject to the draft. If they could find financial assistance, selected men and women could leave the camps to pursue studies at interior universities and colleges. Others sought and obtained work permits to take jobs outside the camps. On Reclamation projects, internees built canals and ditches that brought thousands of acres of land into agricultural production not only to feed the camps but also to help in the war effort. Not to be ignored, however, were incidents of protest and violence by internees at both Heart Mountain and Tule Lake. The relocation "experiment" occurred at tremendous cost to Japanese of the West Coast in the loss of property and possessions, to say nothing of the psychological assaults upon the Japanese as a group. The large Tule Lake Center on the Klamath Project was the last to close on May 5, 1946, when the Bureau of Reclamation officially received it, and its barbed wire fences and guard towers, back from the WRA.[9]

Civilian Public Service Camps

The emergency demands of the Great Depression, and now another world war, made it increasingly clear that public service natural resource bureaus assumed roles far beyond their original purposes. Reclamation adapted to the crises of the times. In the Great Depression, Reclamation built huge industrial-scale dams while it facilitated the establishment of CCC camps and negotiated programs for the resettlement of farmers fleeing drought. After Pearl Harbor, Reclamation assumed the dubious role of providing sites for "relocation or concentration camps" for the West Coast Japanese evacuees. In hindsight, some would say that Reclamation performed a more laudable role when it received "assignees" from the Selective Service System, classified as conscientious objectors (COs), who were allowed to perform public service rather than enter the armed forces. Unlike the Japanese, who were under the authority of the War Relocation Board, the objectors were assigned directly to camps administered by Bureau of Reclamation employees.

The Selective Service System offered to those who could make convincing cases for conscientious objector status the option of becoming Civilian Public Service (CPS) enlistees. In this draft category, men could perform public service unrelated to the war effort. They received no government allowance, with American churches supporting pacifism paying for their upkeep. By late July 1942, the major bureaus within the Department of the Interior that administered public lands – National Park Service, Fish and Wildlife Service, and General Land Office – were put on notice that they should expect to administer Civilian Public Service Camps at selected sites. In the Department of Agriculture, the U.S. Forest Service received similar notification. As with the Japanese, attention again focused on CCC camps (all to be closed by July 1, 1942) for housing the enlistees in Civilian Public Service Camps. Acting Secretary of the Interior Conrad L. Wirth informed the Bureau of Reclamation that four sites were under consideration to be operated with allocations from the Selective Service System. They were Rapid Valley, near Hill City and Rapid City, South Dakota; Provo River, near Heber, Utah; Mancos, near Mancos, Colorado; and Deschutes, near Lapine, Oregon. [10]

7.7. Conrad L. Wirth, director of the National Park Service and sometime acting Secretary of the Interior. Courtesy of the National Park Service.

Reclamation, of course, hoped to benefit from the labor of the Civilian Public Service (CPS) enlistees, as did local farmers, who stood in need of agricultural labor during the war. And unlike the Japanese workers, this labor would not have to be paid. Acting Commissioner Harry Bashore informed Chief Engineer Sinclair O. Harper that the Civilian Public Service Camps would be operated in the same fashion as the CCC camps, with the Bureau of Reclamation hiring superintendents, foremen, and mechanics to carry on work with funds provided by the Selective Service System. Bashore suggested that work plans be drawn up for each of the possible projects where Civilian Public Service Camps might be established, "So that should the camps be occupied on short notice, the Bureau will be prepared to immediately undertake its construction program." By the end of 1942, the North

7.8. Commissioner Harry W. Bashore.

Platte Project on the Wyoming-Nebraska border, containing the Goshen and Pathfinder Irrigation Districts, favored the establishment of a Civilian Public Service Camp there as preferable to a Japanese relocation settlement. They feared sabotage from Japanese laborers who had been used without incident that fall in the harvest of sugar beets. In the case of conscientious objectors, a local official feared that the COs might be exposed to "scorn or mistreatment by soldiers on furlough or the families of soldiers, especially if they are spread around on farms where considerable contact will exist with the general public." But the official quickly added, "This is merely surmise on my part as no actual experience has been had here." He pointed out that when farm work slacked off during the winter, rock riprap work could be done on the Fort Laramie Canal. There was always something to do on reclamation projects.

7.9. Chief Engineer Sinclair O. Harper.

There is evidence that some CPS labor was employed at Grand Coulee Dam; at Riverton Project, Wyoming; and on the Strawberry Valley Project south of Provo, Utah, by the Bureau of Reclamation. Commissioner

Bashore, by the fall of 1943, wrote the Secretary of the Interior's office that more CPS labor could be used on the projects. With labor scarce during the war, the Commissioner appreciated the work of the civilian conscripts in the construction of works needed for the development of water supplies for irrigation which was considered a part of the Department of Interior's war food program. He hoped the office of the Secretary of the Interior "would not only protect the present camp strengths ... but ... increase the number of men assigned to each, if it is possible to do so." Ultimately, the government established Civilian Public Service camps on only three reclamation projects, with only a handful of assignees: Deschutes in Oregon; Mancos in Colorado; and Rapid Valley in South Dakota. The camps never exceeded 160 enrollees each. The U.S. Forest Service and the Soil Conservation Service accommodated the largest number of Civilian Public Service camps. Other conscientious objectors worked in scattered locations on farms, hospitals, and facilities for the mentally ill. Nearly 12,000 men qualified for Civilian Public Service under the Selective Service System in World War II.[11]

The Rapid Valley Project, near Hill City and Rapid City, South Dakota, benefited from Civilian Public Service work. Authorized under the Water Conservation and Utilization Act of 1939, work began in 1942. At first, the project was under the direction of the Farm Security Administration and the CCC, but with the retirement of those agencies from the scene at the

7.10. A Civilian Public Service crew grouting at Deerfield Dam on the Rapid Valley Project, South Dakota.

outset of the war, Reclamation's Civilian Public Service Camp took charge of building the earthen Deerfield Dam and Reservoir. The dam was 137 feet high and 800 feet long and the reservoir was designed to supply water to 12,000 acres of farmland, mostly for sugar beets, and municipal water for Rapid City.[12] The Civilian Public Service enlistees often came under criticism for lethargic work habits, but the complaints did not extend to the Mennonite objectors who received praise for their efficient work habits that exceeded those seen in the CCC camps. In response, *The Reporter*, a publication of the National Service Board of Religious Objectors, defended the CPS camps and the difficulties the men faced. It noted in this instance that Bureau of Reclamation engineers at Lapine, Oregon, on the Deschutes Project, and on the Mancos Project in southwestern Colorado often had little insight into the "plight of c.o's under conscription."[13]

Requests for more labor came from the Commissioner in spite of complaints from the field that CPS camp labor was inefficient and sometimes recalcitrant. At the Lapine, Oregon, camp some "assignees" asked for wages, accident insurance, and aid to their dependents. Near the end of the war, the Selective Service System reported that the Bureau of Reclamation had used CPS labor to work on an earthen dam for water storage that had been started by the Civilian Conservation Corps. The COs also worked in sawmills, gravel pits, rock crushers, and repair shops; constructed roads and buildings; and cleared timber from reservoir sites. For jobs requiring more skill, it was necessary for "the technical agency [the Bureau of Reclamation] to train men. These included engineering; drafting; operating of trucks, tractors and earth moving machinery; building with concrete and steel; and the repairing and maintenance of equipment."[14]

Lastly, the Minidoka Project was the site of one of the many prisoner of war camps scattered throughout the United States during the war. Near Paul, Idaho, Morrison-Knudsen completed a camp that the Army accepted from the Bureau of Reclamation in October 1943. It did not receive prisoners until May 1944, when 3,000 German and Italian soldiers arrived and helped in the harvest of that year. Before the camp closed on October 1, 1946, captured Russians, pressed into service on the Western Front by the Nazis, also saw confinement at this Idaho camp.[15]

7.11. Scenes from the German prisoner of war camp on the Belle Fourche Project, South Dakota.

7.11 continued.

Reorganization and Regionalization

Amidst all these special wartime assignments, the Bureau of Reclamation underwent a major reorganization. Reclamation leadership, beginning with Director Newell and Chief Engineer A. P. Davis in 1907-1908, embraced a quasi-regionalism that gave Division Engineers the responsibility for administering geographical divisions. In 1915, Secretary of the Interior Lane's reforms abolished the Division Engineers, replacing them with the Chief Engineer's Office in Denver. The rising importance of the Denver Office confused Project Supervisors, who saw a divided authority between Denver and Washington. Commissioner Mead's attempt at a regionalized administration followed his 1935 directive. It established an "Operation and Maintenance Division" (for the third time) in the Washington Office and decreed five districts within Reclamation: (1) Montana and North Dakota; (2) Idaho, Oregon, and Washington; (3) Wyoming, South Dakota, and Nebraska; (4) Northern California, Nevada, Utah, and Colorado; (5) Southern California, Arizona, and New Mexico. Mead's order fell short of achieving regional administration. He wanted to give attention to economic problems on the projects and address questions of "excess land-holdings." Mead's death in January 1936 and the heavy work schedule imposed on Reclamation by various New Deal projects detracted from these efforts, which became even more notable after the O&M Division was transferred to Denver in 1942.[16]

The Denver Office remained the centrifugal authority around which field offices revolved, much to the irritation of some in the Commissioner's Office in Washington, D.C., and, eventually, Secretary of the Interior Ickes. As William E. Warne, an assistant commissioner in the Washington, D.C., office, put it, the Commissioner's Office, "at times felt that the satrapy 1,500 miles away from Washington worked to diminish their influence on the program and to frustrate policies they wished to enforce in the field." In the 1930s, emphasis upon river basin planning created concern in the office of the Secretary of the Interior that the Bureau of Reclamation did not have its administrative offices close enough to the local constituencies within various river basins. Secretary Ickes worried that agencies outside the Department of the Interior — the Federal Power Commission, the Corps of Engineers, or independent river basin authorities, i.e., T.V.A. — might move more rapidly to build new projects than Reclamation. Regional offices would place the Bureau of Reclamation in a better competitive position.

On another, perhaps more principled level, Secretary Ickes and his New Deal advisors thought that the Denver Office neglected the social and economic goals of water and power development in the river basins of the West. In its headlong drive to build water delivery systems and dams, the Denver Office seemed little concerned with resettlement issues, farm economics, and the perplexing question of "excess lands." A flare up of this issue was imminent. Ickes was aware that Reclamation was not an ardent enforcer of the 160 acre rule (or 320 acres for a man and wife). It was no secret that reorganization would diminish the power of the Denver Office, create regions with direct communication lines to Washington, and make the regions more responsive to the policies of Commissioner Bashore and the Secretary of the Interior.[17]

On September 9, 1943, under the instigation of Secretary of Interior Ickes, Commissioner Bashore ordered that the organization move toward regional administrative decentralization. Bashore had been an engineer with Reclamation since 1906. He now succeeded Commissioner Page, who left office in early August 1943. At first, the reorganization order designated six regional offices. In 1944, a seventh was added. Designated by numbers with the cities of their regional offices, they included: Region 1, Pacific Northwest, Boise, Idaho; Region 2, Mid-Pacific, Sacramento, California; Region 3, Lower Colorado, Boulder City, Nevada; Region 4, Upper Colorado, Salt Lake City, Utah; Region 5, Southwest, Amarillo, Texas; Region 6, Upper Missouri, Billings, Montana; Region 7, Lower Missouri, Denver, Colorado. Not until 1972 were regions officially referred to by their regional names. Regional office duties included project planning, public and water-user relations, and supervision of project operation and maintenance. They could negotiate power contracts in cooperation with the Interior Department's Power Division and coordinate construction projects with other Bureau of Reclamation operations. In the Denver Office, Commissioner Page's order established four branches. Only the Design and Construction Branch remained under the control of the Chief of Engineering. The other three Denver branches were Project Investigations, Operation and Maintenance, and Fiscal and Administrative Management. A new Assistant Commissioner's office was established in Washington, D.C., to coordinate Reclamation operation in the regions.[18]

River Basin Development Issues

While the reorganization reduced the influence of the Office of the Chief Engineer, Reclamation's Denver Office was much admired worldwide for its civil engineering capabilities and accomplishments. By 1944, various departments within the government made plans for the postwar years. For Reclamation, these included an interest in international conservation and development of resources that engaged the United States as it confronted the challenges of Cold War competition with the Soviet Union and international communism.[19] On the home front, the war had delayed irrigation development. Although Congress had again signalled its approval of an irrigation component for the Columbia Basin Project in 1943, it would not go forward until 1948. Historian Paul Pitzer points out that the focus on power generation at Grand Coulee Dam often obscured the fact that the dam was built to irrigate the nation's largest reclamation project and that much planning had been done before the end of the war. According to Pitzer, "No reclamation project in American history received as much advance study and planning as the Columbia Basin Project." Reclamation official Alfred Golzé noted that the 1943 Columbia Basin Project Act was the first legislation to authorize Reclamation to purchase private lands and divide them into small farms for resale to settlers, "in a pattern similar to homesteading on the public domain."[20]

After approval of the Columbia Basin Project in 1935, Congress, in 1937, made appropriations for surveys, investigations, and appraisal of lands for irrigation works. The act, entitled "Prevention of Land Speculation, Columbia Basin Project," invoked the principle that landowners entering into contracts for government water "as a condition precedent to receiving water from the said irrigation works, … shall agree to dispose of excess holdings … wherein the said landowner also shall confer upon the Secretary of the Interior an irrevocable power of attorney to make any such sale on his behalf." This was another attempt by Congress to prevent speculative land values from plaguing the development of the Columbia Basin Project, as had occurred on other projects. The 1943 Columbia Basin Project Act repealed these restrictions and "returned the project to the standard reclamation format," according to William Warne, one time Director of Information for Reclamation. "Standard format" meant permitting the market to determine land prices offered by private landholders, and, perhaps, lax enforcement of excess land rules that now held to the "standard format" of the 160 acre limitation rule. The 1937 Act had imposed the more stringent rule that landholdings should not exceed forty acres (eighty acres for a man and wife). Above that number, the government considered the holdings as "excess lands." The small parcels were in keeping with the Administration's original hope that the Columbia Basin Project would provide a refuge to farmers who fled the Great Plains. It was also hoped that the project would be the home to 300,000 farmers.

Neither of these goals was realized. The Columbia Basin Project Act of 1943 increased the acreage limitation to the traditional 160 acre parcel per individual landowner. This enabled the registration of property in the names of wives, children, and relatives, thus expanding the size of the working farm unit. Residents welcomed abandonment of the 40 acre limitation in light of the economic pressures and mechanization of modern agriculture. The 160 acre parcel recognized the increased wartime and postwar demands on agricultural production. Still, returning to the 160 acre rule reflected new interest in the issue of farm size by Secretary of the Interior Ickes as well as Commissioner Bashore.[21]

These events did not go unnoticed by the water users in California's Central Valley Project. Like most projects settled long before the arrival of water from a reclamation project, many Central Valley landholdings stood in excess of the acreage limitation provisions in the 1902 Reclamation Act. The Central Valley Project had not been exempted from acreage limitation, as had the Colorado Big-Thompson Project, the Newlands Project, and Imperial Valley. Some of the larger landholders in the Central Valley included the Southern Pacific Railroad, Standard Oil of California, Boswell Company, and DeGiorgio Company, to name but a few. These agribusiness companies, among others, eventually became key opponents of the Interior Department's attempts to enforce acreage limitations and the disposal of excess lands in the post-war period.[22] The 1943 Columbia Basin Project Act's reaffirmation of the acreage limitation signaled future problems for the Central Valley Project's water users. In addition, the 1939 Reclamation Project Act extended generous terms for repayments — fifty years, if necessary, in accordance with land quality and the economic value of crops. The long-term, interest-free loans on construction costs left little doubt that the Department of the Interior had a right to expect adherence to Reclamation Law's acreage limitations on land served by government-subsidized water. The excess lands controversy dominated much of the politics of water in California's Central Valley in the decades after World War II.

By 1948, the Columbia Basin Project was no longer seen as a refuge for down and out farmers from the Great Plains. Reclamation required applicants to have at least $7,500 start up capital, and recommended an optimal amount of $20,000. "The people who farmed the Columbia Basin in the 1950s were not the people pictured by the planners in the late 1930s," noted historian Pitzer in his remarks on the settlement of the Columbia Basin Project in the postwar years. Some in Congress viewed the project as the ideal place for the settlement of veterans. As it did at the end of the previous war, Congress debated the wisdom of granting free farms to veterans on irrigation projects, but defeated an attempt to include the provision in the Servicemen's

Readjustment ("GI Bill of Rights") Act of 1944. Congress merely extended the veteran preferences of WWI to claim a homestead to WWII veterans, and, largely, this occurred at the instigation of Reclamation. [23]

While planning and development for the Columbia Basin Project and the Central Valley Project continued, other river basins drew the attention of Reclamation and the Congress. The concept of multi-resource river basin development attracted the more ardent New Dealers, especially Secretary of the Interior Ickes. Should the federal government possess the political will and determination to develop other river basins, the Tennessee Valley Authority stood as an example of what might occur elsewhere. While TVA was an example, its complications were many. As the TVA demonstrated, the emergence of "new" administrative authorities presented a threat to the older bureaucratic entities such as the Bureau of Reclamation, the Interior Department, and the Corps of Engineers. While Secretary of Interior Ickes embraced the idea of river basin development, he saw no need for new "authorities" that might replace Reclamation in the undertakings. Such new agencies could not bring to the effort Reclamation's proved ability to build dams, hydroelectric facilities, and water delivery systems. Bitter, anti-New Deal criticism already assailed the TVA. It was unlikely that any comprehensive river basin development projects would be forthcoming from the Administration or Congress.

Feared and admired as it was, TVA-style planning was not a model to be repeated. Reclamation's expansion during the 1930s placed it in an advantageous position to assume leadership in any basin authorities created in the West. These included the lower Colorado River Basin, California's Central Valley, the Wasatch Mountains, and the Columbia Basin Project. Bureaucratic competition and a lack of political will within the New Deal curtailed ambitious river basin development plans for both Reclamation and the Corps of Engineers that had shared many earlier undertakings.

Notwithstanding the dim prospects for new TVA-like developments, the Missouri River Basin presented possibilities. In the wake of costly floods on the river in the spring of 1943, Congress's House Flood Committee announced its concern with protecting property in the lower Missouri Basin from Sioux City, Iowa, to the river's mouth at St. Louis, Missouri. Since flood control was the province of the U.S. Army Corps of Engineers, the Committee directed the Board of Engineers for River and Harbors to study and identify flood control problems and countermeasures on the lower

Missouri River. Colonel Lewis A. Pick, Missouri River Division Engineer with the Corps of Engineers, conducted the study. Not surprisingly, he looked first at the earlier 308 Report on the Missouri River Basin. In addition to his investigations, he conducted public hearings and met with the House Committee in June 1943. Colonel Pick was a twenty-six year veteran with the Corps, and he had assisted Herbert Hoover with relief efforts in the 1927 Mississippi River flood.[24]

On August 10, 1943, Pick presented to the House Flood Committee a short report on the Missouri River that went far beyond flood control. He addressed the entire question of the multiple-purpose development in the Missouri River Basin. The topics included flood control, hydroelectric potential, wildlife and recreation, domestic water supplies, and river transportation. The Pick report (now a plan) conceded that levees were not sufficient to control the river. Dams would be necessary on the Missouri and its tributaries. The endorsement of dams and hydroelectric facilities, of course, opened the door to the possibility that the Bureau of Reclamation should play a role in any comprehensive development of the Missouri River. It was an invitation to Reclamation to join the Corps, but only as a subordinate partner. Commissioner of Reclamation Harry Bashore was unwilling to accept the Corps' "dominant position" in the development of the Missouri River Basin. The waters of the Missouri west of Sioux City, Bashore maintained, were more applicable to agricultural, domestic, and industrial uses that demanded the leadership of the Bureau of Reclamation. The presentation of the Pick Plan pushed Reclamation to complete its own study and recommendations for the Missouri Basin begun in 1939. W. (William) Glenn Sloan, Assistant Director of Region 6, in Billings, Montana, was the principal investigator.

The Sloan Report of May 1, 1944, represented the interests of Reclamation. It emphasized irrigation and power on the upper Missouri rather than flood control. Irrigation and power development would serve the greater purpose of retaining and expanding the population base in the northern Great Plains. The Pick and the Sloan studies revived a longstanding competition between the Corps and Reclamation. As this rivalry reemerged, several congressmen complicated the situation further by introducing legislation to create a Missouri Valley Authority (MVA) to remove river development entirely from these two agencies. The prospect of an MVA, similar to the TVA and independent of both Reclamation and the Corps alarmed some northern plains congressmen and governors. It also alarmed Reclamation and the Corps. Most worrisome to these agencies and local governments,

President Roosevelt had supported the concept of an MVA. The Missouri River States Committee, formed in 1942 and made up of governors and local interests, saw the MVA as a clear threat to the sovereignty of the states, especially their control over water. With both Reclamation and the Corps at odds, the Missouri River States Committee pushed the agencies to come up with a compromise plan to head off consideration of an MVA for the river's multiple-resource development.

7.12. Glenn Sloan on the far right with Colonel Lewis Pick and others at the site of Yellowtail Dam in Montana.

In October 1944, Reclamation and the Corps worked out a compromise known as the Pick-Sloan Plan. It was incorporated into the Flood Control Act that the president signed in December 1944. Neither agency was to assume a dominant role on the Missouri River. As for President Roosevelt, he had more far reaching visions for the 1944 legislation. In these last months of his presidency and life, he believed the legislation would eventually lead to development of the river under an "authority" similar to the TVA. He, of course, was mistaken, but the legislation provided the foundation for future government work on the Missouri River. The Corps of Engineers had already completed Fort Peck Dam in 1940, and now five more dams and reservoirs were planned downstream on the main channel of the Missouri by the Corps – Garrison, Oahe, Big Bend, Fort Randall, and Gavins Point. Upstream, Reclamation planned as many as nineteen dams in Montana alone on various stems of the river.[25]

7.13. The Secretary of the Interior's *Annual Report* lauded Reclamation's role in World War II.

power in the Missouri Basin States for commodities produced in other parts of the country will be expanded by many millions of dollars annually.

FOOD FOR A FIGHTING NATION

Food is as important in war as men and machines are. The men and women on the farms of America are as truly "war workers" as those employed in shops and factories to turn out planes and tanks. And wherever our fighting men have gone on all the battle fronts they have been the best fed, the best clothed, and the best equipped troops ever mobilized.

To make possible that achievement, the farm families of America have made a record in the production of food and fiber that is without parallel in the Nation's history. They have provided not only for the needs of our fighting forces but for our allies and the civilian population here at home.

The part which farmers on lands irrigated by Federal Reclamation projects have played in that record-breaking achievement is outstanding. The value of crops produced on Bureau-irrigated lands has more than doubled during the war years, partly reflected in increased market prices but also due to greatly increased production of war-essential crops and to additional acreage brought under irrigation.

CROP VALUES HIT RECORD LEVEL

During the past fiscal year, more than 4 million acres of land were served with irrigation water from Bureau projects, representing an increase of 759,000 acres over 1941 and 125,000 acres more than in 1943. The value of crops produced hit an all-time high of $411,226,000, compared to $159,886,000 in the pre-war year of 1941.

In meeting wartime food needs, farmers on Reclamation lands during 1944 produced 57,122,000 bushels of potatoes, 1,941,000 tons of sugar beets, 6,495,000 bushels of onions, 3,674,000 bushels of beans, 36,672,000 bushels of grain, and 49,000,000 bushels of truck crops. To maintain production of meat and dairy products at a high level, they also produced 4,360,000 tons of forage crops, including 3,830,000 tons of alfalfa hay. Production of apples, oranges, grapefruit, peaches, and other fruits and berries totaled 729,000 tons, and crops produced for seed amounted to 4,512,000 bushels.

The average per acre value of crops on Bureau-irrigated lands in 1944 was also the highest on record, an average of $99.27 per acre, compared with a per acre average of $47.30 in 1941. The highest average crop value in 1944 was reported for the Tieton division of the Yakima project at $598.34 per acre, closely followed by the Okanogan project at $592.32 per acre. Both are apple-growing districts in the State of Washington.

673716—46——4

7.13 continued..

with its projects. Revenues received by the United States for use of land in 1944 totaled $195,805. The majority of the leased lands are withdrawn in connection with operation of completed projects. In addition, more than 700,000 acres of reclamation withdrawn land is now administered by the Grazing Service, with transfer of revenues to the Reclamation Fund. Under an agreement entered into February 28, 1945, between the Bureau and the General Land Office, vacant public lands under reclamation withdrawal may be temporarily transferred to the administration of the General Land Office until needed for reclamation purposes. At the end of the fiscal year, approximately 44,000 acres of such lands had been so transferred.

WORLD'S LARGEST POWER PRODUCER

Through a spectacular expansion in its generating facilities to meet the war emergency the Bureau of Reclamation has become the largest power producer in the world. From plants operating on its projects came nearly 14 billion kilowatt-hours of electric energy during the past fiscal year, much of it to war industries for the manufacture of planes and ships, aluminum, magnesium, and other materials and equipment for the fighting forces. Production of electric energy at Bureau projects has quadrupled since Pearl Harbor.

When the war is over this tremendous capacity for power production will be one of the most important factors in the continued industrial and agricultural expansion of the West. It will provide jobs, stimulate the establishment of new industry, aid in developing mineral resources and, in general, serve as the foundation for the establishment of a more balanced economy throughout the West.

POWER FOR WAR

From its beginning in the power field in 1909 with the 6,000-kilowatt Minidoka project plant in Idaho, the Bureau's installed capacity has grown to 2,439,300 kilowatts. This growth was required to keep pace with the needs for electrical power in areas served by reclamation projects and played a vital role in the tremendous expansion of war industries in the West. To meet demands, the installed capacity of Bureau hydroelectric plants was increased since 1941 by nearly a million and a half kilowatts, a gain of nearly 65 percent.

In the fiscal year 1945 the combined output of the plants operating on Bureau projects was approximately 14,000,000,000 kilowatt-hours. Revenues from the sale of energy were in excess of $20,000,000.

Construction of new plant facilities has been virtually halted since 1942 because of the need for diverting critical materials to other war uses. During the past year an additional 82,500-kilowatt unit was installed in the plant at Boulder Dam, which has supplied a steady

7.13 continued.

Reclamation after Two Wars and the Great Depression

The Pick-Sloan Plan was the last major river development authorized by Congress before the end of World War II in 1945. The cooperative partnership with the Corps of Engineers might suggest diminished prospects for Reclamation compared to its growth in the 1930s. On the other hand, it was clear recognition that the United States government had two competent agencies in the field with the mandates and expertise to modify and utilize river resources. While the Corps of Engineers' navigation improvement activities long predated the arrival of the Reclamation Service/Bureau of Reclamation, it made eminent sense to Congress, and even to the leadership of both agencies, to combine their talents in the Missouri River Basin. Neither organization wanted to lose influence and power to an independent river basin authority, so the 1944 Pick-Sloan legislation was a "preemptive strike." Corps and Reclamation objections aside, the future of river valley authorities quickly waned during the Cold War backlash against "creeping socialism" and threats to state sovereignty posed by powerful centralized government.

For Reclamation, the end of World War II renewed prospects for congressional spending on domestic projects and also worldwide aid to underdeveloped nations. The United States, as the dominant power in the postwar world, possessed capital, expertise, and a spirit of enlightened self-interest to reconstruct war-torn Europe and assist the emergence of a postcolonial world. Dams, hydroelectric facilities, and even irrigation were components of economic growth, and engineers from South America, Asia, and the Middle East visited the Denver Office of the Chief Engineer. Reclamation's experience with building great public works projects during the Great Depression in the economically underdeveloped American West gave it the credentials to undertake similar operations throughout the world. Now, a new international field lay open for Reclamation, should Congress and the State Department agree to its participation and cooperation, in foreign development projects designed to raise world living standards in the struggle against the spread of communism.

In the American West, many old problems persisted. The shortage of farmers and farm labor during the war encouraged agricultural mechanization and a larger capital investment, which, in turn, required larger land units. This had been the trend for years, interrupted temporarily by the economic stagnation and retrenchment of the depression years, but the war accelerated the need for mechanization. Yet, the pace of technological change and the memory of the Great Depression raised questions about what the future held for American agriculture. After all, despite its major departure into multi-purpose projects involving hydroelectricity and providing water supplies for urbanization, the origins of Reclamation lay with its "softer side," a commitment to the development of small farms. Still, the rapid changes in agricultural production demanded much larger farms on many of the projects. And nowhere was the question more urgent than in California's Central Valley Project.

Already, farm owners in the Imperial Valley enjoyed exemption from the 160 acre limitation on lands watered from the Colorado River via the All-American Canal. It came as no surprise, therefore, when Congressman Alfred J. Elliott, of California, attached an amendment to the Rivers and Harbors Bill of 1944 to exempt the Central Valley Project from the 160 acre limitation. The provision attracted sufficient opposition in the Senate to defeat the entire Rivers and Harbors bill and appropriation for 1944.[26] The Senate's rejection demonstrated the continued appeal of small farm ideology and the assumption that Reclamation should honor the commitment, written into its original charter, to serve only the interests of small farmers, even in the increasingly agribusiness-oriented economy of California's Central Valley. The defeat of a Rivers and Harbors Bill brought legislative history full circle back to 1901, when a filibuster in the Senate killed a Rivers and Harbors Bill in retaliation for congressional failure to consider a reclamation bill. If the West could not have a national reclamation program, western senators stood ready to deny river and harbor improvements to the Great Lakes, the Mississippi Valley, and the East, West, and Gulf Coasts. The following year Congress approved the 1902 Reclamation Act and also secured passage of a Rivers and Harbors Bill.

While the scuttling of the Rivers and Harbors Bill in 1944 was a sharp reminder of the link between Rivers and Harbors legislation and western reclamation, the manifest issue in this instance was the Jeffersonian ideal of independent small farmers and a persistent antagonism to land monopolies. The spirit of the original Reclamation Act seemed to be alive and well. In 1944, the power of these ideals in the Senate refused to allow government water to be delivered to the Central Valley Project in California to serve "excess lands." Ideals aside, practical politics probably played the larger role in shaping national reclamation in the West. Tradeoffs and bargains with other regions were and are major vehicles for obtaining legislation friendly to a particular region under the American system of federal and representative government.

The machinations of Congress, however, should not obscure the importance and power of ideals. The long campaign for national reclamation infused the program with both a practical and an idealistic side. Ultimately, the ideals were too difficult to sustain in the face of changing agricultural conditions, the lure of urban life, and the bane of chronic land speculation. The early Reclamation projects simply did not and could not live up to the romantic visions of rural life that inspired them.

In more practical terms, congressional passage of the Reclamation Act was an investment by the national government in regional development. Initially, the fact that the Reclamation Fund came from western land sales disguised the agricultural subsidy and the transfer of wealth from East to West. For some critics in Congress, it was all too apparent — and unacceptable. Yet, Congress refused to renege on its commitment to reclamation and divest itself of its huge investment in water management. The power of western senators and representatives in Congress insured continuation of the program, but it also received support because the ideals that inspired it were national in scope and appeal.

Franklin D. Roosevelt and his New Deal administrators understood this all too well. The dams and irrigation works of the first three decades of the twentieth century served as a precedent, both socially and economically, for the dramatic expansion of Reclamation's programs during the Great Depression. This new era took Reclamation far beyond its early ideals. During the 1930s and 1940s, Reclamation transformed itself into a modern, multipurpose resource bureaucracy committed to hydroelectricity and water for a rapidly urbanizing American West. Although much can be made of this transformation, Reclamation continued, rhetorically, to place irrigation at the top of its agenda.

Congress, too, did not give up on the ideology of the family farm. No doubt its attitudes hardened with the publication, in 1944, of Alfred Goldschmidt's study of two communities in California's Central Valley. Goldschmidt was an anthropologist in the Bureau of Agricultural Economics at the University of California who studied Dinuba and Arvin, California. Surrounding Dinuba, the average sized farm was sixty acres. In this community there were more schools, parks, social amenities, and equality of income than in Arvin where farms averaged over five hundred acres. There were five times the number of farms surrounding Dinuba than in Arvin, with fewer migrant farm laborers. The Goldschmidt study concluded that the "differences in the communities may properly be assigned confidently and overwhelmingly to the scale-of-farming factor." After the 1930s, many debates over the social and economic wisdom of acreage limitations in California began with the Goldschmidt report.[27]

While the Bureau of Reclamation held fast to its irrigation credentials, at the end of World War II it embarked upon a plan to share river basin development with its long time rival, the Corps of Engineers, under the Pick-Sloan Plan. Of course, Reclamation already shared the Columbia River with Corps of Engineers' dams downstream from Grand Coulee Dam. On other fronts, it was unclear where the future would lead in the postwar years. Was Reclamation to remain primarily an organization that worked in the West or did it have a future, as suggested at times, in other parts of the nation or even internationally? Much depended upon the ability and will of Congress to support new projects and continue its promises to those already undertaken.

Much also depended upon the postwar economy – whether economic growth would occur or the nation's economy would revert to the bleak years of the Great Depression.[28] Either way, Reclamation stood to prosper, if its experiences during the Great Depression were any indication. As the Federal Government's major water resource and development arm in the West, at the end of World War II, Reclamation proudly displayed a long list of accomplishments that boded well for its future, hard times or boom times.

Endnotes

1 Christine Pfaff, "Safeguarding Hoover Dam during World War II," *Prologue* 35 (Summer 2003), 10-21; Pitzer, *Grand Coulee*, notes that while the United States probably had plenty of power for the war effort without Grand Coulee Dam, it did make the production of aluminum and the atomic bomb effort in the Pacific Northwest possible without inconveniencing the civilian population. He notes that the war justified the dam's construction, "Although Grand Coulee did not win the war, the reverse is true. The rapid industrial expansion and population growth that the war evoked in the Pacific Northwest made a success out of Grand Coulee Dam," and then he cites Gerald Nash, *World War II and the West: Reshaping the Economy* (Lincoln: University of Nebraska Press, 1990), 1, "the war did for the West in four years what might otherwise have taken forty, rapidly building its cities and its industry," 249-50.

2 Pitzer, *Grand Coulee*, 248-49.

[3] Carey McWilliams, "Japanese Evacuation: Policy and Perspectives," *Common Ground*, (Summer 1942), 1-8; Daniel Geary, "Carey McWilliams and Antifascism, 1934-1943," *Journal of American History,* 90 (December 2003), 929-930; Carey McWilliams, *Brothers Under the Skin* (Boston: Little Brown and Company 1948); Peter Richardson, *American Prophet: The Life and Work of Carey McWilliams* (Ann Arbor: University of Michigan Press, 2005).

[4] Janice Dempsy, Circulation Manager *Common Ground*, to Bayshore [*sic*], June 9, 1942, RG 115, Entry 7, Box 748, NARA, Denver.

[5] Memorandum for the Assistant Secretary of the Interior from the Bureau of Reclamation, March 6, 1942; Page to R.S. Call and, District Engineer, Sacramento, California, March 6, 1942, RG 115, Entry 7, Box 748, NARA, Denver; Roger Daniels, *Concentration Camps USA: Japanese Americans and World War II* (New York: Holt, Rinehart and Winston, Inc., 1972), 83; Mike Mackey, "Introduction: Life in America's Concentration Camps," *Journal of the West* 38 (April 1999), 9-13.

[6] Harry Bashore to Burlew, March 18, 1942, RG 115, Entry 7, Box 748; Henry L. Stimson, Secretary of War, to Harold Ickes, April 14, 1942, RG 115, Entry 7, Box 748, NARA, Denver; Daniels, *Concentration Camps*, 96; Jeffery F. Burton, Mary M. Farrell, Florence B. Lord, and Richard W. Lord, *Confinement and Ethnicity: An Overview of World War II Japanese American Relocation Sites* (Seattle: University of Washington Press, 2002), 203-14, 279-324, 129-48; Mamoru Inouye, "Heart Mountain High School, 1942-1945," *Journal of the West* 38 (April 1999), 56-64.

[7] See annual histories of the Minidoka and Boise Projects for the war years (1941-1945) in RG 115, NARA, Denver; Acting Commissioner Bashore to the Secretary of the Interior, June 22, 1943, RG 115, Entry 7, Box 748, NARA, Denver.

[8] Burton, et al., *Confinement and Ethnicity*, 205, 131, 282-4.

[9] Headlines in the Fallon *Standard* (Fallon, Nevada) for May 13, 1942; In 1988, the Congress acted to compensate Japanese Americans for their hardships with token payments of $20,000 each for those still living who had suffered internment in the camps. More significant was the recognition by Congress that an injustice had been committed in those stressful times.

[10] Conrad L. Wirth to Department of Interior First Assistant Secretary with attachment to Golzé, July 27, 1942, RG 115, Entry 7, Box 751; Acting Commissioner Bashore to CCC Advisory Council, July 24, 1942, RG 115, Entry 7, Box 751, NARA, Denver.

[11] S.O. Harper, Chief Engineer, to Construction Engineer, Terry, Montana; Bashore Memorandum for the Departmental Representative on the Advisory Council to CCC, July 24, 1942; Commissioner, Bureau of Reclamation to Chief Engineer, Denver, September 18, 1942; Floyd M. Roush, Superintendent of Goshen Irrigation District, to C. F. Gleason, Superintendent of Power for Bureau of Reclamation at Guernsey, Wyoming, December 15, 1942; Gleason to Commissioner of Bureau of Reclamation, December 19, 1942, RG 115, Entry 7, Box 751, NARA, Denver; "Selective Service System Directory Civilian Public Service Camps," July 7, 1943, RG 115, Entry 7, Box 751, NARA, Denver; *Selective Service in Wartime: Second Report of the Director of Selective Service* (Washington, D.C. 1942), 268; *Selective Service as the Tide of War Turns: The 3rd Report of the Director of Selective Service, 1943-1944* (Washington, D.C.: 1944), 353-4; *Selective Service and Victory: The 4th Report of the Director of Selective Service* (Washington, D.C.: 1945), 690-1.

[12] Albert N. Keim, *The CPS Story: An Illustrated History of Civilian Public Service* (Intercourse, Pennsylvania: Good Books, 1990), 54.

[13] Benson, *The Reporter*, Supplement (November 1, 1944) as quoted in Mulford Q. Sibley and Philip E. Jacob, *Conscription of Conscience: The American State and the Conscientious Objector, 1940-1947* (Ithaca, New York: Cornell University Press, 1952), 253, en. 29.

[14] Benson, *The Reporter*, 253, en. 29; "Selective Service System Directory Civilian Public Service Camps," July 7, 1943, RG 115, Entry 7, Box 751, NARA, Denver; Harry Bashore to Assistant to the Secretary in Charge of Land Utilization, October 11, 1943, RG 115, Entry 7, Box 751, NARA, Denver; "Reform Party Victory," *The Reporter*, August 15, 1944. "Camp Operations Selective Service System: Projects and Service of Conscientious Objectors," typescript, March 1, 1945, RG 115, Entry 7, Box 751, NARA, Denver; Robert Autobee, "The Deschutes Project," Denver: Bureau of Reclamation History Program, typescript; Christopher J. McCune, "Rapid Valley Project," Denver: Bureau of Reclamation History Program, typescript.

[15] Eric A. Stene, "The Minidoka Project," typescript, Denver: Bureau of Reclamation History Program, 1997; also see annual Minidoka Project histories, 1942-1947, RG 115, NARA, Denver.

[16] "Bureau of Reclamation Historical Organizational Structure," 73.

[17] Watkins, *Righteous Pilgrim*, 777; Warne, *The Bureau of Reclamation*, 25-6; "Bureau of Reclamation Historical Organizational Structure," 75; Charles Coate, "The New School of Thought: Reclamation and the Fair Deal, 1945-1953," *Journal of the West* 22 (April 1983), 58-9.

[18] Warne, *Bureau of Reclamation*, 25; "Bureau of Reclamation Historical Organizational Structure," 76.

[19] See chapter, "Conservation and Containment," in Björn-Ola Linnér, *The Return of Malthus: Environmentalism and Post-war Population – Resource Crises* (Isle of Harris, United Kingdom: White Horse Press, 2003), 32-60.

[20] U.S. Department of the Interior, Bureau of Reclamation, Federal *Reclamation and Related Laws Annotated*, volume II, 1943-1958, edited by Richard K. Pelz (Washington, D.C.: Government Printing Office, 1972), 729-39; Pitzer, *Grand Coulee*, 267, 365; Golzé, *Reclamation*, 107.

[21] Pelz, *Reclamation and Related Laws Annotated*, volume I, 556-7; Pitzer, *Grand Coulee*, 273; Warne, *Bureau of Reclamation*, 84; Watkins, *Righteous Pilgrim*, 776-80; "The Columbia Basin Anti-Speculation Bill," *Reclamation Era,* 32 (March 1942), 71; S. T. Harding, *Water for California* (Palo Alto, California: N-P Publications, 1960), 176-7.

[22] Watkins, *Righteous Pilgrim*, 777.

[23] Pitzer, *Grand Coulee,* 289, 272; Pelz, *Federal Reclamation and Related Laws Annotated,* volume II, 728; Brian Q. Cannon, in "Creating a 'New Frontier of Opportunity…,'" cites U.S. Congress, House, Committee on Irrigation and Reclamation, *Settlement of Returning Veterans on Farms in Reclamation Projects*, 90-2, 97.

[24] U.S. Army Corps of Engineers, *The Federal Engineer Damsites to Missile Sites: A History of the Omaha District, U.S. Army Corps of Engineers* (Washington, D.C.: Government Printing Office, 1985), 75.

[25] "Fort Peck Dam Impounds Immense Reservoir," *Reclamation Era*, 32 (March 1942), 67, notes that Fort Peck's reservoir was the longest reservoir in the nation at 189 miles covering 245,000 acres, but its storage volume of 19,000,000 acre-feet falls short of Lake Mead's at 32,359,274; *Damsites to Missile Sites*, 76-80; Marion Clawson, *New Deal Planning: The National Resources Planning Board* (Baltimore: The Johns Hopkins University Press, 1981), 118; Morris L. Cooke, "Who Shall boss the M.V.A.?" *The New Republic*, 112 (April 16, 1945), 498-99; Armstrong, editor, *History of Public Works in the United States*, 312.

[26] Warne, *Bureau of Reclamation*, 76.

[27] As quoted in Hundley, *Great Thirst*, 266; Robert de Roos, *The Thirsty Land: The Story of the Central Valley Project* (New York: Greenwood Press, Publishers, 1968), 88-90; "Small Business and the Community: A Study in Central Valley of California on Effects of Scale of Farm Operations," Senate Special Committee to Study Problems of American Small Business, *Committee Print 13,* 79th Cong., 2nd sess. (1946); Walter Goldschmidt, *As You Sow* (New York: Harcourt, Brace & Co., 1947); Richard Kirkendall, "Social Science in the Central Valley of California: An Episode," *California Historical Society Quarterly,* 43 (September 1964), 195-218.

[28] Philip W. Warken, *A History of the National Resources Planning Board, 1933-1943* (New York: Garland Publishing, Inc., 1979), 183.

APPENDIX A

BUREAU OF RECLAMATION TIMELINE FOR VOLUME 1

This timeline, like the body of the narrative, cannot address the details of each and every project or law related to Reclamation's history up to 1945. At best, the timeline must touch the high points of Reclamation's early history. There are, however, a number of readily available sources that can provide additional, more detailed information. Try these:

For laws related directly to Reclamation:

- United States Department of the Interior, Bureau of Reclamation. *Federal Reclamation and Related Laws Annotated, Volume I of IV, through 1942*. ed. Richard K. Pelz. Washington, D.C.: U.S. Government Printing Office, 1972.

- __________. *Federal Reclamation and Related Laws Annotated, Volume II of IV, 1943-1958*. ed. Richard K. Pelz. Washington, D.C.: U.S. Government Printing Office, 1972.

For brief historical and engineering information on Reclamation's projects:

- United States Department of Interior, Water & Power Resources Service. *Project Data*. Denver: U.S. Government Printing Office, 1981.

- Go to the Dataweb at the Bureau of Reclamation's homepage, http://www.usbr.gov/dataweb/, and select the type of information in which you are interested. Where they exist, project history narratives are included under the "Project" heading.

For more extensive historical information on projects where books have been written:

- Go to Reclamation's history program, http://www.usbr.gov/history/, and look for the link to the history bibliography.

DATE	ITEM
1862	Homestead Act set standard homestead size at 160 acres
1875	*Exploration of the Colorado River of the West,* a book by John Wesley Powell is published.
1877	Desert Land Act linked irrigation to grants of public lands.
1878	*Report on the Lands of the Arid Region of the United States, with a More Detailed Account of the Lands of Utah* by John Wesley Powell.
1879	U.S. Geological Survey created, March 3. Studied geology and water supply.
1888-1892	USGS undertakes irrigation survey at request of Congress
1893	Economic depression stimulated westerners to agitate for a federal irrigation program in the West
1894	Carey Act to use public lands for state-based irrigation developments
1897	Hiram Chittenden's U.S. Army Corps of Engineers *Preliminary Report on Examination of Reservoir Sites in Wyoming and Colorado*, supported federal construction of irrigation reservoirs.
1901	Senator Thomas Carter of Wyoming filibusters 1901 Rivers and Harbors bill because of Congress's failure to fund federal irrigation in the West.
	Theodore Roosevelt becomes President of the United States, September 14
1902	Reclamation Act (Newlands Act) becomes law on June 17.
	Secretary of the Interior Ethan Allen Hitchcock appoints Charles D. Walcott the first director of the U.S. Reclamation Service while continuing as director of the U.S. Geological Survey
	John Wesley Powell dies September 23
1903	Five Reclamation projects authorized in a single letter by Secretary of the Interior Ethan Allen Hitchcock on March 14: Milk River Project, Salt River Project, Sweetwater (North Platte) Project; Truckee-Carson (Newlands) Project, and Gunnison (Uncompahgre) Project.
	Construction on Theodore Roosevelt Dam, Salt River Project, Arizona, 1903-1911

1904	Shoshone Project, Wyoming, authorized, February 10
	Minidoka Project, Idaho, authorized, April 23
	Minidoka Dam, Minidoka Project, Idaho, constructed 1904-1906
	Panama Canal construction by the United States occurred between 1904 and 1914. That work was preceded by a failed French attempt between 1881 and 1888
1905	Boise Project, Idaho, authorized, March 27
	Pathfinder Dam construction on the North Platte Project, Wyoming, 1905-1909
	Klamath Project authorized May 15.
	On June 17, the Derby Dam was dedicated on the Newlands Project. Derby Dam and the Truckee Canal were Reclamation's Specification Number 1.
	Rio Grande Project, New Mexico and Texas, authorized, extending the Reclamation Act to Texas for purposes of this project, December 2
	Umatilla Project authorized, December 4.
	Yakima Project authorized, December 12. Construction 1909-1933.
	Buffalo Bill (Shoshone) Dam construction on the Shoshone Project, 1905-1910
	Gunnison Tunnel construction, Uncompahgre Project, Colorado, 1905-1909
1906	Kinkaid Act offered 640 acre homesteads in Nebraska Sand Hills
	Town Sites and Power Development Act, April 16. Approved development of townsites and hydroelectric on projects
	Texas made a Reclamation state by Congress, June 12
1907	Charles D. Walcott resigns as Director of the USRS to become Secretary of the Smithsonian Institution
	Frederick H. Newell becomes Director of the USRS, March 9
	Kansas v. Colorado affirmed state water rights and threatened development of some Reclamation projects
	USRS and Office of Indian Affairs agree that Reclamation will build Indian irrigation projects

1908	Upper Spanish Fork Powerplant, Strawberry Valley Project, Utah, initial operation
	Reclamation Record established. Name changed to *NewReclamation Era* in 1924
	Winters Doctrine of reserved water rights established in *Winters v. United States* for Indian reservations
1909	Minidoka Powerplant, Minidoka Project, Idaho, initial operation Theodore Roosevelt Powerplant, Salt River Project, Arizona, initial operation
1910	In response to critics, a five member U.S. Army Corps of Engineers committee established to oversee Reclamation spending
1911	Warren Act, February 21, allowed sale of surplus water to private lands off Reclamation projects
	Arrowrock Dam construction, 1911-1915
1912	Boise River Diversion Powerplant went into operation
	Elephant Butte Dam construction, 1912-1916
1913	Reclamation Commission with up to five members to supervise Reclamation, May 1913-1917
1914	Arthur Powell Davis becomes director of the USRS at insistence of Secretary of the Interior Franklin Lane. Frederick H. Newell departs Reclamation in May of 1915
	Reclamation Extension Act, August 13 • Authorized six agricultural experiment stations on Reclamation projects • Congress must henceforth approve projects and appropriations from Reclamation Fund – ending Reclamation's semi-independence from political processes • Extended repayment to twenty years • Settlers must pay operation and maintenance
1915	Chief Engineer's office located in Denver. At the time the position was known as the Chief of Construction
	Panama-Pacific Exposition in San Francisco
1916	Federal Farm Loan Act
1920	Federal Water Power Act
	Revenues added to Reclamation Fund from Oil Leasing Act and Federal Water Power Act

1922	Fall-Davis Report on the Colorado River
	Wyoming v. Colorado – states could assert prior appropriation water rights claims on interstate rivers
	Under the leadership of Herbert Hoover, the Colorado River Compact was negotiated at Bishop's Lodge north of Santa Fe, New Mexico, and signed at the Palace of the Governors in Santa Fe on November 24
1923	Secretary of the Interior Hubert Work renames U.S. Reclamation Service the U.S. Bureau of Reclamation and appoints David W. Davis as the Commissioner of Reclamation.
	Arthur Powell Davis leaves Reclamation after dissolution of the position of director by Secretary Hubert Work
	Fact Finders' Commission appointed in September
1924	Elwood Mead becomes Commissioner of Reclamation, April 3
	Fact Finders' Report submitted to Congress, April 21
	Los Angeles formally filed a claim for 1500 cubic feet per second of the Colorado River water (about 550,000 acre feet/ year)
	Fact Finders' Act (Second Deficiency Appropriation Act for 1924) passed by Congress, December 5
	Congress directed Reclamation to study development of farm communities in the South
	Weymouth "Report on the Problems of the Colorado Basin" provided to Congress
1926	Repayment obligations of the Buford-Trenton Project cancelled by Congress
	Owyhee Project authorized October 12
	Omnibus Adjustment Act, May 25 • Adjusted many repayment provisions on projects and implemented an "ability to pay" approach • Permitted qualifications requirements for settlers • Tried to address issue of speculation on projects • Extended repayment time up to forty years
1927	Mississippi River Flood

1928	St. Francis Dam failed in California, March 12
	Secretary of the Interior orders safety of dam reviews
	Metropolitan Water District of Southern California formed late in the year to facilitate completion of the Colorado River Aqueduct
	Boulder Canyon Project Act (Swing-Johnson Act) became law December 21. • Authorized construction of Hoover Dam • Authorized construction of the All-American Canal • Ratified the Colorado River Compact
	Construction of Owyhee Dam, 1928-1932.
1929	Great Depression begins and lasts until World War II
1931	*Arizona v. California* – Reclamation permitted to build Hoover Dam under Congress's power over navigable waters of the United States
	Construction begins on Hoover Dam. Ends in 1936
1933	Civilian Conservation Corps created, March 31
	Name of Hoover Dam changed to Boulder Dam by Secretary of the Interior Harold Ickes
	Construction begins on Grand Coulee Dam with National Industrial Recovery Act funds
1934	Construction began on Parker Dam, the source of southern California's Colorado River Aqueduct. Parker is a Reclamation facility but was paid for by the Metropolitan Water District of Southern California.
	Construction on the Imperial Division of the All-American Canal, 1934-1940
1935	Hoover Dam dedicated by President Franklin Delano Roosevelt and others
	Columbia Basin Project authorized by Congress August 30
	Central Valley Project initial authorization, December 2
1936	Imperial Dam, at headworks of All-American Canal, construction, 1936-1938
	Commissioner Elwood Mead dies, January 26
	Year of initial hydroelectric generation at Hoover Dam
1937	Colorado-Big Thompson Project authorized August 9

1938	Construction on Green Mountain Dam, Colorado-Big Thompson Project, Colorado, 1938-1943
	Construction on Coachella Division of the All-American Canal, California, 1938-1948
	Construction of Shasta Dam, Central Valley Project, California, 1938-1945
1939	Friant Dam construction, Central Valley Project, California, 1939-1942
	Reclamation Project Act of 1939, August 4 • Expanded purposes for which Reclamation could build projects, e.g., Municipal and Industrial (M&I) • 40 year repayment period becomes standard • In electricity sales "preference shall be given to municipalities and other public corporations or agencies; and also to cooperative and other nonprofit organizations…"
1940	Alva B. Adams Tunnel construction on Colorado-Big Thompson Project begins. Ended 1947
1941	December 7, Japan attacked Pearl Harbor and the United States enters World War II
	Grand Coulee initial hydroelectric generation.
1942	Japanese internment camps established.
1944	Treaty with Mexico regarding waters of the Colorado, Rio Grande, and Tijuana Rivers (ratified February 3) • Allocated waters • Changed International Boundary Commission's name to the International Boundary and Water Commission
	Flood Control Act of 1944, December 22, initially established the Pick-Sloan Missouri Basin Program
	Arizona ratifies Colorado River Compact and begins to pursue interpretive disputes with California

APPENDIX B:

DIRECTORS/COMMISSIONERS BUREAU OF RECLAMATION THROUGH 1945

Charles Doolittle Walcott July 8, 1902-1907	Director, U.S. Geological Survey[1] and Director, U.S. Reclamation Service
Frederick H. Newell. March 9, 1907-1914	Director, U.S. Reclamation Service[2]
Arthur Powell Davis December 10, 1914-1923	Director [Also Chief Engineer from March 1907 to May 1920]
David W. Davis July 1, 1923 – 1924.	Commissioner, U.S. Bureau of Reclamation[3]
Elwood Mead April 3, 1924 – January 26, 1936.	Commissioner
Mae A. Schnurr. 1930 – 1936.	Acting Commissioner at various times.
John Chatfield Page. January 27, 1936 – August 2, 1943.	Acting Commissioner and then Commissioner beginning January 25, 1937
Harry W. Bashore. August 3, 1943 – 1945.	Commissioner.
Michael W. Straus. 1945 – 1953	Commissioner

BIOGRAPHICAL SKETCHES OF THE COMMISSIONERS OF RECLAMATION THROUGH 1945

Charles Doolittle Walcott 1902-1907

He became the first director of the U. S. Reclamation Service in 1902. Frederick Newell became the first chief engineer at the same time.

Walcott was a self-educated geologist and paleontologist, an expert on trilobites, who came to Reclamation indirectly. He started as an amateur fossil collector who educated himself and became the assistant of the state geologist of New York in 1876. Then in 1879 he took an assistant geologist position as one of the original staff with the U.S. Geological Survey. He served on the staff at the Geological Survey until 1894 when he succeeded John Wesley Powell as its director. In 1902, while still director of the Geological Survey, Walcott was appointed director of the new U.S. Reclamation Service by the Secretary of the Interior. This seemed logical at the time since the Geological Survey was responsible for the measurement of streamflow and from 1888 to 1892 had conducted the irrigation survey in the West.

Walcott dutifully took on his new responsibilities and devoted part of his summer field seasons visiting proposed Reclamation projects and inspecting proposed areas of work. It was Walcott who recommended the earliest Reclamation projects for approval by the Secretary of the Interior. However, by late 1906 Walcott was thoroughly frustrated by the dual responsibilities of the Geological Survey and Reclamation Service when he wrote: "The period of organization and of allotment of funds to projects has largely passed and the second step, that of construction is well underway… . it is clear to me that the Reclamation Service and the Geological Survey should have separate administrative heads… .I therefore have the honor to request that I be relived [*sic*] of the directorship of the Reclamation Service at an early date." A few weeks later he also wrote: "Am thoroughly tired of Reclamation service as it interferes with Geol. Survey administration." It seems clear that the scientific study and data collection missions of the Geological Survey and the utilitarian construction and water delivery mission of the Reclamation Service were conflicting with one another.

While Walcott was director of the Reclamation Service, he was also a founder of the Carnegie Institution of Washington, and, in 1907, he resigned both as director of the Reclamation Service and of the Geological Survey to become the secretary of the Smithsonian Institution. He led the Smithsonian until his death in 1927. After leaving Reclamation and the Geological Survey, Walcott continued active in geological and palaeontological field studies, and he was active in many professional organizations, including the National Academy of Sciences and the National Advisory Committee for Aeronautics. He was also a member of the Cosmos Club. He received many honorary doctorates recognizing his palaeontological work.

Frederick H. Newell, Director, Reclamation Service, 1907 – 1914

Born in Pennsylvania in 1863, Frederick H. Newell graduated from the Massachusetts Institute of Technology in 1885. His degree was in mining engineering, and he worked in Colorado, Pennsylvania, and Virginia for the next several years. Then, in 1888, the United States Geological Survey appointed Newell an assistant hydraulic engineer. He began to work on the Irrigation Survey supported by Congress from 1888 to 1892 and established an experimental camp at Embudo, New Mexico, on the Rio Grande River in 1888. There he was interested in studying both measurement of the quantity and quality of water. From 1894 to 1902 Newell was hydrographer-in-charge at the USGS, and from 1902 to 1906 he was chief hydrographer. In addition, from 1902 to 1907 he held the position of Chief Engineer of the U. S. Reclamation Service which was established July 17, 1902.

Newell was a protégé of Gifford Pinchot and one of the conservationists who had the ear of President Theodore Roosevelt. In his autobiography Roosevelt remembered that after President William McKinley's death, and even before he proceeded to the White House, he received Pinchot and Newell in Washington, D.C., to discuss the issue of establishing the Reclamation Service. In 1907, when Charles D. Walcott left the dual positions of director of the USGS and the USRS to become secretary of the Smithsonian Institution, Newell was appointed director of the Reclamation Service. At the time Newell was appointed director, the Reclamation Service was removed completely from its administrative affiliation to the USGS and made an independent bureau within the Department of the Interior.

However, Newell's leadership of the Reclamation Service was troubled by several fiscal and psychological problems.

After leaving Reclamation Newell became the head of the Department of Civil Engineering at the University of Illinois. While at the university he was very active in the professionalization and licensing of engineers as well as in developing organizations and systems to assist public entities with engineering problems.

During his long career Newell served as a founder and first secretary of the National Geographic Society 1892-1893 – an office he held again in 1897-9. He received the Cullum Medal from the American Geographical Society in 1918 with the inscription "He carried water from a mountain wilderness to turn the waste places of the desert into homes for freedom" on the obverse. He also served as president of the American Association of Engineers in 1919, and Newell was a member of the American Society of Civil Engineers and the Cosmos Club of Washington, D. C. At least two towns on Reclamation projects have been named for him. Newell died in 1932 at the age of seventy, and during World War II a liberty ship was named for him.

Arthur P. Davis, Director, Reclamation Service, 1914 – 1923

The nephew of John Wesley Powell, Arthur Powell Davis administered many of the plans his uncle first conceived concerning water use in the West. Born in 1861 in Decatur, Illinois, Davis earned a Bachelor of Science degree in Civil Engineering from Columbian (now George Washington) University in 1888. Through his uncle's help, Davis worked as a topographer for U.S. Geological Survey from 1884 to 1894. Davis was promoted to hydrographer in charge of government stream measurements in 1895. Three years later, in 1898, he oversaw the hydrographic examination of Nicaragua and Panama Canal routes. In 1907, Davis became chief engineer of the Reclamation Service, a position he held until his appointment to director on December 10, 1914. During his tenure as Director, Reclamation outlined the development of the Colorado River Basin before Congress in 1922. Davis was the first to recommend construction of multipurpose dams whose powerplants would amortize costs of the total project. The name of the Reclamation Service changed to Bureau of Reclamation on June 18, 1923. Davis retired the following day. He moved to Oakland to become director of the East Bay Municipal Utility District. Davis Dam on the Colorado River is named for him as was a liberty ship during World War II. He died in Oakland, California, in 1933.

David W. Davis, Commissioner, Bureau of Reclamation, 1923 – 1924

Born in Wales, 1873, Davis and his family immigrated to the United States in 1875. Davis went to work in a coal mine in Iowa at age 12 to support his widowed mother, and is the only Reclamation Commissioner who never received a college degree. He moved to Idaho in the 1890s to improve his health after spending his youth down the mines. Davis founded the First National Bank of American Falls, Idaho. Highlights of his public life include his election as a member of the Idaho State Senate, 1912–1914, and serving one term as Idaho's Governor from 1919 to 1923. On March 16, 1923, Secretary of the Interior Albert Fall appointed Davis special Assistant Secretary of the Interior to supervise the United States Reclamation Service. On June 18, 1923, the office of the Director of the U.S. Reclamation Service was abolished and Davis was named the first Commissioner of the Bureau of Reclamation on July 1. While Commissioner, the Fact Finders Commission held hearings in Salt Lake City in January 1924. Davis left office on April 2, 1924, and died in 1959.

Elwood Mead, Commissioner, Bureau of Reclamation, 1924 – 1936

Mead came to Reclamation with experience in the academic, legislative, and practical fields of irrigation. Born in 1858 in Patriot, Indiana, Mead graduated from Purdue with a Bachelor of Science degree in 1882. He earned his Doctorate in Civil Engineering from Iowa State College in 1883, and taught Mathematics at Colorado Agriculture College from

1883-84 and 1886-88. As territorial and state engineer of Wyoming, Mead devised the state's first water laws during the period 1888 to 1899. He also worked as chief of investigation and drainage irrigation for the U.S. Department of Agriculture from 1897 to 1907. In 1907, Mead accepted the position of Chairman of State Rivers and Water Supply Commission in Victoria, Australia. Eight years later, he returned to the United States as professor of Rural Institutions, University of California, and chairman of the California Land Settlement Board, 1915-1924. Appointed Commissioner of Reclamation in 1924, Mead directed the development of many monumental Reclamation undertakings, notably Hoover Dam, Grand Coulee, and Owyhee. Mead died in office on January 26, 1936. Lake Mead behind Boulder Dam named in his honor.

Mae A. Schnurr, (times vary through 1936) Acting Commissioner, Bureau of Reclamation, 1930 – 1936

Mae Schnurr accepted her first position with the Federal government as a stenographer in 1915. She began her long career with Reclamation as a clerk in October 1923. She soon became the Secretary to the Commissioner, then Elwood Mead. Over time he entrusted more and more responsibility to Schnurr as she proved her outstanding capabilities. In 1929 Mead's request to create a new position of Assistant to the Commissioner was approved and Schnurr was given the post. She was cited as the first woman to be given so high an administrative post in the Department of the Interior. In February 1930, Mead designated Schnurr as Acting Commissioner whenever he and the Assistant Commissioner were absent. As far as is known, this was the first time that a woman had acted as head of a major federal bureau normally directed by a man. Following Mead's death in 1936, Schnurr's title and responsibilities gradually diminished. She eventually transferred to the Secretary of the Interior's office where she became Assistant to the Secretary. She retired from the Federal service in 1944.

John C. Page, Commissioner, Bureau of Reclamation 1936 – 1943

Born in Syracuse, Nebraska, in 1887, Page received his Bachelor of Science degree in civil engineering from University of Nebraska in 1908. In 1909, Page's first job for the USRS was as a topographer surveying canal sites in Colorado. He joined the Grand Junction, Colorado, city engineer's office as an assistant in 1911. In 1925, he became superintendent of the Grand Valley Project. In 1930, Page went to work as office engineer and chief administrative assistant for the Boulder Canyon Project. By 1935, Page advanced to the head of the Engineering Division. The following year, Page filled the acting Commissioner vacancy, and assumed the Commissioner's office on January 25, 1937. Highlights of his term include President Franklin D. Roosevelt signing the Reclamation Project Act of 1939, and the first power generated at the Grand Coulee powerplant. Page died in 1955. The town of Page, Arizona, near Glen Canyon Dam, memorializes the sixth person to lead Reclamation.

Harry W. Bashore, Commissioner, Bureau of Reclamation, 1943 – 1945

A Missouri native, Bashore graduated from Missouri State University in 1906 with a Bachelor of Science in Civil Engineering. He joined Reclamation out of college as a junior engineer on the North Platte Project. From 1927 to 1933, Bashore served as construction engineer at Lake Minatare, before assuming the superintendent's job on the North Platte Project. Bashore left North Platte to work as a construction and investigative engineer in Oregon, Washington, and California. Bashore returned to Wyoming as an engineer on the Kendrick Project from 1933 to 1939. In 1942, Bashore became Assistant Commissioner, before his appointment as Commissioner on August 3, 1943. During his term, Reclamation's bureaucracy grew with the birth of the regional office system in 1943. In 1944, Congress passed the working agreement between Reclamation and the Corps of Engineers for development of the Missouri River Basin (Pick-Sloan Missouri Basin Program).

Michael W. Straus, Commissioner, Bureau of Reclamation, 1945 – 1953

Straus was born in Chicago in 1897. He attended the University of Wisconsin in pursuit of a chemical engineering degree. Instead of engineering, Straus began his career as a newspaper reporter, eventually working his way to the managing editor's job of two of the nation's largest newspapers, and Washington bureau chief of the International News Service. He entered government as an Assistant to Secretary of the Interior Harold Ickes. Ickes appointed Straus first assistant secretary of the department in 1943. As commissioner, the Upper Colorado River Basin Compact was signed, and Grand Coulee Dam, Shasta Dam, and Reclamation's Engineering Research Center at Denver were all dedicated. Straus died August 9, 1970.

Endnotes

[1] In July of 1902, the Reclamation Service was established within the U.S. Geological Survey of the Department of the Interior. The director of the USGS also served as director of the U.S. Reclamation Service.

[2] This table is misleading because from June of 1913 to February of 1920 a number of people shared power with Frederick Newell and Arthur Powell Davis. In fact, for a time, five officials of the Reclamation Service constituted a board or "commission" for the purpose of considering all questions of administrative policy and management and recommending action thereon to the Secretary of the Interior.

November 1913-December 1914: Frederick H. Newell, director; Arthur Powell Davis, chief engineer; Will R. King, chief counsel; William A. Ryan, comptroller; Ignatius D. O'Donnell, supervisor of irrigation. Frederick H. Newell left Reclamation in spring of 1915

The Department of the Interior reduced the quintet to a trio in June of 1915.

June 1915-August 1917: Arthur Powell Davis, director and chief engineer; Will R. King, chief counsel; William A. Ryan, comptroller.

Reclamation Record omits Ryan from the trio in August 1917. From that date to February of 1920, Davis and King remained as director and chief counsel, respectively.

August 1917-February 1920: Arthur Powell Davis, director and chief engineers; Will King, chief counsel. In May of 1920 the chief engineer title went to Frank E. Weymouth.

[3] On June 20, 1923, the name of the Reclamation Service was changed to the Bureau of Reclamation, and the position of Commissioner of Reclamation was established.

APPENDIX C:

CHIEF ENGINEERS BUREAU OF RECLAMATION THROUGH 1945

Frederick Haynes Newell[1] June 1902-March 1907	Chief Engineer
Arthur Powell Davis[2] April 1907-May 1920	Chief Engineer
Sydney B. Williamson January-June 1915	Chief of Construction – Washington, D.C.[3]
Sydney B. Williamson June 1915-March 1916	Chief of Construction – Denver[4]
Frank E. Weymouth March 1916-May 1920	Chief of Construction – Denver
Frank E. Weymouth May 1920-October 1924	Chief Engineer – Denver[5]
On June 20, 1923,	Redesignated chief engineer, upon name change to Bureau of Reclamation,
Raymond F. Walter November 1924-June 1940	Chief Engineer
Sinclair O. Harper July 1940-December 1944	Chief Engineer
Walker (Brig) R. Young January 1945-July 1948	Chief Engineer

Endnotes:

[1] Between 1907 and 1914 also served as director of Reclamation
[2] Between 1914 and May of 1923, he served as both director of Reclamation and chief engineer.
[3] In 1914 the position of "director and chief engineer" was created for A. P. Davis, and the office of chief of construction was established to head the construction work of Reclamation
[4] In June 1915, executive offices were established in Denver under the direction of the Chief of Construction.
[5] Amid a flurry of reorganization, the position of director and chief engineer was split, leaving A. P. Davis with the title of director while the chief of construction assumed the title of chief engineer. There was no change in duties.

APPENDIX D:

SECRETARIES OF THE INTERIOR, DEPARTMENT OF THE INTERIOR, 1902 THROUGH 1945

Secretary	President(s)
Ethan Allen Hitchcock February 20, 1899-March 4, 1907	William McKinley and Theodore Roosevelt
James R. Garfield March 5, 1907-March 5, 1909	Theodore Roosevelt
Richard A. Ballinger March 6, 1909-March 12, 1911	William Howard Taft
Walter L. Fisher March 13, 1911-March 5, 1913	William Howard Taft
Franklin K. Lane March 6, 1913-February 29, 1920	Woodrow Wilson
John Barton Payne March 15, 1920-March 4, 1921	Woodrow Wilson
Albert B. Fall March 5, 1921-March 4, 1923	Warren G. Harding
Hubert Work March 5, 1923-July 24, 1928	Warren G. Harding and Calvin Coolidge
Roy O. West July 25, 1928-March 4, 1929	Calvin Coolidge
Ray Lyman Wilbur March 5, 1929-March 4, 1933	Herbert C. Hoover
Harold L. Ickes March 4, 1933-February 15, 1946	Franklin D. Roosevelt and Harry S. Truman

APPENDIX E:

NAME AND STATUS CHANGES U.S. RECLAMATION SERVICE/BUREAU OF RECLAMATION THROUGH 1945

United States Reclamation Service (USRS) — Created in 1902 and placed under the jurisdiction of the U.S. Geological Survey with a shared director — Charles D. Walcott.

United States Reclamation Service (USRS) – From March 9, 1907, until June 20, 1923, an independent bureau within the Department of the Interior

United States Bureau of Reclamation (BR or USBR) – On June 20, 1923, renamed the Bureau of Reclamation by the Secretary of the Interior and David W. Davis was appointed "commissioner." There was no position of director in the newly named bureau.

BIBLIOGRAPHY

PRIMARY SOURCES (PAPERS AND MANUSCRIPTS)

Arthur Powell Davis collection [cited as Davis Papers]. American Heritage Center, University of Wyoming, Laramie.

Bureau of Reclamation Records (Record Group 115). National Archives and Records Administration, National Archives and Records Center, Denver, Colorado.

Elwood Mead Papers [cited as Mead Papers]. Bancroft Library, University of California, Berkeley.

Frederick H. Newell Papers. American Heritage Center, University of Wyoming, Laramie.

Morris Bien Papers, 1879-1932. American Heritage Center. University of Wyoming, Laramie.

SECONDARY SOURCES (PUBLISHED MATERIALS)

Abernathy, M. Glenn, Dilys M. Hill, and Phil Williams, editors. *The Carter Years: The President and Policy Making*. New York: St. Martin's Press, 1984.

"Activities of Water Users' Associations." *Reclamation Record*, 5 (February 1914), 44-8.

Adams, Henry. *The Education of Henry Adams*. New York: Random House Modern Library ed., 1931; first published Massachusetts Historical Society, 1907.

Ady, Abel. "The Extension Bill." *Reclamation Record*, 5 (September 1914), 306.

Alexander, Thomas G. *A Clash of Interests: The Interior Department and the Mountain West, 1863-1896*. Provo, Utah: Brigham Young University Press, 1977.

———. "The Powell Irrigation Survey and the People of the Mountain West." *Journal of the West*, 7 (January 1968), 48-54.

———. "Stewardship and Enterprise: The LDS Church and the Wasatch Oasis Environment, 1847-1930." *Western Historical Quarterly,* 25 (Fall 1994), 340-64.

Allen, Barbara. *Homesteading the High Desert.* Salt Lake City: University of Utah Press, 1987.

"American Ambulance Hospital in Paris (Neuilly)." *Reclamation Record*, 8 (November 1917), 503-4.

"Arizona's Roosevelt Dam to be Rededicated." *U.S. Water News Online.* (March 1996).

Armstrong, Ellis L., editor. *History of Public Works in the United States, 1776-1976.* Chicago: American Public Works Association, 1976.

Arrington, Leonard J. *Great Basin Kingdom: An Economic History of the Latter-Day Saints, 1830-1900.* Cambridge: Harvard University Press, 1958.

Austin, Mary. *Land of Little Rain.* Boston and New York: Houghton Mifflin, Inc., 1903.

Autobee, Robert. "The Deschutes Project." Denver: Bureau of Reclamation History Program, typescript.

Bailey, L.H. *The Country-Life Movement in the United States.* New York: The Macmillan Company, 1911.

Baldridge, Kenneth W. "Reclamation Work of the Civilian Conservation Corps, 1933-1942." *Utah Historical Quarterly,* 39 (Summer 1971), 265-85.

Bancroft, George. *History of the United States of America: From the Discovery of the Continent.* New York: Appleton, 1882-1886; originally, Boston: Little, Brown and Company, 1854.

Banks, F.A. "Significance of Grand Coulee Dam." *Reclamation Era*, 26 (December 1936), 278-80.

Bannister, L. Ward. "National Resources Planning Board: How It should be Constituted." *Reclamation Era*, 30 (January 1940), 7.

Barry, John M. *Rising Tide: The Great Mississippi Flood of 1927 and How It Changed America.* New York: Simon & Schuster, 1997.

Bartlett, Richard A. *Great Surveys of the American* West. Norman: University of Oklahoma Press, 1962.

Bassett, R. E., Secretary of Immigration, Elephant Butte Water Users Association. "Advertising and Immigration Work." *Reclamation Record*, 5 (October 1914), 374-5.

Beard, Daniel. Oral History Interview, Daniel Beard: 1993-1995. Denver: Bureau of Reclamation Oral History Program, 1995.

Billington David, Donald C. Jackson, and Martin V. Melosi. *The History of Large Federal Dams: Planning, Design, and Construction.* Denver: Bureau of Reclamation, 2005

.

Blanchard, C. J. *Call of the West: Homes are Being Made for Millions of People in Arid West*. Washington, D.C.: Press of Judd & Detweiler, Inc., 1909.

———. "Co-operation on Reclamation Projects." *Reclamation Record*, 5 (February 1914), 42.

———. "Current Comments from the Projects: Idaho, Minidoka." *Reclamation Record*, 5 (March 1914), 80.

———. "Give Our Fighting Men Their Opportunity on the Land." *Reclamation Record*, 10 (February 1919), 50-1.

———. *Home-making by Government: Account of 11 Immense Irrigation Projects to be Opened in 1908*. Washington, D.C.: Press of Judd & Detweiler, Inc., 1908.

Bledstein, Burton J. *The Culture of Professionalism: The Middle Class and the Development of Higher Education in America.* New York: Norton, 1976.

Bloch, Ivan. "The Columbia River Salmon Industry." *Reclamation Era*, 28 (February 1938), 26-30.

Bogener, Stephen. *Ditches Across the Desert: Irrigation in the Lower Pecos Valley*. Lubbock: Texas Tech University, 2003.

Bonner, Robert E. "Local Experience and National Policy in Federal Reclamation: The Shoshone Project, 1909-1953." *Journal of Policy History,* 15 (Summer 2003), 301-23.

"Bonneville to Market Grand Coulee Power." *Reclamation Era*, 30 (November 1940), 322.

Bonnifield, Mathew P. *The Dust Bowl: Men, Dirt, and Depression.* Albuquerque: University of New Mexico Press, 1979.

"Boulder Canyon Project Act." *New Reclamation Era*, 20 (May 1929), inside front cover.

Bowers, William L. *The Country Life Movement in America, 1900-1920.* Port Washington, NewYork: Kennikat Press, 1974.

Branson, E.C. "Planned Colonies of Farm Owners," in *Economic Problems of Reclamation.* Department of the Interior, Bureau of Reclamation Elwood Mead, Commissioner. Washington, D.C.: Government Printing Office, 1929.

Breisach, Ernest A. *American Progressive History: An Experiment in Modernization.* Chicago, Illinois: University of Chicago, 1993.
Brookings Institution. Institute for Government Research. *The U.S. Reclamation Service: Its History, Activities and Organization.* New York: D. Appleton and Company, 1919.

Broadhead, Michael J. *David J. Brewer: The Life of a Supreme Court Justice, 1837-1910.* Carbondale: Southern Illinois University Press, 1994.

Brown, Bruce. *Mountain in the Clouds: A Search for the Wild Salmon.* New York: Simon & Schuster, 1982.

Bryson, Michael A. *Visions of the Land: Science, Literature, and the American Environment from the Era of Exploration to the Age of Ecology.* Charlottesville: University Press of Virginia, 2002.

Burton, Jeffery F., Mary M. Farrell, Florence B. Lord, and Richard W. Lord. *Confinement and Ethnicity: An Overview of World War II Japanese American Relocation Sites.* Seattle: University of Washington Press, 2002.

Campbell, Hardy W. *Campbell's 1907 Soil Culture Manua*l. Lincoln: Nebraska State Printer, 1907.

Campbell, Robert B. "Newlands, Old Lands," *Pacific Historical Review,* 71 (May 2002), 203-38.

Cannon, Brian Q. "Creating a 'New Frontier' Opportunity: World War II Veterans and the Campaign for Western Homesteads," unpublished manuscript read at the Pacific Coast Branch of the American Historical Association, Honolulu, Hawaii, August 2003.

———. "Water and Economic Opportunity: Homesteaders, Speculators, and the U.S. Reclamation Service, 1904-1924." *Agricultural History*, 76 (Spring 2002), 188-207.

Carlson, Leonard A. *Indians, Bureaucrats, and the Land: The Dawes Act and the Decline of Indian Farming.* Westport, Connecticut: Greenwood Press, 1981.

———. "Federal Policy and Indian Land: Economic Interests and the Sale of Indian Allotments." *Agricultural History,* 57 (January 1983), 33-45.

Carlson, Martin E. "William E. Smythe: Irrigation Crusader." *Journal of the West*, 7 (January 1968), 41-7.

Carpenter, Daniel P. *The Forging of Bureaucratic Autonomy: Reputations, Networks, and Policy Innovation in Executive Agencies, 1862-1928.* Princeton, New Jersey: Princeton University Press, 2001.

Carver, T. N., Director of the Rural Organization Service of the Department of Agriculture. "The Probable Effects of the European War on American Agriculture." *Reclamation Record*, 5 (November 1914). 405-6.

"CCC Program for Federal Reclamation, Fiscal Year 1939." *Reclamation Era*, 28 (August 1938), 163.

"CCC Work to Continue on Reclamation Projects." *Reclamation Era*, 28 (April 1938), 74.

Chandler, A. E. *Elements of Western Water Law.* San Francisco, California: Technical Publishing Co., 1918.

Chittenden, Hiram M. "Examination of Reservoir Sites in Wyoming and Colorado." Washington, D.C.: U.S. Army Corps of Engineers, 1898.

———. "Government Construction of Reservoirs in Arid Regions." *North American Review*, 174 (1902), 245-58.

Clarke, Jeanne Nirnabar. *Roosevelt's Warrior: Harold L. Ickes and the New Deal*. Baltimore, Maryland: Johns Hopkins University Press, 1996.

Clawson, Marion. *New Deal Planning: The National Resources Planning Board.* Baltimore: The Johns Hopkins University Press, 1981.

Clements, Kendrick A. *Hoover, Conservation and Consumerism: Engineering the Good Life.* Lawrence; University Press of Kansas, 2000.

Coate, Charles. "The New School of Thought." *Journal of the West,* 22 (April 1983), 58-63.

Cole, D. W. "Extension of Time for Payment of Water Charges." *Reclamation Record*, 9 (July 1918), 329-31.

Cole, Donald B. "Transmountain Water Diversion in Colorado." *Colorado Magazine,* 25 (March 1948), 49-64.

"The Columbia Basin Anti-Speculation Bill." *Reclamation Era*, 32 (March 1942), 71.

"Comparative Cost of Public and Private Irrigation Projects." *Reclamation Record*, 5 (February 1914), 56.

Cone, William S., Superintendent of Construction. "The Salt River Valley Power Situation." *Reclamation Record*, 5 (May 1914), 175-6.

Conkin, Paul K. *Tomorrow a New World: the New Deal Community Program.* Ithaca, New York: Cornell University Press, 1959.

———. "The Vision of Elwood Mead." *Agricultural History,* 34 (1960). 88-97.

"Construction of Boulder Dam Begun on July 7, 1930." *New Reclamation Era*, 21 (August 1930), 146-7.

Cooke, Morris L. "Who Shall boss the M.V.A.?" *The New Republic*, 112 (April 16, 1945), 498-9.

"Current Comments from Projects: Idaho, Minidoka." *Reclamation Record*, 5 (March 1914), 80.

"Dams Completed During Past Five Years." *Reclamation Era*, 29 (February 1939), 38.

Dana, Marshall N. "Dr. Elwood Mead." *Reclamation Era*, 26 (May 1936), 115.

Danbom, David B. *Born in the Country: A History of Rural America.* Baltimore, Maryland: Johns Hopkins University Press, 1995.

Daniels, Roger. *Concentration Camps USA: Japanese Americans and World War II.* New York: Holt, Rinehart and Winston, Inc., 1972.

Davis, Arthur P. "Results of National Irrigation," *Reclamation Record*, 10 (December 1919), 546-7.

Davis, Louis S. "E.C.W. Activities on Reclamation Projects." *Reclamation Era*, 26 (December 1936), 298-9.

Davison, Stanley Roland. *The Leadership of the Reclamation Movement, 1875-1902.* New York: Arno Press, 1979.

Dawdy, Doris Ostrander. *Congress in Its Wisdom: The Bureau of Reclamation and the Public Interest.* Boulder, Colorado: Westview Press, 1989

Deakins, Alfred. *Progress Report: Irrigation in Western American, So Far as Its Relationship to the the Circumstances of Victoria.* Melbourne, Australia: Government Printer, 1885.

Debler, E. B. "Stabilization by Irrigation." *Reclamation Era*, 30 (November 1940), 309-11.

deBuys William, with photographs by Joan Myers. *Salt Dreams: Land & Water in Low-Down California.* Albuquerque: University of New Mexico Press, 1999.

de Crevecoeur, J. Hector St. John. *Letters from an American Farmer.* London: Printed for Thomas Davies, 1783.

"Defense Resources Committee Appointed." *Reclamation Era*, 30 (September 1940), 258.

"Definite Federal Policy on the Colorado River Urged." *New Reclamation Era*, 15 (April 1924), 51-2.

Dent, T. W. "Community Ditches: Their Rights and Power." *Reclamation Record*, 6 (August 1915), 371-4.

Denver *Post,* September 20, 1924.

de Roos, Robert. *The Thirsty Land: The Story of the Central Valley Project.* New York: Greenwood Press, Publishers, 1968.

de Stanley, Mildred. *The Salton Sea: Yesterday and Today*. Los Angeles: Triumph Press, Inc., 1966.

"Description of Reclamation Exhibit, Panama-Pacific International Exposition, San Francisco, Cal." *Reclamation Record*, 6 (June 1915), 257-8.

de Tocqueville, Alexis. *Democracy in America*, translated by Harvey C. Mansfield and Delba Winthrop. Chicago, Illinois: University of Chicago Press, edition of 2000.

Dodds, Gordon B. *Hiram Martin Chittenden: His Public Career*. Lexington: University of Kentucky Press, 1973.

Dominy, Floyd E. Oral History Interview: April 6, 1994: Boyce, Virginia. Denver, Colorado: Bureau of Reclamation Oral History Project.

Dunar, Andrew J. and Dennis McBride. *Building Hoover Dam: An Oral History of the Great Depression*. Reno: University of Nevada Press, 1993.

Dunbar, Robert G. *Forging New Rights in Western Waters*. Lincoln: University of Nebraska Press, 1983.

"The Earth Movers I." *Fortune*, 28 (August 1943), 119-24.

"Eighteen U.S. Bureau of Reclamation Projects to Cost $218,440,000." *Western Construction News and Highways Builder*. (January 1934), 5-11.

Elliot, Russell R. *Servant of Power: A Political Biography of Senator William M. Stewart*. Reno: University of Nevada Press, 1983.

Espeland, Wendy Nelson. *The Struggles for Water: Politics, Rationality, and Identity in the American Southwest.* Chicago: University of Chicago, 1998.

Eveden, Matthew D. *Fish Versus Power: An Environmental History of the Fraser River.* Cambridge, U.K.: Cambridge University Press, 2004.

Executive Order 8526 "Coordinating the Electrical Facilities of Grand Coulee Dam Project and Bonneville Project." *Federal Register*, 3390. Volume 5. August 29, 1940.

"F. A. Banks Discusses Columbia Development." *Reclamation Era*, 26 (January 1936), 3-5.

"F. E. Weymouth Leaves Bureau of Reclamation." *New Reclamation Era*, 15 (October 1924), 160.

Fallon *Standard* (Fallon, Nevada), May 13, 1942.

Fiege, Mark. *Irrigated Eden: The Making of an Agricultural Landscape in the American West.* Seattle: University of Washington Press, 1999.

Flack, J. Kirkpatrick. *Desideratum in Washington: The Intellectual Community in the Capital City, 1870-1900.* Cambridge, Massachusetts: Schenkman Publishing Company, Inc., 1975.

Flores, Dan. "Zion in Eden: Phases of Environmental History of Utah." *Environmental Review* 7 (Winter 1983), 325-44.

Fogg, P. M. "A History of the Minidoka Project, Idaho, to 1912 Inclusive." Typescript. Boise: Idaho State Library.

"Fort Peck Dam Impounds Immense Reservoir." *Reclamation Era*, 32 (March 1942), 67.

Fradkin, Philip L. *A River No More: The Colorado River and the West.* Berkeley: University of California Press, 1996, paperback edition [originally New York: Knopf, 1981].

Frederick, Kenneth D., and Roger A. Sedjo, editors. *America's Renewable Resources: Historical Trends and Current Challenges.* Resources for the Future: Washington, D.C., 1991.

Fredrickson, George M. *The Inner Civil War: Northern Intellectuals and the Crises of the Union.* New York: Harper & Row, Publishers, 1965.

Fund for Reclamation of Arid Lands, Message from the President of the United States Transmitting a Report of the Board of Army Engineers in Relation to the Reclamation Fund, 61st Congress 3rd Session, HR, Doc. 1262, January 6, 1911. Washington, D.C.: Government Printing Office, 1911.

Gammack, Thomas H. "Hydroelectric Myths." *World's Work*, 58 (May 1929), 118-22.

Ganoe, John T. "The Beginnings of Irrigation in the United States." *Mississippi Valley Historical Review,* 25 (June 1938), 59-78.

Gates, Paul Wallace. *History of Public Land Law Development.* Washington, D.C.: Zenger Publishing Co., Inc., 1968.

Geary, Daniel. "Carey McWilliams and Antifascism, 1934-1943." *Journal of American History,* 90 (December 2003), 912-34.

Gentry, Carolyn, editor. *The Roosevelt Dam.* Produced by Roosevelt Memorial Association Film Library, located at the Library of Congress, American Memory, http://memory.loc.gov/ammem/collections/troosevelt_film/

Glad, Paul W. *McKinley, Bryan, and the People.* Philadelphia: Lippincott, 1964.

Glass, Mary Ellen. *Water for Nevada: The Reclamation Controversy, 1855-1902.* Reno: University of Nevada Press, 1964.

Glennon, R.J., and P.W. Culp. "The Last Green Lagoon: How and Why the Bush Administration Should Save the Colorado River Delta." *Ecology Law Quarterly,* 28 (October 2001), 903-92.

Goetzmann, William H. *Exploration and Empire: The Explorer and Scientist in the Winning of the American West.* New York: Knopf, 1966.

Golzé, Alfred R. "Civilian Conservation Corps Accomplishments on Federal Reclamation Projects: Fiscal Year 1938." *Reclamation Era*, 28 (September 1938), 190-2.

———. *Reclamation in the United States.* Caldwell, Idaho: The Caxton Printers, 1961.

Greenland, Powell. *Hydraulic Mining in California: A Tarnished Legacy.* Spokane, Washington: The Arthur Clark Company, 2001.

Gressley, Gene. "Arthur Powell Davis, Reclamation, and the West." *Agricultural History,* 42 (July 1968), 241-57.

Hall, Marcus. "Repairing Mountains: Restoration, Ecology, and Wilderness in Twentieth-Century Utah." *Environmental History,* 6 (October 2001), 584-610.

Hall, Patricia Kelly, and Steven Ruggles. "'Restless in the Midst of Their Prosperity': New Evidence on the Internal Migration of Americans, 1850-2000." *Journal of American History,* 91 (December 2004), 829-46.

Hall, William Hammond. *Irrigation in California (Southern): the Field, Water-Supply, and Works, Organization and Operation in San Diego, San Bernardino, and Los Angeles Counties: the Second Report of the State Engineer of California on Irrigation and the Irrigation Question.* Sacramento: State Office, J. D. Young, Superintendent, State Printing, 1888.

Harding, S. T. *Water for California.* Palo Alto, California: N-P Publications, 1960.

Hargreaves, Mary W. M. *Dry Farming on the Northern Great Plains: Years of Readjustment, 1920-1990.* Lawrence: University Press of Kansas, 1993.

Harvey, Mark W. T. *A Symbol of Wilderness: Echo Park and the American Conservation Movement.* Albuquerque: University of New Mexico Press, 1994.

Hays, Samuel P. *Conservation and the Gospel of Efficiency: The Progressive Conservation Movement, 1890-1920.* Cambridge, Massachusetts: Harvard University Press, 1959.

Hoffman, Abraham. *Vision or Villainy: Origins of the Owens Valley-Los Angeles Water Controversy.* College Station: Texas A&M University Press, 1981.

Hofstadter, Richard. *The Age of Reform: From Bryan to F.D.R.* New York: Alfred A. Knopf, 1955.

Holbrook, Stewart H. *The Columbia.* New York: Rinehart and Co., Inc., 1956.

———. *Far Corner.* New York: Macmillan, Co., 1952.

"Honorable Harold L. Ickes, Secretary of the Interior Delivers Address at the Dedication of Boulder Dam." *Reclamation Era,* 25 (November 1935), 209-10.

"The Hospital at Neuilly, France." *Reclamation Record*, 8 (September 1917), 406.

House, E. B. "The Duty of Water." *Reclamation Record*, 5 (October 1914), 355.

"How National Forest Administration Benefits Water Users: The Relation of Grazing to Stream Flow." Prepared by the Forest Service. *Reclamation Record*, 7 (May 1916), 223-4.

Huffman, Roy E. *Irrigation Development and Public Water Policy*. New York: The Ronald Press Company, 1953.

Hughes, Thomas P. *Networks of Power: Electrification in Western Societies, 1880-1930.* Baltimore, Maryland: The Johns Hopkins University Press, 1983.

Hulse, James. "Indian Farm or Inland Fishery? The Pyramid Lake Reservation." *Halcyon,* 15 (1993), 131-48 [a publication of the Nevada Humanities Committee].

Hundley, Norris, Jr. *The Great Thirst: Californians and Water, A History*. Berkeley: University of California Press, originally published 1992, revised edition, 2001.

———. *Water and the West: The Colorado River Compact and the Politics of Water in the American West.* Berkeley: University of California Press, 1975.

Hunter, Louis C. and Lynwood Bryant. *A History of Industrial Power in the United States, 1780-1930.* Volume 3. Cambridge, Massachusetts: Massachusetts Institute of Technology, 1991.

Hurt, R. Douglas. *American Agriculture: A Brief History*. Ames: Iowa State University Press, 1994.

———. *The Dust Bowl: An Agricultural and Social History.* Chicago, Illinois: Nelson-Hall, 1981.

Hyde, Anne F. *An American Vision: Far Western Landscape and National Culture, 1820-1920.* New York: New York University Press, 1990.

Ickes, Harold L. "Friant Dam Construction Started, Central Valley Project, California." *Reclamation Era*, 29 (1939), 289-90.

———. "Our Rights to Power." *Colliers*, 102 (November 12, 1938).

———. *The Secret Diary of Harold L. Ickes. Volume I.* New York: Simon and Schuster, 1953.

"Impending Reclamation Disaster May Be Averted." *New Reclamation Era*, 15 (May 1924), 67-9.

"Imperial Valley, California," *Reclamation Era*, 26 (February 1936), 45-6.

"Increase the Duty of Water," *Reclamation Record*, 5 (May 1914), 160.

"Increased Food Production on Reclamation Projects." *Reclamation Record*, 8 (December 1917), 547-50.

Inouye, Mamoru. "Heart Mountain High School, 1942-1945." *Journal of the West,* 38 (April 1999), 56-64.

Jackson, Donald C. *Building the Ultimate Dam: John S. Eastwood and the Control of Water in the West.* Lawrence: University Press of Kansas, 1995.

———. "Engineering in the Progressive Era: A New Look at Frederick Haynes Newell and the U.S. Reclamation Service." *Technology and Culture,* 34 (July 1993), 539-74.

———, and Norris Hundley, Jr., "Privilege and Responsibility: William Mulholland and the St. Francis Dam Disaster." *California History,* 82, 3 (2004), 8-47.

Jackson, W. Turrentine, Rand F. Herbert, and Stephen R. Wee, introduction. *Engineers and Irrigation: Report of the Board of Commissioners on the Irrigation of the San Joaquin, Tulare and Sacramento Valleys of the State of California, 1873.* Fort Belvoir, Virginia: Office of History of the U.S. Army Corps of Engineers, 1990. Reprint published as "Engineer Historical Studies" Number 5.

James, George Wharton. *Reclaiming the Arid West: The Story of the United States Reclamation Service.* Washington, D.C.: Government Printing Office, 1917.

Johnson, Alvin S. "Economic Aspects of Certain Reclamation Projects." In U.S. Department of the Interior, Bureau of Reclamation, Elwood Mead, Commissioner, *Economic Problems of Reclamation.* Washington, D.C.: Government Printing Office, 1929.

———. *Pioneer's Progress: An Autobiography.* New York: Viking Press, 1952.

Josephson, Paul R. *Industrialized Nature: Brute Force Technology and the Transformation of the Natural World.* Washington, D.C.: Island Press, 2002.

Josephy, Alvin, Jr. "Here in Nevada: A Terrible Crime." *American Heritage*, 21 (June 1970), 93-100.

Kahrl, William L. *The Conflict Over Los Angeles' Water Supply of the Owens Valley*. Berkeley: University of California Press, 1982.

Keim, Albert N. *The CPS Story: An Illustrated History of Civilian Public Service*. Intercourse, Pennsylvania: Good Books, 1990.

Kelley, Robert L. *Battling the Inland Sea: American Political Culture, Public Policy, & the Sacramento Valley, 1850-1986*. Berkeley: University of California Press, 1989.

Kennedy, David. *Freedom from Fear: The American People in the Depression and War, 1929-1945*. New York: Oxford University Press, 1999.

Keys, John W. III. "Address of the Commissioner Presented at the Bureau of Reclamation Centennial Celebration at Hoover Dam on the Colorado River." June 17, 2002.

Kleinsorge, Paul L. *The Boulder Canyon Project: Historical and Economic Aspects*. Palo Alto, California: Stanford University Press, 1941.

Kluger, James R. *Turning Water on with a Shovel: The Career of Elwood Mead.* Albuquerque: University of New Mexico Press, 1992.

Knack, Martha C., and Omar C. Stewart. *As Long as the Rivers Shall Run*. Berkeley: University of California Press, 1984.

Knight, Oliver. "Correcting Nature's Error: The Colorado-Big Thompson Project." *Agricultural History,* 30, 4 (1956), 157-69.

Kolupaila, Steponas. "Early History of Hydrometry in the United States." *Journal of the Hydraulics Division, Proceedings of the American Society of Civil Engineers* 86 (January 1960), 1-51.

Koppes, Clayton R. "Public Water, Private Land: Origins of the Acreage Limitation Controversy, 1933-1953." *Pacific Historical Review,* 47 (November 1978), 607-36.

Lamar, Howard R. *The New Encyclopedia of the American West.* New Haven, Connecticut: Yale University Press, 1998.

Lampen, Dorothy. *Economic and Social Aspects of Federal Reclamation.* University Studies in Historical and Political Science. Series 98, Johns Hopkins, Baltimore, Maryland: The Johns Hopkins Press, 1930.

Lang, William. "1949: Year of Decision on the Columbia River," *Columbia: The Magazine of Northwest History*, 19 (Spring 2005), 8-17.

Las Vegas *Age*. October 4 and 11, 1935.

Las Vegas *Evening Review Journal*. October 1, 1935.

Layton, Edwin T., Jr. *The Revolt of the Engineers: Social Responsibility and the American Engineering Profession.* Cleveland, Ohio: The Press of Case Western University, 1971.

Layton, Stanford J. *The Rationalization of Homesteading and Rural Life in the Early Twentieth Century American West.* Provo, Utah: Brigham Young University, 1987.

Lee, George W. "Cooperation on the Okanogan Project." *Reclamation Record,* 7 (December 1916), 557-9.

Lee, Lawrence B. "Introduction." In William E. Smythe. *The Conquest of Arid America*. Seattle: University of Washington, reprint 1969.

———. *Reclaiming the American West: An Historiography and Guide* (Santa Barbara, California: ABC – Clio, 1980.

———. "William Ellsworth Smythe and the Irrigation Movement: A Reconsideration." *Pacific Historical Review*, 41 (August 1972), 289-311.

Leslie, Jacques. *Deep Water: The Epic Struggle over Dams, Displaced People, and the Environment.* New York: Farrar, Straus and Giroux, 2005.

Lichatowich, Jim. *Salmon without Rivers: A History of the Pacific Salmon Crisis.* Washington, D.C.: Island Press, 1999.

Liebling, A.J. "Lake of the Cui-Ui Eaters." *The New Yorker*, 30 (January 1, 8, 15, 22, 1955).

Lilley, William, III, and Lewis L. Gould. "The Western Irrigation Movement, 1878-1902: A Reappraisal." In *The American West: A Reorientation.* Gene Gressley, editor. Laramie: University of Wyoming Publications, 1966.

Limerick, Patricia. *Desert Passages: Encounters with the American Deserts.* Albuquerque: University of New Mexico Press, 1985.

———. *Something in the Soil: Legacies and Reckonings in the New West.* New York: W.W. Norton & Company, 2000.

Linenberger, Toni Rae. *Dams, Dynamos, and Development: The Bureau of Reclamation's Power Program and Electrification of the West.* Denver: Bureau of Reclamation, 2002.

Linnér, Bjorn-Ola. *The Return of Malthus: Environmentalism and Post-war Population – Resource Crisis.* Isle of Harris, United Kingdom: White Horse Press, 2003.

Lowitt, Richard. *The New Deal in the West.* Bloomington: Indiana University Press, 1984.

Lowitt, Richard, and Judith Fabry, editors. *Henry A. Wallace's Irrigation Frontier: On the Trail of the Corn Belt Farmer, 1909.* Norman: University of Oklahoma Press, 1991.

Lumpee, Henry L. "The Civilian Conservation Corps: Irrigators." *Reclamation Era*, 26 (November 1936), 268-9.

Lyman, William Denison. *The Columbia River: Its History, Its Myths, Its Scenery, Its Commerce.* Portland, Oregon: Binfords & Mort, Publishers, 1963 (fourth edition), orig. pub. 1909.

Mackey, Mike. "Introduction: Life in America's Concentration Camps." *Journal of the West,* 38 (April 1999), 9-13.

Malone, Michael P. and Richard Etulain. *The American West: A Twentieth Century History.* Lincoln: University of Nebraska Press, 1989; second paperback edition, 1990.

Maxwell's Talisman, 6 (July 1906).

McCarthy, Michael G. *Hours of Trial: The Conservation Conflict in Colorado and the West, 1891-1907.* Norman: University of Oklahoma Press, 1977.

McConnell, Grant. *The Decline of Agrarian Democracy.* New York: Atheneum, 1969.

———. *Private Power & American Democracy.* New York: Alfred Knopf, 1966.

McCool, Daniel. *Command of the Waters: Iron Triangles, Federal Water Development, and Indian Water.* Tucson: University of Arizona Press, 1994.

———. *Native Waters: Contemporary Water Settlements and the Second Treaty Era.* Tucson: University of Arizona Press, 2002.

McCullough, David. *The Johnstown Flood: The Incredible Story Behind One of the Most Devastating Disasters American Has Ever Known.* New York: Simon and Schuster, 1987.

McCully, Patrick. *Silenced Rivers: The Ecology and Politics of Large Dams.* London and New York: Red Books, 1996.

McCune, Christopher J. "Rapid Valley Project." Bureau of Reclamation History Program, typescript.

McPhee, John. *Encounters with the Archdruid.* New York: Farrar, Straus & Giroux, 1971.

McWilliams, Carey. *Brothers Under the Skin.* Boston: Little Brown and Co., 1948.

———. "Japanese Evacuation: Policy and Perspectives." *Common Ground,* (Summer 1942), 1-8.

Mead, Elwood. *Agricultural Development in Palestine: Report to the Zionist Executive.* London: Zionist Executive, 1924.

———. "Construction of Irrigation Works and Selection of Settlers Directed by the Bureau of Reclamation." *U.S. Daily,* (September 17, 1928).

———. *Helping Men Own Farms: A Practical Discussion of Government and Aid in Land Settlement.* New York: Macmillian Company, 1920.

Merritt, Raymond H. *Engineering in American Society, 1850-1875.* Lexington: University of Kentucky Press, 1969.

Miles, Nelson A. "Our Unwatered Empire." *North American Review*, 150 (March 1890), 370-81.

Miller, Char. *Gifford Pinchot and the Making of Modern Environmentalism.* Washington, D.C.: Island Press, 2001.

Moeller, Beverley B. *Phil Swing and Boulder Dam.* Berkeley: University of California Press, 1971.

Morgan, Murray. *The Columbia.* Seattle, Washington: Superior Publishing Co., 1949.
———. *The Dam.* New York: Viking, 1954.

"Morris Bien Dies in Takoma Park," Washington *Evening Star.* (July 29, 1932).

Morrissey, Katherine G. *Mental Territories: Mapping the Inland Empire.* Ithaca, New York: Cornell University Press, 1997.

Mumford, Lewis. "Theory and Practice of Regionalism." *Sociological Review*, 20 (January 1928), 18-33.

Nadeau, Remi A. *The Water Seekers.* Garden City, New York: Doubleday & Co., 1950.

Nash, Gerald. *State Government and Economic Development: A History of Administrative Policies in California, 1849-1933.* Berkeley: University of California Press, 1964.

———. *World War II and the West: Reshaping the Economy.* Lincoln: University of Nebraska Press, 1990.

"National Forest Reserve." *The Nation*, 37 (September 6, 1883), 201.

Nelson, Gordon E. *The Lobbyist: The Story of George H. Maxwell, Irrigation Crusader*. Bowie, Maryland: Headgate Press, 2001.

Neuberger, Richard Lewis. *Our Promised Land.* New York: The Macmillan Company, 1938.

———. *Our Promised Land,* edited by David L. Nicandri (Moscow: University of Idaho Press, 1989).

Newell, Frederick Haynes. *Irrigation in the United States*. New York: Thomas Y. Crowell & Company, 1902.

———. "Reclamation Law." In U.S. Department of the Interior, U.S. Geological Survey. *Third Annual Report of the Reclamation Service, 1903-4*. Second Edition. Washington, D.C.: Government Printing Office, 1905.

———. "Statistics of Irrigation and Water Supply." *The Independent,* 45 (May 4, 1893), 595-6.

Newlands, Francis G. "National Irrigation Works." *Forestry and Irrigation,* 8 (February 1902), 63-6.

The New York *Times*. (May 28, 1871); (October 1, 1935).

Noggle, Burl. *Teapot Dome: Oil and Politics in the 1920s*. New York: Norton, 1965.

Norcross, Fred N. "Genesis of the Colorado-Big Thompson Project," *Colorado Magazine* 30 (January 1953), 29-37.

Nunis, Doyce Blackman, editor. *The Saint Francis Dam Disaster Revisited.* Los Angeles: Ventura Historical Society of Southern California; Ventura County Museum of History, 1995.

Nye, David E. *American Technological Sublime*. Cambridge, Massachusetts: The MIT Press, 1994.

———. *Electrifying America: Social Meanings of a New Technology, 1890-1940*. Cambridge, Massachusetts: The MIT Press, 1990.

Ogden, Gerald R., compiler. *The Excess Lands Provisions of Federal Reclamation Law: A Bibliography and Chronology*. Davis, California: Agricultural History Center, University of California, Davis, July 1980.

Olson, James C. *History of Nebraska*. Lincoln: University of Nebraska Press, 1955.

Olson, James S., editor. *Historical Dictionary of the New Deal: From Inauguration to Preparation for War*. Westport, Connecticut: Greeenwood Press, 1985.

"Operations on Indian Projects Transferred." *Reclamation Record*, 15 (February 1924), 31.

Oregonian (Portland, Oregon), (November 2, 1993), A10.

Orsi, Jared. *Hazardous Metropolis: Flooding and Urban Ecology in Los Angeles*. Berkeley: University of California Press, 2004.

Orsi, Richard J. *Sunset Limited: The Southern Pacific Railroad and the Development of the American West, 1850-1930*. Berkeley: University of California Press, 2005.

Outland, Charles. *Man-made Disaster: The Story of the St. Francis Dam*. Glendale, California: Arthur C. Clark Company, 1977.

Page, John C. "Also One-Third is Ill-Watered." *Reclamation Era,* 28 (January 1938), 2-4.

———. "Don't Forget the Drought." *Reclamation Era*, 31 (November 1941), 282.

———. "Water Conservation and Control." *Reclamation Era*, 27 (March 1937), 46-8.

Parfit, Michael. "The Floods that Carved the West." *Smithsonian,* 26 (April 1995), 48-76.

Pelz, Richard K., editor. U.S. Department of the Interior, U.S. Bureau of Reclamation, *Federal Reclamation and Related Laws Annotated, Volume 1, Through 1942*. Washington, D.C.: Government Printing Office, 1972.

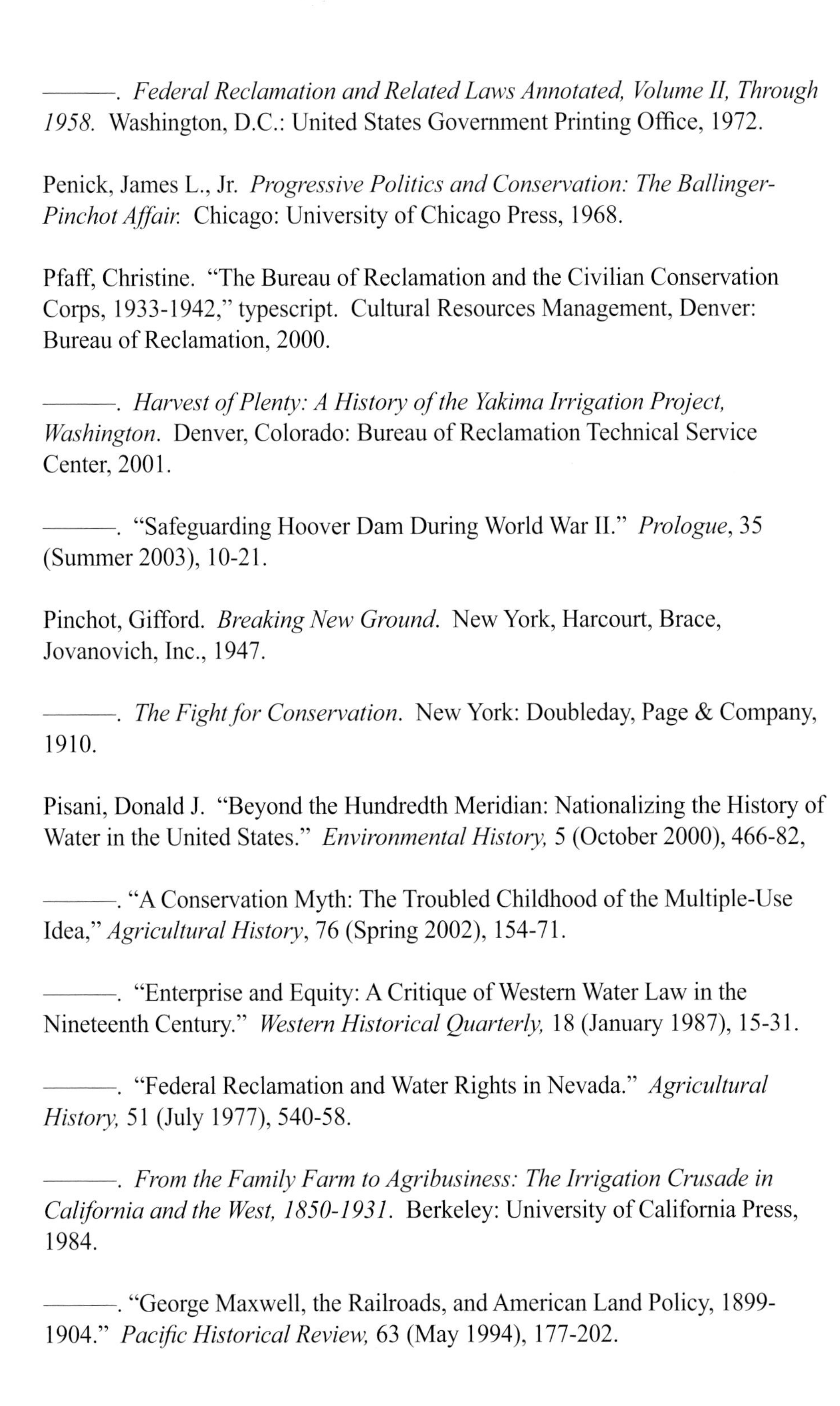

———. *Federal Reclamation and Related Laws Annotated, Volume II, Through 1958.* Washington, D.C.: United States Government Printing Office, 1972.

Penick, James L., Jr. *Progressive Politics and Conservation: The Ballinger-Pinchot Affair.* Chicago: University of Chicago Press, 1968.

Pfaff, Christine. "The Bureau of Reclamation and the Civilian Conservation Corps, 1933-1942," typescript. Cultural Resources Management, Denver: Bureau of Reclamation, 2000.

———. *Harvest of Plenty: A History of the Yakima Irrigation Project, Washington.* Denver, Colorado: Bureau of Reclamation Technical Service Center, 2001.

———. "Safeguarding Hoover Dam During World War II." *Prologue*, 35 (Summer 2003), 10-21.

Pinchot, Gifford. *Breaking New Ground.* New York, Harcourt, Brace, Jovanovich, Inc., 1947.

———. *The Fight for Conservation.* New York: Doubleday, Page & Company, 1910.

Pisani, Donald J. "Beyond the Hundredth Meridian: Nationalizing the History of Water in the United States." *Environmental History,* 5 (October 2000), 466-82,

———. "A Conservation Myth: The Troubled Childhood of the Multiple-Use Idea," *Agricultural History*, 76 (Spring 2002), 154-71.

———. "Enterprise and Equity: A Critique of Western Water Law in the Nineteenth Century." *Western Historical Quarterly,* 18 (January 1987), 15-31.

———. "Federal Reclamation and Water Rights in Nevada." *Agricultural History,* 51 (July 1977), 540-58.

———. *From the Family Farm to Agribusiness: The Irrigation Crusade in California and the West, 1850-1931.* Berkeley: University of California Press, 1984.

———. "George Maxwell, the Railroads, and American Land Policy, 1899-1904." *Pacific Historical Review,* 63 (May 1994), 177-202.

———. "Irrigation, Water Rights, and the Betrayal of Indian Allotment." *Environmental Review,* 10 (Fall 1986), 157-76.

———. "Reclamation and Social Engineering in the Progressive Era." *Agricultural History,*57 (January 1983), 46-63.

———. "State vs. Nation: Federal Reclamation and Water Rights in the Progressive Era." *Pacific Historical Review,* 51 (August 1982), 265-82.

———. *To Reclaim a Divided West: Water, Law, and Public Policy, 1848-1902.* Albuquerque: University of New Mexico Press, 1992.

———. *Water and American Government: The Reclamation Bureau, National Water Policy and the West, 1902-1935.* Berkeley: University of California Press, 2002.

Pitzer, Paul C. *Building the Skagit: A Century of Upper Skagit Valley History, 1870-1970.* Portland, Oregon: The Galley Press, 1978.

———. *Grand Coulee: Harnessing a Dream.* Pullman: Washington State University Press, 1994.

Pomeroy, John Norton. *A Treatise on the Law of Water Rights.* St. Paul, Minnesota: West Publishing Company, 1893.

———. "Riparian Rights" and "Water Courses" in *Johnson's New Universal Cyclopedia: A Scientific and Popular Treasury of Useful Knowledge.* Four volumes. Frederick A. P. Bernard and Arnold Guyot, editors in chief. New York: A. J. Johnson and Company, 1881.

Powell, John Wesley. "History of Irrigation," *The Independent,* 45 (May 4, 1893), 593-5.

———. "The New Lake in the Desert." *Scribner's Magazine*, 10 (October 1891), 463-8.

———. "The Irrigable Lands of the Arid Region," *Century Magazine*, 39 (March 1890), 766-77.

———. *Report on the Lands of the Arid Region.* See bibliographic listing at U.S. Congress, House. Committee on Appropriations. J. W. Powell, *Report on the Lands...*

"President Talks at Boulder Dam." New York *Times*. October 1, 1935. 2:2.

Pringle, Henry F. *Theodore Roosevelt: A Biography*. New York: Harcourt, Brace and Company, 1931.

"Progress of Work, October, 1909." *Reclamation Record*, 1 (November 1909), 103-8.

Prucha, Francis Paul. *The Great Father: The United States Government and Indians*. Volumes 1 and 2. Lincoln: University of Nebraska Press, 1984.

Pursell, Carroll. *The Machine in America: A Social History of Technology*. Baltimore: John Hopkins University Press, 1995.

Rabbitt, Mary C. *Minerals. Lands, and Geology for the Common Defence and General Welfare, 1879-1904,* volume II. Washington, D.C.: Government Printing Office, 1980.

Randolph, C. B. "Cooperative Farm Organizations." *Reclamation Record*, 5 (December 1914), 470.

Rasmussen, Wayne D. *Taking the University to the People: Seventy-Five Years of Cooperative Extensions*. Ames: Iowa State University Press, 1989.

Rauchway, Eric. *Murdering McKinley: The Making of Theodore Roosevelt's America*. New York: Hill & Wang, 2003.

Reclamation Record, 1 (March 1909), 25-36.

"Recommendations for New Projects." *New Reclamation Era*, 15 (May 24, 1924), 77.

"Reform Party Victory." *The Reporter*, (August 15, 1944) [periodical of the National Service Board for Concientious Objectors].

Reisner, Marc. *Cadillac Desert: The American West and Its Disappearing Water*. New York: Viking, 1986.

"Report of the Central Board of Review." *Reclamation Record*, 7 (July 1916), 298.

"Report of the State Engineer." *Appendix to the Journals of the Senate and Assembly of the Twenty-Second Session of the Legislature of the State of Nevada, 1905*. Carson City, Nevada: State Printing Office, 1905.

"Report of the State Engineer." *Appendix to the Journals of the Senate and Assembly of the Twenty-Second Session of the Legislature of the State of Nevada, 1909*. Volume 2. Carson City, Nevada: State Printing Office, 1909.

The Reporter, Supplement. November 1, 1944 [periodical of the National Service Board for Concientious Objectors]..

Richardson, Elmo R. *The Politics of Conservation: Crusades and Controversies, 1897-1913.* Berkeley: University of California Press, 1962.

Richardson, Peter. *American Prophet: The Life and Work of Carey McWilliams.* Ann Arbor: University of Michigan Press, 2005.

Righter, Robert W. *The Battle over Hetch Hetchy: America's Most Controversial Dam and the Birth of Modern Environmentalism*. New York: Oxford University Press, 2005.

Robbins, William G. "'At the End of the Cracked Whip:' The Northern West." *Montana: The Magazine of Western History,* 38 (Winter 1988), 2-11.

———. *Colony and Empire: The Capitalist Transformation of the American West.* Lawrence: University Press of Kansas, 1994.

———, editor. *The Great Northwest: The Search for Regional Identity*. Corvallis: Oregon State University Press, 2001.

Robinson, Michael C. *Water for the West*. Chicago, Illinois: Public Works Historical Society, 1979.

Rocca, Al M. *America's Master Dam Builder: The Engineering Genius of Frank T. Crowe*. New York: University Press of America, 2001.

Rocha, Guy Louis. "The IWW and the Boulder Dam Project: The Final Death Throes of American Syndicalism." *Nevada Historical Society Quarterly,* 21 (Spring 1978), 3-24.

Rogers, J. David. "A Man, a Dam and a Disaster: Mulholland and the St. Francis Dam." *Southern California Quarterly,* 77 (Spring/Summer 1995), 1-109.

———. "Reassessment of the St. Francis Dam Failure." *Engineering Geology Practice in Southern California,* edited by B.W. Pipkin and R.J. Proctor. *Association of Engineering Geologists, Special Publication No. 4* (1992), 639-66.

"Roosevelt Dam Reaches New Heights." *Civil Engineering*, 66 (May 1996), 10.

Roosevelt, Theodore. *An Autobiography*. New York: The Macmillan Company, 1914.

———. *State Papers as Governor and President, 1899-1909*. New York: Charles Scribner's Sons, 1926.

———. *The Winning of the West*. 4 volumes. New York: G. P. Putnam's Sons, 1889-1896.

Rowley, William D. "The Newlands Project: Crime or National Commitment?" *Nevada Public Affairs Review,* 1 (1992), 39-43.

———. *Reclaiming Arid America: The Career of Francis G. Newlands.* Bloomington: Indiana University Press, 1996.

———. *U.S. Forest Service Grazing and Rangelands*. College Station: Texas A&M University Press, 1985.

Rusco, Elmer R., editor. *A Reporter at Large: Dateline—Pyramid Lake, Nevada.* Reno: University of Nevada Press, 2000.

Scheiber, Harry N. and Charles W. McCurdy. "Eminent-Domain Law and Western Agriculture, 1849-1900." *Agricultural History,* 49 (January 1975), 112-30.

Schmitt, F. E., and John W. Haw. "Report on Survey of Federal Reclamation in the West." *Reclamation Era*, 25 (February 1935), 29.

Schuyler, James D. *Reservoirs for Irrigation, Water Power and Domestic Water Supply*. New York: Wiley and Sons, 1909.

Selective Service as the Tide of the War Turns: The 3rd Report of the Director of Selective Service. Washington, D.C.: Government Printing Office, 1944.

Selective Service and Victory: The 4th Report of the Director of Selective Service. Washington, D.C.: Government Printing Office, 1945.

Selective Service in Wartime: Second Report of the Director of Selective Service. Washington, D.C.: Government Printing Office, 1942.

Shallat, Todd. *Structures in the Stream: Water, Science, and the Rise of the U.S. Army Corps of Engineers.* Austin: University of Texas Press, 1994.

Sherow, James Earl. *Watering the Valley: Development along the High Plains Arkansas river, 1870-1950.* Lawrence: University Press of Kansas, 1990.

Shumway, G.L. "Give Settlers a Free Hand." *The Irrigation Age*, 22 (March, 1907), 145-7.

Shurts, John. *Indian Reserved Water Rights: The Winters Doctrine in Its Social and Legal Context, 1880s-1930s.* Norman: University of Oklahoma Press, 2000.

Sibley, Mulford Q. and Philip E. Jacob. *Conscription of Conscience: The American State and the Conscientious Objector, 1940-1947.* Ithaca, New York: Cornell University Press, 1952.

Sitkoff, Harvard. *A New Deal for Blacks: The Emergence of Civil Rights as a National Issue.* New York: Oxford University Press, 1978.

"Six New Projects Are Recommended." *New Reclamation Era*, 15 (May 1924), 69.

Smith, Henry Nash. "Clarence King, John Wesley Powell, and the Establishment of the U.S. Geological Survey." *Mississippi Valley Historical Review, 34* (June 1947), 37-58.

———. "Rain Follows the Plow: The Notion for Increased Rainfall for the Great Plains." *Huntington Library Quarterly*, 10 (1947), 175-88.

———. *Virgin Land: The American West as Symbol and Myth.* Cambridge, Massachusetts: Harvard University Press, 1950.

Smith, Karen. *The Magnificent Experiment: Building the Salt River Reclamation Project, 1890-1917.* Tucson: University of Arizona Press, 1982.

Smith, Norman. *A History of Dams.* Secausus, New Jersey: Citadel Press, 1972.

———. *Man and Water: A History of Hydro-Technology.* New York: Charles Scribner's Sons, 1975.

Smythe, William E. *The Conquest of Arid America.* New York: Harper & Brothers Publishers, 1900.

———. "An International Wedding: The Tale of a Trip on the Borders of Two Republics," *Sunset* (October 1900).

———. "The 20th Century West." *Land of Sunshine* [predecessor of *Out West*], 15 (June-December 1901), 61-4.

Starr, Kevin. *Material Dreams: Southern California Through the 1920s.* New York: Oxford University Press, 1990.

Steen, Harold K., editor. *The Conservation Diaries of Gifford Pinchot.* Durham, North Carolina: Forest History Society Pinchot Institute for Conservation, 2001.

———. *The U.S. Forest Service: A History.* Seattle: University of Washington Press, 1976.

Stegner, Wallace. *Beyond the Hundredth Meridian: John Wesley Powell and the Second Opening of the West.* Lincoln: University of Nebraska Press, 1953.

Steinberg, Theodore. *Nature Incorporated: Industrialization and the Water of New England.* New York: Cambridge University Press, 1991.

———. "'That World's Fair Feeling:' Control of Water in Twentieth Century America." *Technology and Culture.* 34 (April 1993) 401-9.

Steinel, Alvin T. *History of Agriculture in Colorado.* Fort Collins, Colorado: State Agricultural College, 1926.

Stene, Eric A. "The Minidoka Project." Typescript. Denver: Bureau of Reclamation History Program, 1997.

Sterling, Everett W. "The Powell Irrigation Survey, 1888-1893." *Mississippi Valley Historical Review* 27 (December 1940), 421-34.

Stevens, Joseph E. *Hoover Dam: An American Adventure.* Norman: University of Oklahoma Press, 1988.

Stoutemyer, B. E. "Irrigation Districts: Their Relation to the Reclamation Service." *Reclamation Record*, 5 (October 1914), 861-3.

Stradling, David, editor. *Conservation in the Progressive Era.* Seattle: University of Washington Press, 2004.

Strom, Claire. *Profiting from the Plains: The Great Northern Railway and Corporate Development of the American West.* Seattle: University of Washington Press, 2003.

Strong, Douglas H. *Tahoe: An Environmental History.* Lincoln: University of Nebraska Press, 1984.

Sutter, Paul S. *Driven Wild: How the Fight Against the Automobile Launched the Modern Wilderness Movement.* Seattle: University of Washington Press, 2002.

Swain, Donald C. "The Bureau of Reclamation and the New Deal, 1933-1940." *Pacific Northwest Quarterly,* 61 (July 1970), 137-46.

———. *Federal Conservation Policy, 1921-1933.* Berkeley: University of California Press, 1963.

Taylor, Joseph E., III. *Making Salmon: An Environmental History of the Northwest Fisheries Crisis.* Seattle: University of Washington Press, 1999.

Teele, Raymond P. *Irrigation in the United States.* New York: D. Appleton and Co., 1915.

———. "Land Reclamation Policies in the United States." USDA *Bulletin* No. 1257, August 23, 1924. Washington, D.C.: Government Printing Office, 1924.

Teisch, Jessica B. "Great Western Power, 'White Coal' and Industrial Capitalism in the West." *Pacific Historical Review* 70 (May 2001), 221-53.

Thomas, George. *The Development of Institutions under Irrigation with Special Reference to Early Utah.* New York: The Macmillan Company, 1920.

Thomson, David. *In Nevada: The Land, the People, God, and Chance.* New York: Vintage Books, 1999.

Townley, John M. "Reclamation and the Red Man: Relationships on the Truckee-Carson Project." *The Indian Historian,* 11, 1 (1978), 21-8.

———. "Soil Saturation Problems on the Truckee-Carson Project," *Agricultural History,* 52 (April 1978), 280-91.

———. *Turn this Water into Gold: The Story of the Newlands Project.* Reno: Nevada Historical Society, 1977.

Tugwell, Rexford G. "Reflections on Farm Relief." *Political Science Quarterly,* 43 (December 1928), 481-97.

Turner, Frederick Jackson. "Pioneer Ideals and the State University." in *The Frontier in American History*. New York: Holt, Rinehart and Winston, 1962, originally published 1920.

———. "The Problem of the West." *Atlantic Monthly*, 78 (September 1896), 289-97.

Tyler, Daniel, *The Last Water Hole in the West: The Colorado-Big Thompson Project and the Northern Colorado Water Conservancy District.* Boulder: University Press of Colorado, 1992.

———, *Silver Fox of the Rockies: Delphus E. Carpenter and Western Water Compacts*. Norman: University of Oklahoma Press, 2003.

Tyson, H. G., Jr. "A Reply to Comptroller Ryan's Article." *Reclamation Record,* 5 (June 1914), 211-3.

U.S. Army Corps of Engineers. "Examination of Reservoir Sites in Wyoming and Colorado." By Hiram M. Chittenden. Washington, D.C.: U.S. Army Corps of Engineers, 1898.

———. *The Federal Engineer Damsites to Missile Sites: A History of the Omaha District, U.S. Army Corps of Engineers.* Washington, D.C.: Government Printing Office, 1985.

U.S. Congress, House. Committee on Appropriations. Hearings on the Reclamation Service before the Committee on Appropriations. 67th Congress, 2nd Session, December 23, 1921. Washington, D.C.: Government Printing Office, 1922.

———. Committee on Appropriations. J. W. Powell, *Report on the Lands of the Arid Region of the United States, with a More Detailed Account of the Lands of Utah. With Maps.* Ex. Doc. No. 73. 45th Congress, 2nd Session.

———. Committee on Irrigation of Arid Lands. Hearing before the Committee on Irrigation of Arid Lands. February 1909.

———. Committee on Irrigation and Reclamation, *Settlement of Returning Veterans on Farms in Reclamation Projects*, 79 Cong., 1st Session, April 12, 1945.

———. J. W. Powell, "Geographical and Geological Surveys West of the Mississippi," Report No. 612, 43rd Congress, 1st Session.

U.S. Congress, Senate. *Federal Reclamation by Irrigation.* Senate Document No. 92, 68th Congress, 1st Session. Serial Number 8238. (1924).

———. *Investigation of the Department of the Interior and the Bureau of Forestry.* Volume 1. 61st Congress, 3rd Session, Senate Document 719.

———. "Letter of the Commissioner of Agriculture" in answer to a Senate resolution of February 6 1874, transmitting a paper prepared by Hon. George P. Marsh, on the subject of irrigation, "Irrigation: Its Evils, the Remedies, and the Compensations." 43rd Cong., 1st Session, S. Misc. Doc. No. 55.

U.S. Department of Agriculture, Economic Research Service. *A History of Federal Water Resources Programs, 1961-1970*, Miscellaneous Publication No.1379, by Beatrice Hort Holmes. Washington, D.C.: U.S. Government Printing Office, 1979.

U.S. Department of the Interior, U.S. Bureau of Reclamation. "Bureau of Reclamation Historical Organizational Structure." Denver: Bureau of Reclamation History Program, 1997, very limited printing.

———. Elwood Mead, Commissioner, Alvin Johnson. "Economic Aspects of Certain Reclamation Projects." in *Economic Problems of Reclamation.* Washington, D.C.: Government Printing Office, 1929.

———. Elwood Mead, Commissioner. E. C. Branson, "Planned Colonies of Farm Owners," in *Economic Problems of Reclamation.* Washington, D.C.: Government Printing Office, 1929.

———. *Federal Reclamation and Related Laws Annotated,* Volume I, Through 1942. Richard K. Pelz, editor. Washington, D.C.: U.S. Government Printing Office, 1972.

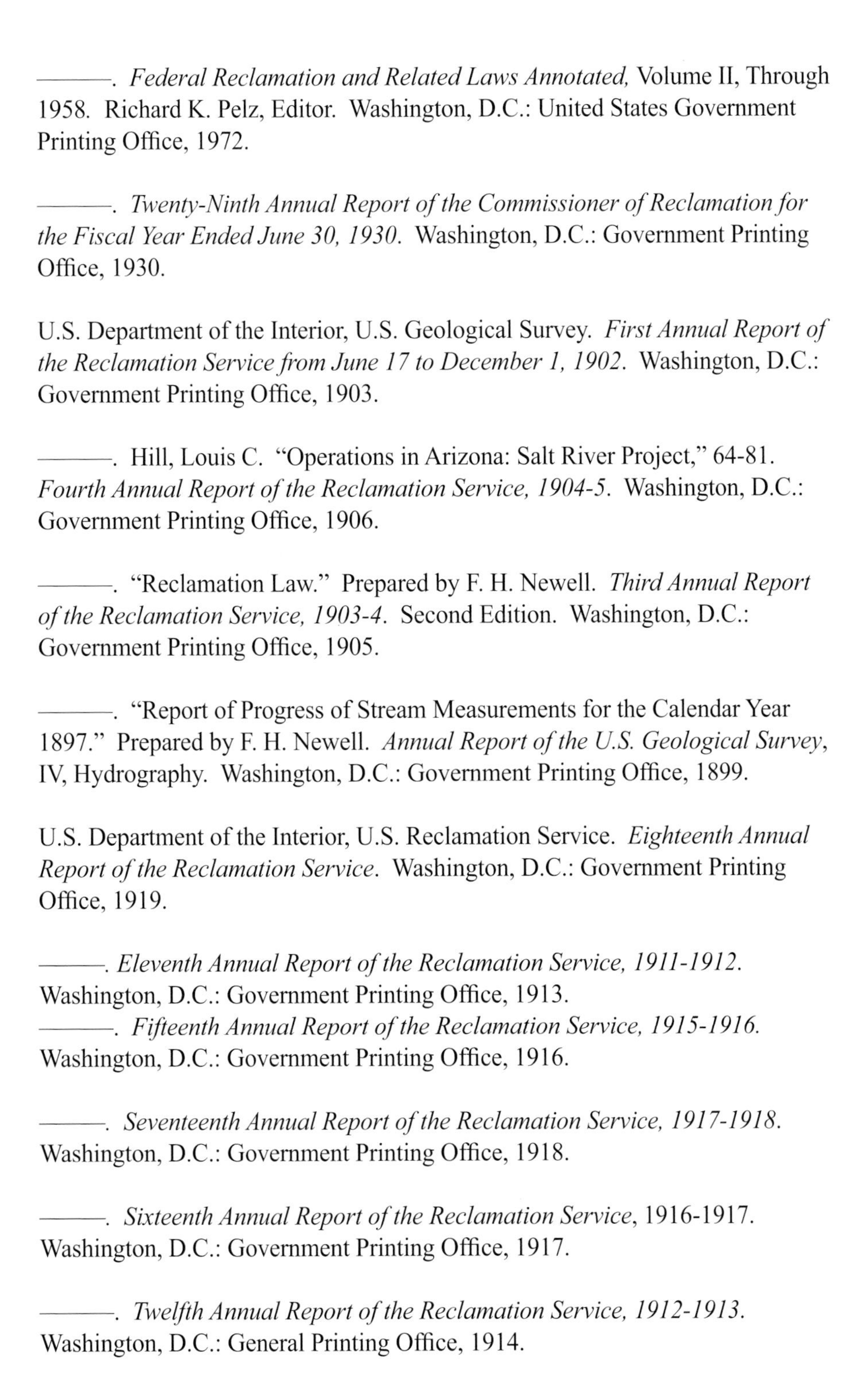

———. *Federal Reclamation and Related Laws Annotated,* Volume II, Through 1958. Richard K. Pelz, Editor. Washington, D.C.: United States Government Printing Office, 1972.

———. *Twenty-Ninth Annual Report of the Commissioner of Reclamation for the Fiscal Year Ended June 30, 1930.* Washington, D.C.: Government Printing Office, 1930.

U.S. Department of the Interior, U.S. Geological Survey. *First Annual Report of the Reclamation Service from June 17 to December 1, 1902.* Washington, D.C.: Government Printing Office, 1903.

———. Hill, Louis C. "Operations in Arizona: Salt River Project," 64-81. *Fourth Annual Report of the Reclamation Service, 1904-5.* Washington, D.C.: Government Printing Office, 1906.

———. "Reclamation Law." Prepared by F. H. Newell. *Third Annual Report of the Reclamation Service, 1903-4.* Second Edition. Washington, D.C.: Government Printing Office, 1905.

———. "Report of Progress of Stream Measurements for the Calendar Year 1897." Prepared by F. H. Newell. *Annual Report of the U.S. Geological Survey,* IV, Hydrography. Washington, D.C.: Government Printing Office, 1899.

U.S. Department of the Interior, U.S. Reclamation Service. *Eighteenth Annual Report of the Reclamation Service.* Washington, D.C.: Government Printing Office, 1919.

———. *Eleventh Annual Report of the Reclamation Service, 1911-1912.* Washington, D.C.: Government Printing Office, 1913.
———. *Fifteenth Annual Report of the Reclamation Service, 1915-1916.* Washington, D.C.: Government Printing Office, 1916.

———. *Seventeenth Annual Report of the Reclamation Service, 1917-1918.* Washington, D.C.: Government Printing Office, 1918.

———. *Sixteenth Annual Report of the Reclamation Service*, 1916-1917. Washington, D.C.: Government Printing Office, 1917.

———. *Twelfth Annual Report of the Reclamation Service, 1912-1913.* Washington, D.C.: General Printing Office, 1914.

———. *Twentieth Annual Report of the Reclamation Service, 1920-1921.* Washington, D.C.: Government Printing Office, 1921.

U.S. Department of Interior, Water and Power Resources Service. *Project Data.* Denver, Colorado: Government Printing Office, 1981.

United States Presidents. *A Compilation of the Messages and Papers of the Presidents, 1789-1918.* Prepared under the Direction of the Joint Committee on Printing, of the House and Senate, pursuant to an Act of the Fifty-Second Congress of the United States (with additions and encyclopedia index by private enterprise). Compiled by James D. Richardson. New York: Bureau of Literature, c.[1897-1924].

Von Nordhoff, Ellen. "The American Frontier as a Safety Valve: The Life, Death, Reincarnation, and Justification of a Theory." *Agricultural History,* 36 (Fall 1962), 123-49.

Walcott, Charles Doolittle. "Vast Extent of Arid Lands." *National Magazine*, 15 (1902), 571-2.

Ward, Evan R. *Border Oasis: Water and the Political Ecology of the Colorado River Delta, 1940-1975.* Tucson: University of Arizona Press, 2003.

———. "The Twentieth Century Ghost of William Walker: Conquest of Land and Water as Central Themes in the History of the Colorado Delta." *Pacific Historical Review,* 70 (August 2001), 359-85.

Warken, Philip W. *A History of the National Resources Planning Board, 1933-1943.* New York: Garland Publishing, Inc., 1979.

Warne, William E. *The Bureau of Reclamation.* New York: Praeger Publishers, 1973.

Watkins, T.H. *Righteous Pilgrim: The Life and Times of Harold L. Ickes, 1874-1952.* New York: Henry Holt and Company, 1990.

Webb, Walter Prescott. "The American West Perpetual Mirage." *Harper's Magazine*, 214 (May 1957), 25-31.

Wegmann, Edward. *The Design and Construction of Dams. Including Masonry, Earth, Rock-fill, Timber, and Steel Structures, Also The Principal Types of Movable Dams,* fourth edition. New York: John Wiley & Sons, 1904. First edition published in 1888 with the title *The Design and Construction of Masonry Dams.*

"The Week." *New Republic*, 33 (January 17, 1923), 185.

Weinberg, Albert K. *Manifest Destiny: A Study of Nationalist Expansion in American History*. Baltimore: Johns Hopkins Press, 1935.

White, Richard. "The Altered Landscape: Social Change and the Land in the Pacific Northwest." *Regionalism and the Pacific Northwest*, edited by William G. Robbins, Robert J. Frank, Richard E. Ross. Corvallis: Oregon State University Press, 1983.

———. *"It's Your Misfortune and None of My Own," A History of the American West*. Norman: University of Oklahoma Press, 1991.

———. *The Organic Machine: The Remaking of the Columbia River*. New York: Hill & Wang, 1995.

———. *The Roots of Dependency: Subsistence, Environment and Social Change among the Choctaws, Pawnees, and Navajos*. Lincoln: University of Nebraska Press, 1983.

Wickens, James F. *Colorado in the Great Depression*. New York: Garland Publishing, Inc., 1979.

Wickersham, C. P. "Advice from an Optimist." *Reclamation Record*, 5 (September 1914), 323-4.

Widtsoe, John A. *Success on Irrigation Projects*. New York: John Wiley & Sons, Inc., 1928.

"Will Not Deprive Striving Settlers of Land, Fall Says." Yakima *Daily Republic*, February 6, 1922.

Wilson, Edmund. "Hoover Dam." *New Republic*, 68 (September 2, 1931), 66-9.

Wilson, M. L. "Land Utilization." Economic Series Lecture #25, April 16, 1932. Chicago, Illinois: University of Chicago Press, 1932.

Wisner George T. and Edgar T. Wheeler. "Investigations of Stresses in High Masonry Dams of Short Spans." *Engineering News*, 54 (August 10, 1905), 141-4.

Wolf, Donald E. *Big Dams and Other Dreams: The Six Companies Story*. Norman: University of Oklahoma Press, 1996.

Worster, Donald. *Dust Bowl: The Southern Plains in the 1930s*. New York: Oxford University Press, 1979.

———. *Rivers of Empire: Water and Aridity in the Growth of the American West*. New York: Oxford University Press, 1985.

———. *A River Running West: The Life of John Wesley Powell*. New York: Oxford University Press, 2001.

Wright, James E. *The Politics of Populism: Dissent in Colorado*. New Haven, Connecticut: Yale University Press, 1974.

Wrobel, David. *Promised Lands: Promotion, Memory, and the Creation of the American West*. Lawrence: University Press of Kansas, 2002.

Yochelson, Ellis L. *Charles Doolittle Walcott, Paleontologist*. Kent, Ohio: Kent State University Press, 1998.

Young, J.M. "River Planning in the Missouri Basin." *The Military Engineer*, 22 (March-April 1930), 152-8.

Zevin, B. D., editor. *Nothing to Fear: The Selected Addresses of Franklin Delano Roosevelt, 1932-1945*. Boston, Massachusetts: Houghton Mifflin Company, 1946.

INDEX